EXS 96

Cancer: Cell Structures, Carcinogens and Genomic Instability

Edited by Leon P. Bignold

Birkhäuser Verlag
Basel · Boston · Berlin

Editor

Leon P. Bignold
Division of Tissue Pathology
Institute of Medical and Veterinary Science
PO Box 14
Rundle Mall, SA 5001
Australia

Library of Congress Cataloging-in-Publication Data
Cancer : cell structures, carcinogens and genomic instability / edited by Leon P. Bignold.
p. ; cm. -- (EXS ; 96)
Includes bibliographical references and index.
ISBN-13: 978-3-7643-7156-2 (alk. paper)
ISBN-10: 3-7643-7156-0 (alk. paper)
1. Carcinogenesis--Molecular aspects. 2. Cancer cells. 3. Cancer--Pathophysiology.
4. Cancer--Genetic aspects. I. Bignold, Leon P., 1947- II. Series
[DNLM: 1. Neoplasms--genetics. 2. Carcinogens-adverse effects. 3. Gene Expression Regulation, Neoplastic--drug effects. 4. Genomic Instability. 5. Neoplasms-etiology. QZ 202 C21518 2005]
RC268.5.C28 2005
616.99'4071--dc22

2005048329

Bibliographic information published by Die Deutsche Bibliothek
Die Deutsche Bibliothek lists this publication in the Deutsche Nationalbibliografie; detailed bibliographic data is available in the Internet at <http://dnb.ddb.de>.

ISBN 10: 3-7643-7156-0 Birkhäuser Verlag, Basel – Boston – Berlin
ISBN 13: 978-3-7643-7156-2

Printed on acid-free paper produced from chlorine-free pulp. TCF ∞
Cover illustration: Bottom right: Somatic translocations and mis-targeting of HATs and HDACs (see Fig. 2, p. 27, detail). Bottom left: Proposed dose-response relationship for radiation-induced effects (see Fig. 1, p.168). Top right: Bioactivation of aromatic amines or amides (AA) towards ultimate DNA-reactive sulphate or acetoxy esters (see Fig. 3, p. 71, detail). Top left: A typical metaphase spread from a fifth generation (G5) mTERC$^{-/-}$ mouse fibroblast showing a fusion product (arrow) and 41 chromosomal arms, an example of aneuploidy (see Fig. 3, p. 56).

Printed in Germany
ISBN 10: 3-7643-7156-0
ISBN 13: 978-3-7643-7156-2
9 8 7 6 5 4 3 2 1

e-ISBN: 3-7643-7378-4

www. birkhauser.ch

Contents

List of Contributors

Vinay Arora, Apoptosis Research Centre, Children's Hospital of Eastern Ontario, Research Institute, 401 Smyth Road, Ottawa, Ontario K1H 8L1, Canada; e-mail: vinay@mgcheo.med.uottawa.ca

Benjamin J. Arthurs, Chemical Structure & Dynamics, Pacific Northwest National Laboratory, MS K8-88, 3335 Q Ave, Richland, WA 99354, USA; e-mail: barthurs@wsu.edu

Bharati Bapat, Samuel Lunenfeld Research Institute, Department of Pathology and Laboratory Medicine, Mount Sinai Hospital, 600 University Avenue, 9th Floor, Room 992B, Toronto, Ontario M5G 1X5, Canada; and Department of Laboratory Medicine and Pathobiology, University of Toronto, Toronto, ON, M5S 1A1, Canada; e-mail: bapat@mshri.on.ca

Leon P. Bignold, Division of Tissue Pathology, Institute of Medical and Veterinary Science, PO Box 14, Rundle Mall, SA 5001, Australia; e-mail: leon.bignold@adelaide.edu.au

Herman H. Cheung, Apoptosis Research Centre, Children's Hospital of Eastern Ontario, Research Institute, 401 Smyth Road, Ottawa, Ontario K1H 8L1, Canada; e-mail: herman@mgcheo.med.uottawa.ca

B. L. D. Coghlan, Centre for European Studies and General Linguistics, University of Adelaide, Adelaide, SA 5005, Australia

William B. Coleman, Department of Pathology and Laboratory Medicine, UNC Lineberger Comprehensive Cancer Center, University of North Carolina School of Medicine, Campus Box 7525, Room 515, Brinkhous-Bullitt Building, Chapel Hill, NC 27599, USA; e-mail: william.coleman@pathology.unc.edu

James R. Davie, Manitoba Institute of Cell Biology, 675 McDermot Avenue, Winnipeg, Manitoba, R3E 0V9 Canada; e-mail: davie@cc.umanitoba.ca

Michael J. Davies, The Heart Research Institute, 145 Missenden Road, Camperdown, Sydney, NSW 2050, Australia; e-mail: m.davies@hri.org.au

Bojan Drobic, Manitoba Institute of Cell Biology, 675 McDermot Avenue, Winnipeg, Manitoba, R3E 0V9 Canada; e-mail: bdrobic@gmail.com

Anette Duensing, Molecular Virology Program, University of Pittsburgh Cancer Institute, Hillman Cancer Center, Research Pavillion, Suite 1.8, 5117 Centre Avenue, Pittsburgh, PA 15213, USA; e-mail: aduensing@pitt.edu

Stefan Duensing, Molecular Virology Program, University of Pittsburgh Cancer Institute, Hillman Cancer Center, Research Pavillion, Suite 1.8, 5117 Centre Avenue, Pittsburgh, PA 15213, USA; e-mail: duensing@pitt.edu

Katherine L. Dunn, Manitoba Institute of Cell Biology, 675 McDermot Avenue, Winnipeg, Manitoba, R3E 0V9 Canada; e-mail: umdunnkl@hotmail.com

Troy R. Durham, Centre for Cellular and Molecular Biology, School of Biological and Chemical Sciences, Deakin University, 221 Burwood Highway, Burwood, Victoria 3125, Australia; e-mail: trd@deakin.edu.au

Assam El-Osta, The Alfred Medical Research and Education Precinct (AMREP), Baker Medical Research Institute, Epigenetics in Human Health and Disease Laboratory, Second Floor, Commercial Road, Prahran, Victoria 3181, Australia; e-mail: assam.el-osta@baker.edu.au

Paula S. Espino, Manitoba Institute of Cell Biology, 675 McDermot Avenue, Winnipeg, Manitoba, R3E 0V9 Canada; e-mail: umespino@cc.umanitoba.ca

Brian J. Estes, Chemical Structure & Dynamics, Pacific Northwest National Laboratory, MS K8-88, 3335 Q Ave, Richland, WA 99354, USA; e-mail: bestes@gonzaga.edu

Jay Harper, Vascular Biology Program, Children's Hospital Boston, Karp Research Building 12.214, 300 Longwood Avenue, Boston, MA 02115, USA; and Department of Surgery, Harvard Medical School, Boston, MA 02115, USA; e-mail: jay.harper@childrens.harvard.edu

Hiroyuki Hayakawa, The Channing Laboratory, Room 861, Brigham and Women's Hospital, 181 Longwood Avenue, Boston, MA 02115, USA; e-mail: hiroyuki_hayakawa@hms.harvard.edu

Okio Hino, Department of Pathology, Juntendo University School of Medicine, 2-1-1 Hongo, Bunkyo-ku, Tokyo 113-8421, Japan; e-mail: ohino@med.juntendo.ac.jp

H.P.A. Jersmann, Department of Medicine, University of Adelaide, Adelaide, SA 5005, Australia; e-mail: hubertus.jersmann@adelaide.edu.au

Toshiyuki Kobayashi, Department of Pathology, Juntendo University School of Medicine, 2-1-1 Hongo, Bunkyo-ku, Tokyo 113-8421, Japan; e-mail: koba396@msd.junterdo.ac.jp

Robert G. Korneluk, Apoptosis Research Centre, Children's Hospital of Eastern Ontario, Research Institute, 401 Smyth Road, Ottawa, Ontario K1H 8L1, Canada; e-mail: bob@mgcheo.med.uottawa.ca

Andreas Luch, Massachusetts Institute of Technology, Center for Cancer Research, 77 Massachusetts Avenue, E17-132, Cambridge, MA 02139, USA; e-mail: luch@mit.edu

Nadja V. Melquiot, The Channing Laboratory, Room 861, Brigham and Women's Hospital, 181 Longwood Avenue, Boston, MA 02115, USA; e-mail: nmelquiot@rics.bwh.harvard.edu

William F. Morgan, Radiation Oncology Research Laboratory, BRB 7-011, and Greenebaum Cancer Center, University of Maryland, Baltimore, Baltimore, MD 21201-1551, USA; e-mail: wfmorgan@som.umaryland.edu

Marsha A. Moses, Vascular Biology Program, Children's Hospital Boston, Karp Research Building 12.214, 300 Longwood Avenue, Boston, MA 02115, USA; and Department of Surgery, Harvard Medical School, Boston, MA 02115, USA; e-mail: marsha.moses@childrens.harvard.edu

Carmel Mothersill, Medical Physics and Applied Radiation Sciences Unit, McMaster University, Hamilton, Ontario L8S 4K1, Canada; e-mail:

mothers@mcmaster.ca
Karl Münger, The Channing Laboratory, Room 861, Brigham and Women's Hospital, 181 Longwood Avenue, Boston, MA 02115, USA; e-mail: kmunger@rics.bwh.harvard.edu
Asha S. Multani, Department of Molecular Genetics, Unit # 011, The University of Texas M. D. Anderson Cancer Center, 1515 Holcombe Blvd., Houston, TX 77030, USA
Christine L. Nguyen, The Channing Laboratory, Room 861, Brigham and Women's Hospital, 181 Longwood Avenue, Boston, MA 02115, USA; e-mail: christine_nguyen@student.hms.harvard.edu
Kazuo Okimoto, Toxicology Group, Safety Research Laboratories, Dainippon, Pharmaceutical Co., Ltd., 33-94, Enoki-cho, Suita, Osaka 564-0053, Japan; e-mail: kazuo_okimoto@dainippon-pharm.co.jp
Sen Pathak, Department of Molecular Genetics, Unit # 011, The University of Texas M. D. Anderson Cancer Center, 1515 Holcombe Blvd., Houston, TX 77030, USA; e-mail: spathak@mail.mdanderson.org
David I. Pattison, The Heart Research Institute, 145 Missenden Road, Camperdown, Sydney, NSW 2050, Australia; e-mail: d.pattison@hri.org.au
Stavroula Raptis, Samuel Lunenfeld Research Institute, Department of Pathology and Laboratory Medicine, Mount Sinai Hospital, 600 University Avenue, 9th Floor, Room 992B, Toronto, Ontario M5G 1X5, Canada; and Department of Laboratory Medicine and Pathobiology, University of Toronto, Toronto, ON, M5S 1A1, Canada; e-mail: raptis@mshri.on.ca
Marianne Sowa, Chemical Structure & Dynamics, Pacific Northwest National Laboratory, MS K8-88, 3335 Q Ave, Richland, WA 99354, USA; e-mail: marianne.sowa@pnl.gov
C. B. Seymour, Medical Physics and Applied Radiation Sciences Unit, McMaster University, Hamilton, Ontario L8S 4K1, Canada
Elizabeth T. Snow, Centre for Cellular and Molecular Biology, School of Biological and Chemical Sciences, Deakin University, 221 Burwood Highway, Burwood, Victoria 3125, Australia; e-mail: esnow@deakin.edu.au
Gregory J. Tsongalis, Laboratory of Molecular Pathology, Department of Pathology, Dartmouth-Hitchcock Medical Center, Lebanon, NH 03756, USA

Preface

This volume began with an invitation from the publishers to edit a volume of EXS on Cancer. This invitation undoubtedly derived from my articles in Cellular and Molecular Life Sciences in 2002 and 2003 on the relationships between the morphology, aetiology and pathogenesis of tumours, especially in relation to genetic instability. After many years of teaching the theories of cancer in undergraduate medical school courses, it seemed to me that the variably chaotic histopathologic features of tumours parallel in some way, the variably unstable genomes of tumour cells, which were being discovered in the 1990s. Thus the title of the volume has come to include morphology, carcinogenesis and genetic instability.

The invitation came while I was working with Herrn Dr. med. Hubertus Jersmann (MD Düsseldorf, PhD, now Senior Lecturer in Medicine of the University of Adelaide) and Professor Brian Coghlan (Emeritus Professor of German, the University of Adelaide), on the work of the nineteenth century cancer pathologists, especially David Paul von Hansemann (1858–1920). With the delivery of the manuscripts from the authors of the chapters, it became obvious that a background chapter for the volume could include some of the material which we had "uncovered" together. Because of this, chapter 1 is authored by the three of us, and the "new" material figures prominently.

I am extremely grateful to the contributors of the chapters, without whom, of course there could be no volume at all, and to Dr. Beatrice Menz, the supervising editor at Birkhäuser, who has been unfailingly helpful and patient.

The staff of the Barr-Smith Library at the University of Adelaide were very helpful throughout and Mr. Peter Dent of the Photography Department of the Institute of Medical and Veterinary Science, Adelaide, assisted with the design of the cover.

Leon P. Bignold March 2005

Cancer morphology, carcinogenesis and genetic instability: a background

Leon P. Bignold[1], B. L. D. Coghlan[2] and H. P. A. Jersmann[3]

[1] *Division of Tissue Pathology, Institute of Medical and Veterinary Science, PO Box 14, Rundle Mall, SA 5001, Australia*
[2] *Centre for European Studies and General Linguistics, University of Adelaide, Adelaide, SA 5005, Australia*
[3] *Department of Medicine, University of Adelaide, Adelaide, SA 5005, Australia*

Summary. Morphological abnormalities of both the nuclei and the cell bodies of tumour cells were described by Müller in the late 1830s. Abnormalities of mitoses and chromosomes in tumour cells were described in the late 1880s. Von Hansemann, in the 1890s, suggested that tumour cells develop from normal cells because of a tendency to mal-distribution and other changes of chromosomes occurring during mitosis. In the first decades of the 20th century, Mendelian genetics and "gene mapping" of chromosomes were established, and the dominant or recessive bases of the familial predispositions to certain tumour types were recognised. In the same period, the carcinogenic effects of ionising radiations, of certain chemicals and of particular viruses were described. A well-developed "somatic gene-mutational theory" of tumours was postulated by Bauer in 1928. In support of this, in the next three decades, many environmental agents were found to cause mitotic and chromosomal abnormalities in normal cells as well as mutations in germ-line cells of experimental animals. Nevertheless, mitotic, chromosomal, and other mutational theories were not popular explanations of tumour pathogenesis in the first half of the 20th century. Only in the 1960s did somatic mutational mechanisms come to dominate theories of tumour formation, especially as a result of the discoveries of the reactivity of carcinogens with DNA, and that the mutation responsible for xeroderma pigmentosum causes loss of function of a gene involved in the repair of DNA after damage by ultraviolet light (Cleaver in 1968). To explain the complexity of tumourous phenomena, "multi-hit" models gained popularity over "single-hit" models of somatic mutation, and "epigenetic" mechanisms of gene regulation began to be studied in tumour cells. More recently, the documentation of much larger-than-expected numbers of genomic events in tumour cells (by Stoler and co-workers, in 1999) has raised the issue of somatic genetic instability in tumour cells, a field which was pioneered in the 1970s mainly by Loeb. Here these discoveries are traced, beginning with "nuclear instability" though mitotic-and-chromosomal theories, single somatic mutation theories, "multi-hit" somatic theories, "somatic, non-chromosomal, genetic instability" and epigenetic mechanisms in tumour cells as a background to the chapters which follow.

Key words: Cancer, carcinogenesis, chromosomes, genetic instability, historical, nuclei.

Introduction

There are several excellent histories of the study of cell biology and of tumours [1–14], which give coverage of most of the various aspects of cells and this disease process. However, no perfectly satisfactory account of the history of investigations of the relationships between the morphological features and aetiological factors of tumours is available, especially in terms of the genetic theories of the pathogenesis of tumour formation. Essentially, the history of the investigation of mutations in tumour cells is characterised by an early attempt by von Hansemann to provide a theory that embraced morphological abnormalities and chromosomal changes. This theory was rejected and largely forgotten, but was followed, over a period of approximately 70 years, by slow recognition, first of single mutations in tumours, then of multiple mutations in tumours. Recently, using molecular methods based on the polymerase chain reaction, so many mutations in tumour cells have been demonstrated that only acquired somatic non-mitotic, non-chromosomal genetic instability, together with alterations of gene expression, appears to provide a likely explanation. This chapter sketches the major milestones of tumour genetics, with detail given mainly when it has not previously been published in English.

The nature of the essential abnormality of cancerous tissues has been discussed since the disease was recognised as a process separate from inflammatory disorders, by the Hippocratic School in the 6th century BC. However, studies of tumours according only to clinical features and macroscopic appearances, by Greek, Roman, Arab, and medieval and Renaissance Europeans led to concepts and schemes of classification of tumours that were largely arbitrary and unhelpful [2, 8, 9, 12].

Beginning in the 16th century, various forms of microscopes were developed in Europe. The compound form is said to have been invented by the Janssen brothers in about 1590 [15, 16]. Early compound microscopes suffered especially from poorly made glass, as well as chromatic and spherical aberration, and were not necessarily more useful than the simple microscopes of the period, for example, those of Leewenhoek (1632–1723) [15, 16]. The progress of microscopic discoveries in the 17th and 18th centuries was achieved mainly through gradual improvements in the making of the glass for lenses. Thus, early in the period, lacteal and lymphatic vessels were described and only later were red cells in blood and "globules" in lymph, as well as fibres in many tissues discovered [4]. One of the last, but important discovery using a simple microscope with a non-achromatic lens (see below) was that of the nucleus by Robert Brown in the early 1830s [17]. Regarding tumours, one view in this period was that tumours derive from particular ("plastic") types of lymph, perhaps through a process of coagulation [8, 12, 18].

Achromatic lenses and the work of Johannes Müller

The development of cell theory probably could not have occurred without the development of better microscopes. These were developed concurrently by microscope makers in England, France, Germany and Italy [15, 16]. Attempts to make lenses with reduced chromatic aberration by combining glass of the crown and flint types began in the 18th century [16]. However, only in the mid-1830s, and because of the theoretical advances in optics provided by J.J. Lister

(1769–1869, father of Lord Lister [13]) in 1830, were useful achromatic lenses manufactured [16]. Microscopes with these lenses could magnify 500 times, and a wealth of scientific discovery followed. Within a few years, it was appreciated that cells are the basic living units of the body, and which, by multiplying and secreting extracellular materials, form all the tissues, i.e. the "Cell theory". Wolff [2] gives priority to Raspail (a French histochemist, 1794–1878 [13]), both for this idea, and for being first to use the word "cell" for the microscopic structures now recognised as such. The same idea, however, was propounded in detail by Schleiden (1804–1881) (for plant cells) in 1838 [19] and by his friend Schwann (1810–1882) (who gave generous credit to Schleiden) for animal cells in 1839 [20].

Schwann did this work while he was an assistant to Johannes Müller (1801–1858), Professor of Anatomy and Physiology at the University of Berlin, who was perhaps the most remarkable scientific teacher in modern history. This is because his students included not only Schwann, His (1811–1887), Henle (1809–1885) and Kölliker (1817–1905) [13], but also Virchow (1821–1902), who was the major proponent of cellular pathology (see below), Helmholtz (1821–1894), who propounded the law of conservation of energy as well as contributing to optics and acoustics, and Wilhelm Wundt (1832–1920), who founded experimental psychology. Another of Müller's students was Brücke (1819–1892), who in turn was a major influence on Sigmund Freud (1856–1939) [21].

Although Müller's research interests were mainly in neurophysiology, in 1838 he published [22] studies of the cells and their nuclei in tumours using a Schiek microscope of the latest type. Müller [22] documented the variable sizes and shapes of cells and nuclei, not only between tumour types, but also between cases of the same tumour type, and also among the cells of individual tumours. In so far as nuclei were understood to probably contain at least some of the hereditary material of cells [23], the finding of variability of nuclear morphology in cells might be considered the original observation of a form of hereditary/genetic instability in tumour cells, even if it was not appreciated as such at the time.

The beginnings of histology, cell biology and the cellular pathology of tumours

The improved techniques for microscopy (above) rapidly resulted in understanding of the general structure of animal and plant tissues, as well as their embryological development, in the current senses [4, 9–12]. These basic observations identified several major issues which were to dominate the study of cell biology for the next 40 years. The views of Virchow, who was Professor of Pathology at Berlin 1856–1902, are given prominence in the following discussion, because he was the most prominent pathologist of the 19th century, and strongly influenced the entire discipline of pathology in that era.

1. How do cells arise? Schwann [20], Müller [22] and Kölliker [24] thought that cells could arise spontaneously in some specific (but invisible) type of interstitial fluid, which they referred to as "blastema". Müller in particular thought that the process involved first crystallisation of nuclear material in the "blastema" and, second, the aggregation of cytoplasm around the crystals [22]. Müller thought cancer cells arose from particular cancerous invisible particles, which he called *"seminum morbii"* [22]. The alternative view, being that cells must always develop from pre-existing cells was espoused by Raspail, Remak (1815–1865) and, most famously, Virchow. For accounts of this well-documented controversy, which lasted into the late 19th century, see especially [5, 8–12] and references therein.
2. How do cells and nuclei divide to generally provide for daughter cells, which are the same as each other and the mother cell? Even with the best achromatic lenses, it was not possible in histological preparations to see any more detail of nuclear division than a division of nuclear material into two parts, followed by division of the cell, and this matter was not resolved in this period (see below).
3. To what degree do the embryological origins of the cells determine their ultimate morphology? Histological studies of embryos had led to suggestions, particularly by Remak and His [1, 3, 4, 8–12], that there are two or three basic embryological layers, from which adult cell types derive. Tracing the developmental "pedigrees" of cell types in adults was a major activity in embryology until the 20th century [1, 3]. A part of the stimulus for these studies was the urge to further investigate earlier observations of Caspar Wolff (1733–1794), von Baer (1792–1876) and Oken (1779–1851) [3, 4, 25, 26], who had shown that the phases of embryonic development which occur in one species resemble, albeit temporarily, the adult forms of "simpler" species that are "lower" or "earlier" in the evolutionary tree. The results of these histological studies generally supported the earlier observations, and were popularised as the saying "Ontogeny recapitulates Phylogeny", especially by Haeckel (1834–1919) [4, 26–28]. These studies formed part of the well-documented struggle (broadly from 1858–1940) to establish scientific evidence for and against Darwin's theory of evolution [29–33].
4. Can cells undergo changes of mature morphological type in adults? Or are they always faithful to their original lineage? Virchow pointed out that chondrocytes and osteocytes associated with the callus of healing of fractures of bone arise from basic connective or "supportive" tissue [34] (*Bindegewebe*: for a useful note on the English translation of this word, see Translator's Notes in [25]). For these changes of cell type, Virchow used the phrase "histological substitution" in the second edition of his "Cellular pathology" (1858) [34], but used "metaplasia" for the same process in the fourth edition of "Cellular pathology" in 1871 [35] and in a later article in 1884 [32] (see also [36]). However, the same authors who supported fixed continuities of embryological layer to adult cell type also tended to support a view of fixity of cell type in adult tissues (see above).

5. As an extension of the fourth issue above, can cells of tumours come from a single "tumour precursor cell" in connective tissue or from the more specialised cells of each type? Again, Virchow insisted that interchanges of cell types are common phenomena and that cancers do not come from epithelial cells, but rather from particular cells of the *Bindegewebe* [34, 35]. Eventually, the opposite view, especially in regards to the derivation of carcinomas from epithelial cells only, came to be most widely held, due especially to the work of Waldeyer (1836–1921) and Thiersch (1822–1895) [8–12]. The present view is that some, but not all, cell types or their local "stem cells" retain some ability to adopt different directions of histological differentiation under particular circumstances [37].
6. What is the stimulus to the excessive growth of tumours? By the mid-19th century, ideas of "blastemas" and "plastic lymph" had been abandoned, and Virchow's [34] suggestion that a chronic local "irritation" must be the first step of tumour formation was popular. Thiersch, in 1865 suggested that a local over-nutrition of tissues might cause excessive growth [2], and Bol, in 1876 proposed that tumour cells derive their growth in some way from some influence of embryonal-like mesenchymal cells at the site of tumour formation [2]. Cohnheim's idea [38], published in 1882, was that the cells of tumours are essentially embryonal in nature, being "left-over" embryonal cells. In this way, carcinomas are epithelial because they arise from "left-over" embryonal epithelial cells. To account for the sudden activation of these "dormant" embryonal cells, Cohnheim suggested a mechanism of local hyper-nutrition [39]. Ribbert (1855–1920) initially invoked embryonal rests as the source of tumour cells [2], but later thought that normal adult cells might be stimulated to grow entirely because of a loss of normal local inhibitory "tissue tension" [39] (reviewed in English [2, 40, 41]).
7. What is the mechanism of the nuclear pleomorphism to tumours? None of the various mechanisms of tumour formation (above), however, provide an explanation for nuclear pleomorphism in tumour cells. Perhaps stimulated in part by this consideration, much effort was expended in the late 19th century on the (intranuclear-) parasitic theories, for which the best evidence obtained was that the intranuclear irregularities of tumour cells resemble either the bodies of, or the effects of, parasites. (Wolff [2] devotes 150 pages to these theories. Shimkin [8] gives a useful table.) These were very popular up to the early 20th century and were mentioned by Ewing as late as 1940 [42], although the evidence in their favour was entirely morphological, and conversion of normal cells to cancer cells by the intranuclear "cancer parasites" was not achieved.
8. In the most general sense, what is the relationship of disease processes to normal physiological processes? The Ancient Greeks and Romans, especially Galen, held that all diseases, excepting trauma and parasitic disorders, are due to "imbalances of (normal physiological) humours" and hence are endogenous in origin [2, 5, 8–12]. Müller [23] held that diseases are "abnormal physiologies" and Virchow repeatedly and strongly stated the same view [9, 43, 44]. To give just two quotations, in 1855 [45], Virchow stated "All pathological for-

mations are either degenerations, transformations, or repetitions of typical physiological structures." In 1877 [46], Virchow repeated the same opinion "... I must strongly emphasise that pathological formations never develop beyond the physiological possibilities of the species." One object of Virchow's position was to distance himself from all non-mechanistic ideas in pathology, such as the involvement of "vital energies" and "special life forces" [43]. An effect of his views, however, may have been to encourage his students to find physiological processes that paralleled any new phenomenon of disease they might wish to describe (see below).

Mitosis, chromosomes and Von Hansemann's theory of cancer

In the next period, however, microscopical techniques were introduced, which allowed the discoveries that form the basis of our much of our current histological and pathological understanding. From the 1870s, a variety of non-optical improvements were made to histological techniques, including new fixatives, paraffin embedding, better section cutting and better stains (aniline dyes, followed by haematoxylin and eosin) [15, 16]. In addition, "optically homogenous" oil immersion, with appropriate new types of lenses, and achromatic substage light condensers were introduced by Abbe (1840–1905) in the late 1870s at the Zeiss factory in Jena [16]. In the 1880s, Abbe invented apochromatic ("away from colour") lenses, which were composed of glass with various novel additives, especially borate. These lenses were released commercially in three series, in 1886, 1888 and 1894 [16], so that optical resolutions (0.25 μm), close to the highest that can be achieved using visible light, were obtained.

These improved methods were applied to the events of cell division, as outlined in [3, 4, 9–14], and the condensation of chromatin into "threads" (also termed "loops", "filaments" and later "chromosomes", or "nuclear segments" [1]) prior to nuclear division was discovered by many authors, including Strassburger, Waldeyer, Flemming, Boveri, van Beneden and others [1, 3, 4, 14].

Von Hansemann's first paper on cancer (1890)

In relation to tumours, several authors (references in [47]) commented on the abnormalities of chromosomes of tumour cells, but only Klebs [48] considered that these might have any pathogenetic significance for tumour formation. Most notably, however, the topic was taken up by von Hansemann (1858–1920) in 1889, who had graduated in Medicine only a few years before (*Staatsexamen* 1886) and was, at the time, the junior (third) *Assistent* in Pathology to Virchow in Berlin. Von Hansemann used all of the new techniques, and produced a remarkable synthesis of the cell biological principles of the time and his own observations to create the first chromosomal theory of tumour formation. The general outline of his concepts appeared in his first paper in 1890 [47], while later articles and two books [49, 50] contained extensions and modifications of his ideas, as well as responses to the frequently negative comments published by other authors.

The first paper (1890) [47] is difficult to understand for two reasons. First, von Hansemann probably felt that, before he could elaborate a notion of cancer, he had to describe a normal biological process which, when mildly abnormal, would produce appearances resembling those of cancer (perhaps under the influence of Virchow, see above). Thus, it was probably not enough, in 1890, for von Hansemann to observe that, if chromosomes carry the genetic material of the cell, and are abnormal in cancer cells, then the abnormal chromosomes are probably the cause of the abnormalities of cancer cells. Put another way, von Hansemann possibly had to satisfy Virchow's somewhat abstract notions of disease pathogenesis (see above) and find a whole analogous system of biological process, which in some way resembled many if not most of the tumourous phenomena. Second, at the time von Hansemann wrote the paper, the differences between the chromosomal replications and divisions in meiosis and mitosis were not recognised, nor were the numbers of chromosomes in human adult cells or gametes known. Furthermore, the individuality of chromosomes was only one theory among many at the time, and "genes" and "gene maps" lay in the future.

At the beginning of this paper [47], von Hansemann discussed ideas of the variability of amounts of chromatin in cells generally. He then noted that injection of the chromatin of a sperm is an important aspect of fertilisation of the egg, and that the amount of chromatin increases and decreases in cells associated with spermatogenesis in testicular tissue. Von Hansemann then observed that in tumour tissue, increases and decreases of chromatin in tumour cells occurs, and that the smallest nuclei appear to become degenerate. This last process that von Hansemann observed, seems to be in some way analogous to the expulsion of the polar bodies (referring to them as did Hertwig [25] as *Richtungskörperchen*, which later came to be used for "centrioles") from the developing egg in the ovary. Next, von Hansemann discussed asymmetric distribution of chromosomes in mitosis as the main mechanism of the formation of small nuclei in some detail. In the next section of the paper, he reviewed theories of cell heredity as they were known in 1890, mentioning especially the ideas of Weismann and Naegeli, in relation to "quantitative" and "qualitative" asymmetries of cell division (not nuclear division) during formation of the blastula. This discussion led to consideration of the progress of differentiation of the cells in the early embryo, with two "pivotal" statements being made, to try to link differentiation, autonomy and growth. They were:

1. "With every further qualitative work division, the cells lose the capability to exist autonomously."
2. "With every new generational phase, a changed growth energy ["nutritional, formative, and functional activity" (Virchow)] takes place which often

manifests itself in a change in direction of growth." (The original was in bold emphasis).

Von Hansemann then went on to discuss the capacities for differentiation of adult cells, as demonstrated by the results of transplantation experiments, before returning to oogenesis. In this part, he used the view of Weismann (references in the paper) that the process of development of the ripe egg in the ovary is one of transition from a cell which is of a differentiated (germinal epithelial) type, to one which has no capacity to subsequently "differentiate" (of course unless fertilised), and thus should be considered as completely undifferentiated. Von Hansemann therefore described the change which the ripening egg has undergone as *Entdifferenzirung*, which is best translated as "dedifferentiation", in the sense of a cell having been differentiated, but being no longer so. (In German, *ent-* is usually used for "the condition following the removal of something", for example when a church is deconsecrated, as opposed to unconsecrated.) Von Hansemann then named the separate phenomenon of the process of differentiation of the egg after fertilisation (i.e. all embyronic development) "prosoplasia" and named the dedifferentiation of the ovarian germinal epithelium to the ripe egg, "anaplasia".

In the remainder of the paper, von Hansemann discussed further the functions and "differentiation" of cells, and the role of changes of chromatin. In one passage, von Hansemann justified all of this background material with the words (which possibly relates to the Virchovian position of "every disease process is an abnormal physiological process', see above):

"Touching on this theme may be justified by the fact that, in so far as I have wished to draw any conclusions from my observations on epithelial cancers, I had to take a position *vis à vis* a sequence of biological questions."

He then cited transplantation experiments with embryonic tissues, to emphasise that tumour cells are not only different to embryonal ones (and thus in opposition to the "embryonal rest" theories of Cohnheim, see above) but also possibly have an egg-formative character. From all of this, von Hansemann's concluding sentence in the 1890 article [47] can be comprehended:

"Thus, as far as anaplastic cells are concerned, they must not be confused with embryonal ones, in fact, there is a clear contrast between the two, and the embryonal cells begin where the anaplasia ends, with the egg."

Von Hansemann's later works

Shortly after von Hansemann's article [43], the true nature of the chromatin ejected from the ripening egg (surplus haploid nuclei) was recognised [3, 4, 51]. Also, the model of differentiation involving quantitative changes of chromatin was abandoned [3, 52], Von Hansemann in later publications [49, 50] gradually abandoned the "egg-formative" "physiological prototype process" part of his theory, but retained the essence of the chromosomal mechanism of tumour formation. He described the phenomena of progressive disturbance of mitoses in tumour cells and of increasingly abnormal morphology and characteristics of chromosomes, including their "lysis", "stickiness" and other changes, in tumour cell nuclei in association with increasing clinical aggressiveness [49, 50]. This cellular process of mitotic and chromosomal instability is currently often referred to as "clastogenesis" [53, 54] and the associated behavioural process as "tumour progression" [55, 56]. He also discussed the relationships of the chromosomes to cell function, and provided numerous other, and still-relevant, insights into the relationship between the morphology of tumours and their pathogenesis.

Contemporaneous responses to von Hansemann's theories; Boveri's theory

At the time, however, his ideas were rejected. Most authors, for example R.C. Whitman [57] (at the University of Colorado, not to be confused with C.H. Whitman, Director of the Woods Hole Marine Biological Institute, Maine) did not understand von Hansemann's ideas, or confused them with the concept of "backward reversal" of embryonic differentiation (as originated by Bol in 1876, see above, and [58]), or rejected them out of hand as impossible [39].

Boveri (1862–1915), a biologist who had previously made great contributions to the understanding of mitosis, published a volume on the origin of tumours in 1914 [59]. In this book, Boveri mainly suggested that quadipolar mitosis might be a significant mechanism in the induction of tumourous behaviour in cells. His theory was poorly considered in terms of pathology, and appears to owe more to von Hansemann than he (Boveri) admitted. Thus, von Hansemann is mentioned by name (pages, 6, 24, 67, and 108 of the English translation of Boveri's volume [59]) but never by citation of Hansemann's articles or books. Moreover, Boveri wrote (pp 23–24 [57]) "The cell of a malignant tumour is accordingly (and here I take up again the idea of Hansemann) a cell with a definite (*sic*) abnormal chromatin complex." However, on p 111 Boveri [59] stated "The essence of my theory, is not abnormal mitosis, but in general, a definite (*sic*) *abnormal chromosome complex* (original italics)."

Because von Hansemann had already, by 1914, described a variety of non-mitotic abnormalities of chromosomes in tumour cells (see above), Hansemann's priority seems to have been overlooked. Boveri [59] seems to have used ideas which were very similar to those of von Hansemann, except for specific reference to quadripolar mitosis.

"Dedifferentiation" and "anaplasia" remain in use

Von Hansemann's terminology of "dedifferentiation" and "anaplasia" were extremely successful, and remain in use today. This is because the classification of tumours prior to 1890, was of "homologous" (like the adjacent normal tissue) and "heterologous" (unlike the adjacent normal tissue). This classification, which dated from Laennec in 1804 [2] and was used by Virchow [34] and others, did not permit any intermediary types, while von Hansemann's concept (dedifferentiation/anaplasia) was of a process which could occur in grades and degrees [2]. Von Hansemann's views were based on actual histopathological phenomena, which were being more and more widely documented in diagnostic histopathology throughout the world from the 1890s onwards, using the new techniques and microscope lenses (see above). "Dedifferentiation" and "anaplasia" entered the medical lexicon, where they remain firmly to this day.

Reappraisal

Perhaps correctly, von Hansemann's notion of "anaplasia" (with its component of either the correct "dedifferentiation" or the incorrect "undifferentiation" concept of the cell) has been discounted. However, from the perspective of the 21st century, we can see that his basic idea of chromosomal disorder as the basis of tumour formation may well be valid, and current aspects of chromatin and chromosomes in cancer are the subjects of chapters 2 and 3 in this volume.

On the basis of all of this, it would appear that von Hansemann may deserve more recognition as a contributor to genetic theories of tumours and oncology generally, than he is currently awarded.

Early 20th century studies of carcinogenesis in relation to the cell biology of cancer

Hereditary factors and Mendelian genetics in tumours

In the 18th and early 19th century, many authors, including John Hunter [18] considered that families can inherit predispositions to cancers, generally in keeping with "humoural/diathesis" concepts of disease [60, 61]. Detailed studies of families to test this were undertaken and continued into the 20th century, for example, by Warthin [62]. A few familial predispositions, however, were known to be to tumours of one type only. For example, von Recklinghausen's neurofibromatosis was known to be a familial disorder as early as the 1880s [63]. Only after the application of Mendelian genetics to human diseases was the nature of these predispositions established. Thus, familial polyposis coli was found to be autosomal dominant, and xeroderma pigmentosum was shown to be autosomal recessive in their respective genetic transmissions by the 1920s [60, 61].

At the beginning of the 20th century, inherited predispositions to tumours were investigated experimentally. Numerous breeding programmes of experimental animals were conducted, especially in the UK and the USA [8], and it was found that, indeed, the inbred offspring of animals with certain tumour types were more liable to the same tumour, but not to tumours generally. Maud Slye (1879–1954) thought that the results of such studies showed that human tumours are of a "recessive" type, but this was not supported by other workers, notably Little (1888–1971) [8, 64].

Another line of investigation was the transplantability of experimental tumours between members of the same species, and across species [8, 65–69]. Initially, transplantation experiments were conducted in the investigation of infectious theories of cancer according to the "Koch's postulates" used for infectious diseases. No transfers of disease to normal recipient cells by tumour tissue occurred. Subsequently, transplantation of tumours was used to study the hereditary factors associated with the susceptibility of the recipient animals to tumour "take", and claims were made that this was dominantly inherited [66]. Later, the effects of immunological reactions to these transplants were recognised, and it transpired that most reactions appeared to be due to the recipients' reactions to the species-related antigens of the donor, rather than any reactions to tumour-specific antigens [67]. Other studies were directed at factors associated with metastasis, which was similar to von Hansemann's feature of "capacity of the tumour cells for independent existence", or "autonomy" (see above). Leo Loeb [68] in 1937 considered that the major determinants of growth of transplanted tumours include immune reactions of the host, but also that the "growth energy" or "growth momentum" of the transplanted tumour tissue is important. Because growth rate of tumours and degree of dedifferentiation are often related, and rapidly growing tumours may access host blood vessels faster than the host tissues can react with fibrosis, this may be an adequate explanation of tumourous "greater capacity for independent existence" (see above).

Despite this unsatisfactory situation concerning the actual significance of heterotypic survival, these transplants of tumours provided useful models of cancer for the study of various aspects of cancer, not the least of which was anti-cancer therapies. The distinction between degree of "autonomy" and "susceptible to immunological rejection by the recipient animal" could not be made easily until the advent of the nude mouse in the 1980s [70].

Chemical agents

Although workers in certain occupations had been known to be susceptible to cancers in the 18th century, the chemical or physical basis of these were not widely considered. This may have been due in part to the fact that these diseases were still considered to be due to some generalised imbalances of humours, and thus direct action of these agents on cells were perhaps not understood to be relevant. Snuff cancer was described in 1761 by John Hill, chimney-sweep's cancer 1775 by Percival Pott and pipe smoker's cancer (of the lower lip) by Soemmering in 1795 [8]. Arsenical compounds were described as causing skin cancers in animals in 1822 and in humans in 1888 [8, 12]. Tar and paraffin cancers were described by von Volkmann in 1875 [2, 5], and mine worker lung cancers were recognised in 1879 [8].

In 1895, aniline dyes were found to cause urinary tract cancers [8]. In the early 20th century, coal tar was proved to be carcinogenic in rabbit skin by Ichikawa and Yamaguchi [8, 12], although Hannau had failed to produce such lesions by repeated application of coal to the scrota of dogs in an experiment in the 1880s [2].

In 1930, the first pure carcinogenic hydrocarbon was isolated from coal tar by Kennaway and his group [71], allowing for detailed studies of the biological effects of these compounds, with so-far-unsuccessful attempts to establish relationships between their chemical structure and carcinogenic potential [72, 73]. The ability of some chemical carcinogens to cause germ-line mutations in experimental animals was shown in the late 1920s and 1930s [6].

Meanwhile, the number of categories of known carcinogens has expanded to include aromatic amines, nitrosamines and alkylating agents [8, 74]. In the 1960s, it was also found that chemical carcinogens can cause strand breaks in DNA in cells [8]. Some problematic inconsistencies between the chemical activities, including degrees of DNA "adduct" formation and the carcinogenic potencies of various chemical agents were documented early in these studies (for a recent discussion see [75]).

Aspects of chemical carcinogenesis are the subject of chapters 4 and 5 of this volume.

Physical agents

Ultraviolet light was discovered in 1801 by Rittner, who noted the ability of a component of sunlight beyond violet light to darken silver chloride [76]. Sunlight was suggested to be the cause of sailor's cancers by Unna in 1894 and ultraviolet light was shown to be able to cause skin cancers in white mice in 1928 [8]. Chapter 6 of this volume deals with current issues in ultraviolet carcinogenesis.

X-rays were discovered in 1895 by Roentgen (1845–1923) [8, 12], and uranium salts were shown to emit gamma rays in 1896 by Becquerel (1852–1908) [8, 12]. The former were understood to cause skin cancers as early as 1902 by Freiben [8, 12] and isotopes taken internally were reported to cause bone cancers in 1925 [8]. Experimental induction of germ-line mutation in *Drosophila* by X-rays was demonstrated in 1928 by H.J. Muller [6], and it was established in the 1930s that irradiation causes chromosomal lesions in cell cultures *in vitro* [77]. Chapters 7 and 11 deal with current aspects of radiation-induced carcinogenesis.

Infectious agents

Numerous parasitic theories of neoplasia were proposed from the 19th century on the basis of structures suggested to be these parasites in the cytoplasm and nuclei of tumour cells (see above). The association of bilharzia and bladder cancer was suggested as early as 1889 [8].

Peyton Roux, in 1911, reported that a tumour of fowls could be due to a transmissible, filterable agent [8, 12], and subsequently Shope reported that a filterable agent could transmit papillomata of the skin of rabbits [78]. Subsequent developments in the field showed that oncogenic viruses may be of either RNA or DNA type [79]. Current aspects of viral oncogenesis in relation to the host genome are discussed in chapter 8 of this volume.

Tissue processes as "targets" in carcinogenesis

While the focus of this chapter has so far been on individual cell, there remain the problems of intra- and intercellular controls of tumour cell behaviour, and the overall concept of cancer as a disorder of a single fundamental biological process. The concept of tumours arising by a disturbance of a normal tissue process has been popular since Virchow in the middle of the 19th century (see above).

Abnormal hyperplasia has been recognised as a frequent preliminary morphological change in tumours since Virchow (see above), and is well documented, for example, in tumours of the human endometrium [80], in experimentally induced lesions of the skin [81] and in the breast of mice [82]. One of the major features of tar-induced experimental skin tumours is that a phase of reversible epidermal hyperplasia occurs, as stressed by several authors [82–84].

Abnormal wound healing was proposed as the basic process of cancer by several authors, for example, Haddow [85], and later workers suggested that local hormones (for example, "chalones" [86]) that control the cell proliferation associated with healing may mediate these abnormal responses.

Abnormal "differentiation" as the fundamental process of carcinogenesis continues to be extensively investigated, although the concepts of modern authors are distinct from the proposed "loss of differentiation" (*Entdifferenzierung*) described by von Hansemann (see above). Harris [87] in 1990, reviewing differentiation and tumour formation, noted the "ancient question" of whether a tumour grows rapidly because it does not differentiate, or does not differentiate because it grows rapidly. Harris [87] conceded that this association could arise if a separate cause has both effects, and the present author has shown that, among the various human tumours, examples are to be found in which lack of differentiation and high growth rate are not correlated at all [88]. For some recent reviews of notions of differentiation as a primary event in cancer see [89–91].

The role of the mesenchyme in tumour formation (foreshadowed by Virchow, and espoused by Ribbert, see above) has had several recent supporters, especially those concerned with epithelial-mesenchymal interactions in normal biology [92–94]. Epithelial-connective tissue lineage infidelity is a separate issue, which has been revived (since Virchow) and referred to as "transdifferentiation" [95, 96], epithelial-mesenchymal "plasticity" [97], epithelio-mesenchymal transformation [98] and epithelial-mesenchymal transition [99].

Another cellular process relevant to tumours, which is currently undergoing intensive current study, is apoptosis [100]. This process is a defence of the body against tumours, in that, under normal circumstances, all cells except the permanent stem cells of tissues, ultimately die and in the case of epithelial cells are shed rather than being resorbed. Thus, cells which suffer mutations during transit amplification of epithelia are eliminated by shedding. These issues are discussed in chapter 9 of this volume.

Nineteenth century ideas of the involvement of blood vessels in the pathogenesis of tumours by way of a "nutritive" growth-stimulating effect are mentioned in chapter 3. The role of angiogenesis in tumour formation is a new concept, and is discussed, in terms of current investigative techniques, in chapter 10 of this volume

Somatic mutation in tumour cells and the number of mutations per cancer cell

"Single-hit" somatic mutation theories of tumours

Although the idea of alteration of hereditary material of adult cells as the basis of cancer is implicit in Virchow's idea of hybridising of somatic cells [2], in parasitic theories [2, 8], and in von Hansemann's concepts of altered chromosomal composition (see above), in the first half of the 20th century the ideas were not universally accepted in relation either to spontaneous human tumours, or to chemically or physically induced experimental tumours.

Nevertheless, with the rediscovery of Mendel's work in general, interest in the alteration of genetic material re-emerged. De Vries, in his "Mutation theory" (1902) [101], reviewed evidence of alteration of the genetic material in somatic cells of plants, using the term "vegetative mutations" of which "bud variations" comprised one type. The involvement of alterations of somatic cells in this variation was recognised by Bateson [102], and the term "somatic mutation" was used by Tyzzer in 1916 [103], and by Whitman in 1919 [57].

Probably the most extensive early discussion of "mutation" in relation to tumours was that of Bauer [60]. Because no translation of the work of this author has apparently been published, some detail will be given here. Bauer [60] provided a well-constructed volume, in which various biological phenomena, experimental results, and human diseases were considered. Based on the work of germ-line mutation in *Drosophila* by Morgan and co-workers [104], Bauer observed:

"Considering the problem of tumours, it is of decisive importance whether such mutations, whose occurrence in the germinal cells has been proved a thousand-fold, can also appear in the body cells. In this respect, one can say *a priori* that it would have to appear uncommonly striking if the fundamental biological process of gene alteration were possible only in the chromosomes of the primordial germ cells (*Urkeimzellen*) and impossible in all the other cells."

Bauer [60] went on to discuss the botanical evidence in favour of somatic mutation as a possible phenomenon, before discussing the possibility that human developmental lesions that occur singly as an uninherited condition (such as isolated exostosis) could be due to somatic mutations during embryonic, foetal or histological development. This, he pointed out, could be particularly true if multiple lesions of the same type occur as an inherited condition (e.g. inherited multiple exostoses). Let Bauer [60] speak for himself:

"In contrast to these systematised forms of disease, there are also locally limited forms of the same fundamental disturbance, e.g. partial albinism of one eye, solitary exostosis, individual bone cysts, unilateral cystic kidneys, etc.

These forms, which in contrast to the generalised forms, are locally limited, are characterised by the fact that: (1) they virtually always appear singly, (2) they are never inherited or hereditable, and (3) morphologically, they are essentially identical to the generalised forms.

In all these and similar cases, medicine found itself embarrassed with respect to their aetiological interpretation. We do not wish to make fun of medievally naïve ideas which explained such systemic illnesses, if they appeared in the multiple manner, by general pressure, and if they appeared solitarily, by local pressure of a constricted amniotic cavity, and explained the fact of hereditability of the ailment by the hereditary tendency to constriction of this cavity. We must, however, reiterate that there was no satisfactory interpretation for the facts just given. Mostly, one avoided the difficulty by contenting oneself with the observation that the relevant disease appears now multiple and hereditary, and now solitary.

Thus, in all these malformations, with their identical essence, which sometimes appear generalised, and sometimes are locally limited, we are dealing with the same process as in the bud mutations of a plant. A mutation when it appears in the germinal cells is hereditable, and in the bearers of the mutation, then spreads to the entire tissue system dependent upon it. The same mutation when it appears in the body cells has effects that are locally limited; only in the cells which are descended from the cell first are mutated. Thanks to the change occurring in the same gene, the mutation causes the same morphological picture, but is naturally never hereditable as it, being a mutation of somatic cells, does not belong to the germinal line."

Bauer [60] made numerous other relevant observations, but this early work has been cited infrequently in recent years, and is currently neglected.

idea by statistical evaluation, on the basis of "population genetics" pioneered by R.A. Fisher [58, 136]. Several authors [137, 138] concluded that only a few somatic mutations might be necessary to cause a cell to become tumourous.

Armitage and Doll [139, 140] added tumour "progression" to the initiation/promotion model as a third stage of neoplasia. Progression later came to be considered to be caused by mutations [141] and by mutations arising from genetic instability in particular (see below).

Non-mitotic, non-chromosomal somatic genetic instability in tumours and replicative infidelity of DNA

From the foregoing, it seems that at least three broad types of "genetic instability" have been identified. First "mitotic instability", described essentially by von Hansemann (see above), results in mal-distribution of otherwise normal chromosomes to daughter cells, creating cells with imbalances of chromosomes. Second, there is "chromosomal instability", by which the chromosome structure (see especially [6, 14, 77]) is compromised, so that the chromosomes become more "sticky" or prone to breakage. This tends to be associated with "mitotic instability". A justification for this separation of "mitotic" and "chromosomal" phenomena is the existence of anti-tumour drugs, especially of the vinblastine group, which specifically disrupt the tubulin of the spindle fibres of mitosis [142]. The third type of genetic instability occurs without significant mitotic or chromosomal disturbance, and is associated with a reduction of the ability of a cell to faithfully reproduce its DNA sequence. The original information in favour of this concept (reviewed [143]) developed from the discoveries of bacteria that mutate more quickly ("mutator strains") than wild types. Later, it was found that, from a single tumour, strains of tumour cells could be grown that had different biological features and different karyotypes, indicating tumour "heterogeneity (see [143]). At approximately the same time, it was discovered that the mutant gene responsible for the increased rates of skin cancer among individuals with xeroderma pigmentosum encoded an enzyme associated with repair of DNA [144]. Subsequently, the idea has been supported by the discovery that a variety of inherited predispositions to tumours are associated with mutations of genes responsible for either repair of DNA or preservation of fidelity of replication (during S phase synthesis) of DNA [145]. In the 1990s, the application of methods based on the polymerase chain reaction (PCR, invented by Mullis [146]) led to the eventual quantification of the number of genomic events in the whole genome of carcinoma cells (by Stoler and co-workers in 1999 [147]). This study used directly *ex vivo* cells, and a new method of inter- (simple repeat sequence) PCR. The results [146] showed that the numbers of genomic events exceeded possible aetiological events and also the number of mitotic and chromosomal aberrations. Thus, the action of an acquired, non-mitotic, non-chromosomal somatic genetic instability seems to be strongly supported. The basic ideas of this mode of carcinogenesis have been outlined and investigated particularly by L.A. Loeb since the 1970s (e.g. [148–151], and reviewed in [75]).

Aspects of genetic instability in tumours are discussed in chapters 11, 12 and 13 of this volume.

Abnormal gene expression in cancer

Despite the above ideas concerning the mutational basis of tumours, the possibility that an abnormality of gene regulatory mechanisms might contribute in a fundamental way to tumour cell pathogenesis remains. At present, many gene regulatory mechanisms are known, including chemical alteration of the DNA itself (especially methylation) and proteins and RNAs that act on (local) promoter regions. In addition, there are regulators of translation of RNA and factors that control the cell cycle (i.e. multiplication) of target cells. Aspects of these issues are covered in chapter 14 of this volume.

Conclusions

The history of the relationships of the morphology of tumour cells and their cellular genetics has involved numerous contributions from many apparently separate fields of biology. The broad cellular morphological observations, in terms of the variability of form, function and behaviour, were established in the 19th century due to the work of Müller, Virchow, von Hansemann and others. In particular, mitotic and chromosomal lesions noted by von Hansemann, although virtually ignored at the time, may find support in more recent karyokinetic studies of chronic myelocytic leukaemia, and some sarcomatous conditions. The genetic observations of most recent times, however, have revealed so much previously unsuspected genomic disturbance in tumour cells that some form of non-mitotic, non-chromosomal instability appears to be involved. Because all of these processes involve nuclear structures, and appear to be provokable by carcinogens, this volume has been designed to bring together chapters which deal with many aspects of these studies, and illuminate the latest aspects of these contemporary issues in cancer research.

Acknowledgements
We are indebted to the National Library of Congress (Washington D.C.) for the gift of a copy of reference [84] and to the staffs of the Barr-Smith Library, University of Adelaide, and the Library of the University of Leipzig for assistance. Ms Elizabeth Goodwin typed many of the translations, and help was rendered also by S. Coghlan, A. Coglan and M.-L. Krone.

References

1 Hertwig O (1892) *The Cell. Outlines of General Anatomy and Physiology*. Translation by M. Campbell, 1895. Swan Sonnenschein & Co, London
2 Wolff J (1907) *The Science of Cancerous Diseases from the Earliest Times to the Present.* Translation by B Ayoub, and with an introduction by S Jarcho, 1990, Science History Publications, Sagamore Beach, MA
3 Wilson EB (1924) *The Cell in Development and Heredity*. 3rd edn., Macmillan, New York
4 Nordenskiöld E (1928) *The History of Biology*. Translation by LB Eyre, Tudor Publishing, New York
5 Cameron GR (1952) *Pathology of the Cell.* Oliver and Boyd, Edinburgh
6 Koller PC (1957) The genetic component of cancer. In: RW Raven (ed.): *Cancer, Vol 1.* Butterworth and Co, London, 335–403
7 Allen G (1978) *Life Science in the Twentieth Century*. Cambridge Univ Press, Cambridge
8 Shimkin MB (1979) *Contrary to Nature*. US Dept Health, Education and Welfare, Publication No (NIH) 79–720, Washington DC
9 Rather LJ (1978) *The Genesis of Cancer. A study in the History of Ideas*. The Johns Hopkins University Press, Baltimore
10 Harris HA (1995) *The Cells of The Body: A History of Somatic Cell Genetics*. Cold Spring Harbor Laboratory Press, Cold Spring Harbor, NY
11 Harris HA (1999) *The Birth of The Cell.* Yale University Press, New Haven
12 Fitzgerald PJ (2000) *From Demons and Evil Spirits to Cancer Genes*. Armed Forces Institute of Pathology, Washington DC
13 Dohm G (2001) *Geschichte der Histopathologie/History of Histopathology*. Springer-Verlag, Berlin
14 Lima-de-Faria A (2003) *One Hundred Years of Chromosome Research, and What Remains To Be Learned.* Kluwer, Dordrecht
15 Carpenter WB (1891) *The Microscope and its Revelations. 7th edn.* J&A Churchill, London
16 Bradbury S (1967) *The Evolution of the Microscope*. Permagon Press, Oxford
17 Brown R (1833) On the organs and mode of fecundation in Orchideae and Asclepiadeae. Transactions Linnean Soc (London) 26: 685–745 (Communicated to the Linnean Society November 1831)
18 Hunter J (1837) *The Works of John Hunter FRS.* Translation by JF Palmer, Longman, London
19 Schleiden (1838) *Contributions to phytogenesis*. Translation by H Smith 1847, The Sydenham Society, London
20 Schwann T (1839) *Microscopical researches into the accordance in the structure and growth of animals and plants*. Translation by H Smith, 1847, The Sydenham Society, London
21 Bringmann WG, Lück HE, Miller R, Early CE (eds) (1997) *A Pictorial History of Psychology.* Quintessence Publishing Co, Carol Stream
22 Müller J (1838) *On the Nature and Structural Characteristics of Cancer and those Morbid Growths which may be Confounded with It*. G. Reimer, Berlin. Translation by C West, 1840, Sherwood, Gilbert and Piper, London
23 Rather LJ, Rather P, Frerichs JB (1986) *Johannes Muller and the Nineteenth-Century Origins of Tumor Cell Theory*. Science History Publications, Canton
24 Kölliker A (1852) *A Manual of Human Microscopic Anatomy.* Translation by G Busk, 1860, Parker, London
25 Hertwig O (1890) *Textbook of the Embryology of Man and Mamals.* Translation by EL Mark, 1899, Swan Sonnenschein & Co, London
26 Gaskin E (1968) *Investigation into Generation 1651–1828*. Hutchinson, London
27 Haeckel E (1907) *The evolution of man: a popular scientific study.* 5th edn, Translation by J McCabe, Watts, London
28 Bolsche W (1906) *Haeckel, his life and work*. With introduction and supplementary chapter by the translator, J McCabe. TF Unwin, London
29 Huxley JS (1958) *Introduction to the Mentor edition of the Origin of Species by C Darwin*, New American Library, New York, ix-xv
30 Barnett S (ed.) (1958) *A Century of Darwin*. Heinemann, London
31 Bowler PJ (1983) *The Eclipse of Darwinism*. Johns Hopkins Univ Press, Baltimore
32 Schwartz JF (1999) *Sudden Origins: Fossils, Genes and the Emergence of Species*. John Wiley and

Sons, New York
33 Ruse M (1982) *Darwinism defended: a guide to the evolution controversies.* Addison-Wesley, Advanced Book Program/World Science Division, Reading, Mass, USA
34 Virchow R (1858) *Cellular Pathology as Based Upon Physiological and Pathological Histology. 2nd edn.* Translation by F Chance, reprinted 1971 by Dover Publications, New York
35 Virchow R (1871) *Cellular Pathology as Based Upon Physiological and Pathological Histology. 4th edn.* A Hirschwald, Berlin
36 Virchow R (1884) Ueber Metaplasie/On metaplasia. *Virchow's Archives* 97: 410–430
37 Bignold LP (2005) Embryonic reversions and lineage infidelities in tumour cells: genome-based models and role of genetic instability. *Int J Exp Pathol* 86: 67–79
38 Cohnheim J (1882) *Lectures in General Pathology.* Translation by AB McKee 1889. New Sydenham Society, London
39 Ribbert H (1911) *Das Karzinom des Menschen/Human Cancer.* Friedrich Cohen, Bonn
40 Adami JG (1908) *The Principles of Pathology, vol 1 General Pathology.* Lea and Febiger, Philadelphia
41 Bashford RF, Murray JA, Cramer W (1905) *The Growth of Cancer under Natural and Experimental Conditions.* Sci Rpts Imperial Cancer Research Fund, No 2, Part II, Taylor and Francis, London
42 Ewing J (1940) *Neoplastic diseases. 4th edn.* Saunders, Philadelphia
43 Rather LJ (1962) Introduction; Harvey, Virchow, Bernard and the Methodology of Science. In: *Disease, Life and Man.* Translation and editing by LJ Rather, Collier, New York, 13–38
44 Ackerknecht EH (1953) *Rudolph Virchow, Doctor Statesman Anthropologist.* University of Wisconsin Press, Madison, USA, 98–105
45 Virchow R (1855) Cellular Pathology. Virchow's Archives, 8:1. Translation by LJ Rather in *Disease, Life and Man.* Collier, New York, 1962, 86–115
46 Virchow R (1877) Standpoints in Scientific Medicine. Virchow's Arch 70: 1. Translation by L J Rather in *Disease, Life and Man*, Collier, New York, 1962, 156–164
47 Von Hansemann DP (1890) On the Asymmetrical Division of Cells in Epithelial Carcinomata and their Biological Importance. *Virchow's Arch* 119: 299–326
48 Klebs E (1889) *Die allgemeine Pathologie/General Pathological. Vol 2.* G Fisher, Jena
49 Von Hansemann D (1893) *Studien über die Spezificität, den Altruismus und die Anaplasie der Zellen mit besonderer Berücksichtigung der Geschwülste/Studies of the specificity, altruism and anaplasia of cells with special reference to tumours.* A Hirschwald, Berlin
50 Von Hansemann D (1897) *Die Mikroscopische Diagnose der Bosartigen Geschwülste/The Microscopic Diagnosis of Malignant Tumours.* (2nd edn, 1902). A Hirschwald, Berlin
51 Dunn LC (1965) *A Short History of Genetics.* McGraw-Hill, New York
52 Farley J (1982) *Gametes and Spores, Ideas about Sexual Reproduction 1750–1914.* Johns Hopkins University Press, Baltimore
53 Moore RC, Bender MA (1993) Time sequence of events leading to chromosomal aberration formation. *Environ Mol Mutagen* 22: 208–213
54 O'Connor PJ, Manning FC, Gordon AT, Billett MA, Cooper DP, Elder RH, Margison GP (2000) DNA repair: kinetics and thresholds. *Toxicol Pathol* 28: 375–381
55 Klein G, Klein E (1984) Oncogene activation and tumor progression. *Carcinogenesis* 5: 429–435
56 Nowell PC (1986) Mechanisms of tumor progression. *Cancer Res* 46: 2203–2207
57 Whitman RC (1919) Somatic mutation as a factor in the production of cancer; a critical review of v. Hansemann's theory of anaplasia in the light of modern knowledge of genetics. *J Cancer Res* 4: 181–202
58 Beneke R (1900) A case of osteoid chrodrosarcoma of the urinary bladder, with comments on metaplasia. *Virchow's Arch* 161: 70–114
59 Boveri T (1914) *Zur Frage der Entstehung der Malignen Tumoren.* G Fischer, Jena. Translation by M Boveri, and published as *Origin of Malignant Tumors*, 1929, Williams and Wilkins, Baltimore
60 Bauer KF (1928) *Mutationstheorie der Geschwülst-Entstehung/Mutation Theory in Tumour formation.* Julius Springer, Berlin
61 Schneider NR, Williams WR, Chaganti RS (1986) Genetic epidemiology of familial aggregation of cancer. *Adv Cancer Res* 47: 1–36
62 Warthin AS (1913) Heredity with reference to carcinoma. *Arch Int Med* 12: 546–555
63 Crowe FW, Schull JW, Neel JV (1956) *A Clinical, Pathological and Genetic study of Multiple Neurofibromatosis.* CC Thomas, Springfield, Illinois

64 Little CC (1928) Evidence that cancer is not a simple Mendelian recessive. *J Cancer Res* 12: 30–46
65 Bashford EF, Murray JA (1904) The transmissibility of malignant new growths from one animal to another. *Sci Rep Imperial Cancer Research Fund* 1: 11–15
66 Tyzzer EE (1909) A series of spontaneous tumours in mice with observations on the influence of heredity on the frequency of their occurrence. *J Med Res* 21: 479–518
67 Woglom WH (1929) Immunity to transplantable tumours. *Cancer Rev* 4: 129–214
68 Loeb L (1937) *The Biological Basis of Individuality*. Charles C Thomas, Springfield, Illinois
69 Triolo VA (1964) Nineteenth century foundations of cancer research, origins of experimental research. *Cancer Res* 24: 4–27
70 Fidler IJ (1986) Rationale and methods for the use of nude mice to study the biology and therapy of human cancer metastasis. *Cancer Metastasis Rev* 5: 29–49
71 Kennaway EL, Hieger I (1930) Carcinogenic substances and their fluorescence spectra. *Br Med J* 1: 1044–1046
72 Badger GM (1954) Chemical constitution and carcinogenic activity. *Adv Cancer Res* 2: 73–127
73 Süss R, Kizel V, Scriber JD (1973) *Cancer experiments and concepts*. Springer-Verlag, New York
74 Lawley PD (1994) From fluorescence spectra to mutational spectra, a historical overview of DNA-reactive compounds. *IARC Sci Publ* 125: 3–22
75 Bignold LP (2004) Carcinogen-induced impairment of enzymes for replicative fidelity of DNA and the initiation of tumours. *Carcinogenesis* 25: 299–307
76 Murrell J (2004) Shedding light on light. *Croat Chemica Acta* 77: 17–30
77 Lea DE (1962) *The Actions of Radiations on Living Cells. 2nd edn*. Cambridge University Press, Cambridge, UK
78 Shope RE (1932) A filterable virus causing a tumor-like condition in rabbits and its relationship to myxomatosis. *J Exp Med* 56: 803–822
79 Gross L (1983) *Oncogenic viruses. 3rd edn*. Permagon Press, Oxford, UK
80 Inoue M (2001) Current molecular aspects of the carcinogenesis of the uterine endometrium. *Int J Gynecol Cancer* 11: 339–348
81 Argyris TS (1985) Regeneration and the mechanism of epidermal tumour promotion. *Crit Rev Toxicol* 14: 211–258
82 Foulds L (1969) *Neoplastic development*. (Vol 2, 1975), Academic Press, London
83 Berenblum I (1958) The study of tumours in experimental animals. In: HW Florey (ed.): *General Pathology, 2nd edn*. Lloyd-Luke, London, 513–549
84 Willis RA (1948) *Pathology of Tumours*, Butterworths, London
85 Haddow A (1972) Molecular repair, wound healing, and carcinogenesis: tumour production a possible overhealing? *Adv Cancer Res* 6: 181–234
86 Apffel CA (1976) Nonimmunological host defenses: a review. *Cancer Res* 36: 1527–1537
87 Harris H (1990) The role of differentiation in the suppression of malignancy. *J Cell Sci* 97: 5–10
88 Bignold LP (2003) The mutator phenotype theory of carcinogenesis and the complex histopathology of tumours: support for the theory from the independent occurrence of nuclear abnormality, loss of specialisation and invasiveness among occasional neoplastic lesions. *Cell Mol Life Sci* 60: 883–891
89 Prasad KN, Hovland AR, Nahreini P, Cole WC, Hovland P, Kumar B, Prasad KC. (2001) Differentiation genes: are they primary targets for human carcinogenesis? *Exp Biol Med (Maywood)* 226: 805–813
90 Yuasa Y (2003) Control of gut differentiation and intestinal-type gastric carcinogenesis. *Nat Rev Cancer* 3: 592–600
91 Lefort K, Dotto GP (2004) Notch signaling in the integrated control of keratinocyte growth/differentiation and tumor suppression. *Semin Cancer Biol* 14: 374–386
92 Verrecchia F, Mauviel A (2002) Transforming growth factor-beta signaling through the Smad pathway: role in extracellular matrix gene expression and regulation. *J Invest Dermatol* 118: 211–215
93 Micke P, Ostman A (2004) Tumour-stroma interaction: cancer-associated fibroblasts as novel targets in anti-cancer therapy? *Lung Cancer*, 45 Suppl 2: S163–175
94 Parmar H, Cunha GR (2004) Epithelial-stromal interactions in the mouse and human mammary gland *in vivo*. *Endocr Relat Cancer* 11: 437–458
95 Zhang Z, Yuan XM, Li LH, Xie FP (2001) Transdifferentiation in neoplastic development and its pathological implication. *Histol Histopathol* 16: 1249–1262

96 Ber I, Shternhall K, Perl S, Ohanuna Z, Goldberg I, Barshack I, Benvenisti-Zarum L, Meivar-Levy I, Ferber S (2003) Functional, persistent, and extended liver to pancreas transdifferentiation. *J Biol Chem* 278: 31950–31957
97 Guarino M, Micheli P, Pallotti F, Giordano F (1999) Pathological relevance of epithelial and mesenchymal phenotype plasticity. *Pathol Res Pract* 195: 379–389
98 Hay ED (1995) An overview of epithelio-mesenchymal transformation. *Acta Anat (Basel)* 154: 8–20
99 Birchmeier W, Birchmeier C (1995) Epithelial-mesenchymal transitions in development and tumour progression. *EXS* 74: 1–15
100 Kerr JF (2002) History of the events leading to the formulation of the apoptosis concept. *Toxicology* 181–182: 471–474
101 De Vries H (1902) *Mutation Theory*. Translation by JB Farmer, AD Darbishire. 1910, Kegan Paul, Trench, Trübner & Co, London
102 Bateson W (1913) *Problems of Genetics*. Yale University Press, New Haven, CT, USA
103 Tyzzer EE (1916) Tumour immunology. *J Cancer Res* 1: 125–155
104 Morgan TH, Sturtevant AH, Muller HJ, Bridges CB (1915) *The Mechanism of Mendelian Heredity*. Constable, London
105 Lawley PD (1994) Historical origins of current concepts of carcinogenesis. *Adv Cancer Res* 65: 17–111
106 Berenblum I, Shubik P (1949) An experimental study of the initiating stage of carcinogenesis, and a re-examination of the somatic cell mutation theory of cancer. *Br J Cancer* 3: 109–118
107 Berenblum I (1967) *Cancer Research Today*. Pergamon Press, New York
108 Burdette WJ (1955) The significance of mutation in relation to the origin of tumours: a review. *Cancer Res* 15: 201–226
109 Earle WR (ed.) (1943) Production of malignancy *in vitro*. *J Natl Cancer Instit* 4: 131–248
110 Sanford KK (1965) Malignant transformation of cells *in vitro*. *Int Rev Cytol* 18: 249–311
111 Hayflick L (1967) Oncogenesis *in vitro*. *Nat Cancer Inst Monograph* 26: 355–385
112 Huebner RJ, Todaro GJ (1969) Oncogenes of RNA tumour viruses as determinants of cancer. *Proc Natl Acad Sci USA* 64: 1087–1094
113 Sporn MB, Roberts AB (1985) Autocrine growth factors and cancer. *Nature* 313: 745–747
114 Rozengurt E (1995) Polypeptide and neuropeptide growth factors: signalling pathways and role in cancer. In: M Peckham, H Pinedo, U Veronesi (eds): *Oxford Textbook of Oncology*, Oxford University Press, Oxford, 12–20
115 Moschos SJ, Mantzoros CS (2002) The role of the IGF system in cancer: from basic to clinical studies and clinical applications. *Oncology* 63: 317–332
116 Bignold LP (2004) Chaotic genomes make chaotic cells: the mutator phenotype theory of carcinogenesis applied to clinicopathological relationships of solid tumors. *Cancer Invest* 22: 338–343
117 Iversen OH (1995) Of mice and men: a critical reappraisal of the two-stage theory of carcinogenesis. *Crit Rev Oncog* 6: 3357–3405
118 McGee J O'D, Isacson PG, Wright NA (eds) (1992) *Oxford Textbook of Pathology*. vol. 1. Oxford University Press, Oxford 636
119 Rubin E, Farber JL (eds) (1992) *Pathology*. 2nd edition. Lippincott, Philadelphia, 144
120 Walter JB, Talbot IC (eds) (1996) *General Pathology*. Churchill Livingstone, Edinburgh, 530
121 Lockhart-Mummery JP (1934) *The Origin of Tumours*. J and A Churchill, London
122 Nichols EM (1969) Somatic variation and multiple neurofibromatosis. *Hum Hered* 19: 473–479
123 Comings DE (1973) A general theory of carcinogenesis. *Proc Natl Acad Sci USA* 70: 3324–3328
124 Knudson AG (2000) Chasing the cancer demon. *Annu Rev Genet* 34: 1–19
125 Knudson AG (2001) Two hits (more or less) to cancer. *Nature Rev Genet* 2: 157–162
126 Bignold LP (2004) The cell-type-specificity of inherited predispositions to tumours: review and hypothesis. *Cancer Letters* 216: 127–146
127 Fearon ER, Vogelstein B (1990) A genetic model for colorectal tumourigenesis. *Cell* 61: 759–767
128 Vogelstein B Kinzler KW (1993) The multistep nature of cancer. *Trends Genet* 9: 138–141
129 Bodmer W (1997) The somatic evolution of cancer. *J Roy Coll Physicians Lond* 31: 82–89
130 Fearnhead NS, Wilding JL, Bodmer WF (2002) Genetics of colorectal cancer: hereditary aspects and overview of colorectal tumourigenesis. *Br Med Bull* 64: 27–43
131 Miller EC, Miller JA (1981) Mechanisms of chemical carcinogenesis. *Cancer* 47 (5 Suppl): 1055–1064

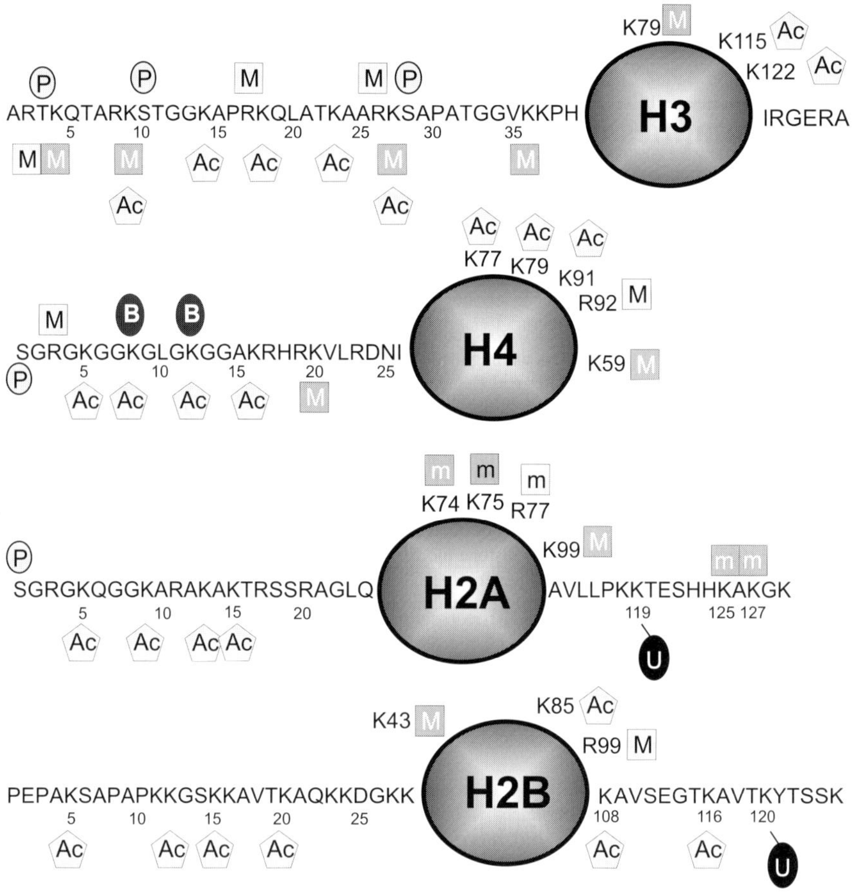

Figure 1. Core histone modifications. The N-terminal and in some cases C-terminal amino acid sequences of human histone are shown. The modifications include methylation (M), acetylation (Ac), phosphorylation (P), ubiquitination (U), and biotinylation (B). Methylation sites that are uncertain are denoted as (m).

nucleosome [7]. H1 and the histone tails stabilize the higher order compaction of chromatin.

The core histones undergo a wide range of post-synthetic modifications, most of which are reversible. These modifications include acetylation, phosphorylation, methylation, ubiquitination, poly ADP ribosylation, and biotinylation [8, 9] (Fig. 1). The majority of modified amino acids reside in the tail domains, but there is an increasing awareness of modified residues occurring in the histone-fold domains [10]. The amino-terminal tails of the four core histones play an important role in chromatin compaction. These tails protrude from the nucleosome, with that of H3 protruding the farthest. The tails of H3, being the longest and positioned in such a way as to allow several contacts

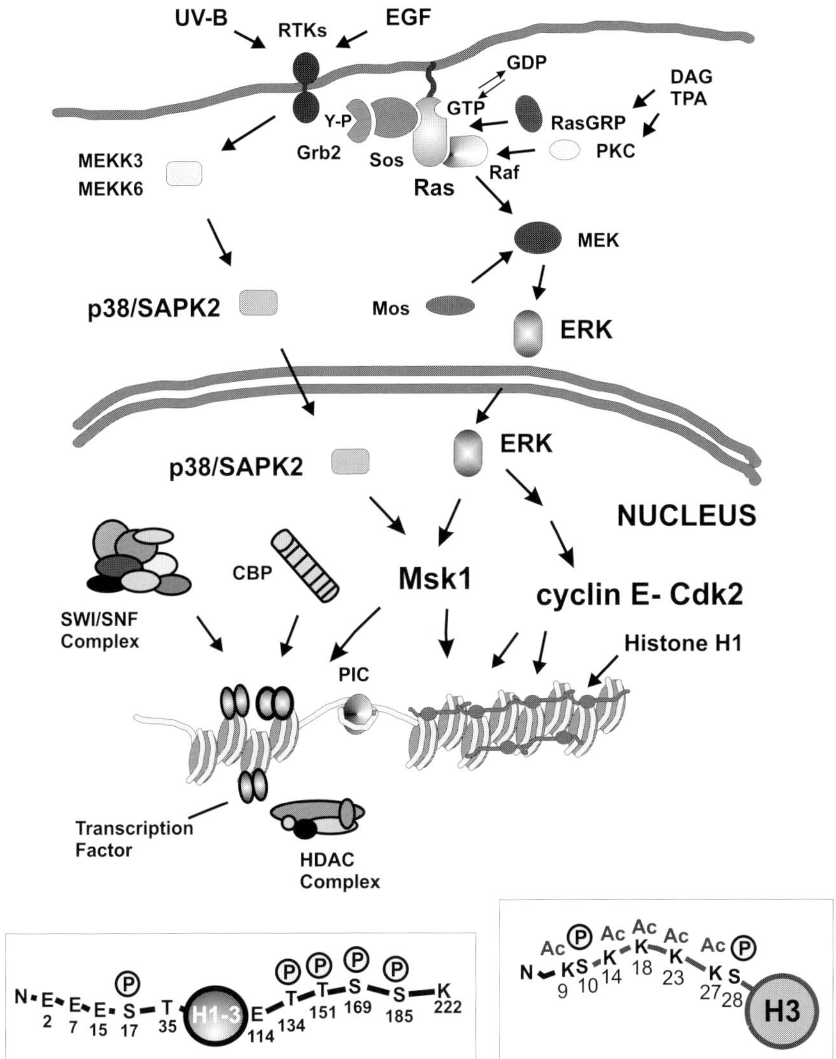

Figure 2. MAPK signal transduction pathways and the modification of chromatin. The Ras-MAPK pathway is activated by EGF and TPA. TPA acts through PKC and/or RasGRP. UV-B activates both the Ras-MAPK and the p38 kinase pathways. Inserts: Left panel, The sites of phosphorylation of H1 subtype H1.3 are located on the N- and C-terminal tails. Right panel, The H3 phosphorylation sites are nestled in a region of acetylation (P, phosphorylation, Ac acetylation). CBP, CREB binding protein (histone acetyltransferase); Cdk2, cyclin-dependent kinase 2; DAG, diacylglycerol; GDP, guanosine diphosphate; GTP, guanosine 5'-triphosphate; HDAC complex, histone deacetylase complex; PIC, preinitiation complex; RasGRP, Ras guanyl nucleotide-releasing protein; RTKs, receptor tyrosine kinases; SOS, Son of Sevenless; TPA, 12-O tetradecanoylphorbol-13-acetate.

with linker DNA, are crucial for the formation of higher order chromatin fibers [11]. Acetylation and phosphorylation of the N-terminal tail of H3 promotes transcriptional activation of local genes by disrupting chromatin compaction.

An understanding of the significance of the histone modification type and position in the nucleosome is starting to emerge. Transcribed regions of the genome are associated with hyperacetylated H3 (including acetylation of K9 and K14), hyperacetylated H4 and H3 hypermethylated at K4 and K79, while inactive genes have poorly acetylated histones and H3 that is hypomethylated at K4 and K79 [12, 13]. Methylation of K9 of H3 is involved in maintaining the condensed structure of heterochromatic regions. The targeting of histone-modifying enzymes to specific regions of chromatin results in the distinct distribution of the modified histone isoforms.

A universal distinguishing feature of transcribed and potentially active (referred to as competent) chromatin is that these regions of the genome have an increased sensitivity to nuclease digestion (for example to DNase I and micrococcal nuclease). The binary distribution of modified histones among the transcribed and silent genes is thought to be one of the factors in conferring nuclease sensitivity onto transcribed chromatin [12, 13].

Activation of an oncogene or deactivation of a tumor suppressor gene results in the decondensation of chromatin [14–16]. Alterations in modification of core and linker histones are thought to be responsible for the more relaxed chromatin structure of these cells. In the following sections, we explore the diverse mechanisms by which cancer cells divert the targeting or alter the activity of histone-modifying enzymes to remodel chromatin structure and adjust gene expression.

Histone acetyltransferases in malignancies

Acetylation of core histones destabilizes histone-DNA interactions as well as histone-histone contacts between adjacent nucleosomes and interactions occurring between histones and specific regulatory proteins. Histone acetylation is catalyzed by a group of enzymes termed histone acetyltransferases (HATs). The highly organized and repressive nature of chromatin stresses the integral role that HATs play in the dynamic changes necessary for gene expression that control cellular processes. For most HATs, acetylation of substrates extend beyond histones and addition of acetyl groups to transcription factors, cell cycle regulators and structural proteins, demonstrating the extensive influence of HATs to normal cellular function and maintenance.

Two of the most extensively studied HATs in transcriptional activation are p300 and its closely related homolog cyclic AMP response element-binding (CREB)-binding protein (CBP). p300 was discovered through its interaction with the adenovirus E1A protein, while CBP was identified through its association with the phosphorylated form of CREB. p300 and CBP act as global transcriptional coactivators, being involved in cellular processes such as cell cycle control, differentiation and apoptosis [17]. p300 and CBP acetylate all four core histones. p300/CBP stimulate transcriptional activation of specific genes through direct or indirect interaction with various promoter-binding transcription factors including CREB, nuclear hormone receptors and oncoprotein activators such as c-Fos, c-Jun and c-Myb. The broad spectrum of p300/CBP interacting proteins provides a general mechanism for integration of several signaling and transcription-response pathways that p300 and CBP modulate [18].

Defects in the expression and function of HATs have been reported in cancer cells. The mis-targeting, ill-timed activation or irregular increase in activity of HATs can lead to expression of genes that allow tumorigenesis [19]. Chromosomal translocations, deletions and mutations that affect genes encoding potent HATs have been associated to the genesis of several malignant conditions, particularly hematological disorders. The following section discusses malignancies that result from aberrant HATs and subsequent consequences to chromatin structure and function.

HAT fusion proteins in leukemias

Hematopoietic disorders such as leukemia frequently manifest chromosomal translocations spanning genes that encode HATs, and result in protein fusions with other transcription factors that lead to defective acetyltransferase activities (Fig. 3) [20]. Fusion proteins that result from these translocations form hybrid transcriptional regulators and chimeric HATs that become dominant over their wild-type counterpart, permitting either a gain or loss of function, and enabling differential expression of target genes. The resulting myelomonocytic neoplasms or other malignancies are attributed to changes in proteins involved in cell cycle control, differentiation and apoptosis. Altered acetylation of histones and protein substrates contribute to defects in chromatin remodeling and allow oncogenesis [21].

The hematological malignancy acute myeloid leukemia (AML) subtype M4/M5 displays frequent somatic translocations that disrupt the 5' end of CBP gene. Commonly, it has been detected that the MYST family putative acetyltransferase monocytic zinc finger (MOZ) gene is juxtaposed beside the CBP gene after (8;16)(p11;p13) translocation, resulting in MOZ-CBP fusion proteins. The chimera retains most of the interaction domains of both parent proteins as well as the MOZ acetyltransferase domain establishing a likely gain of function that is suspected to play a role in leukemogenesis. Although this is so, further studies to directly address this claim and its effects on chromatin organization are lacking [21, 22]. Fusion proteins involving MOZ and p300 have also been observed as rare variants [23]. Furthermore, chromosomal translocations also affect the MOZ-related gene *MORF*, which has been observed to form MORF-CBP and/or CBP-MORF chimeras, and are expressed in juvenile AML subtype M5a and therapy-related myelodysplastic syndrome (MDS) [24, 25]. In some AML patients that present similar clinical conditions as those expressing MOZ-CBP fusions, inversions of chromosome 8 between p11 and q13 allows fusion of the MOZ gene with the nuclear coactivator TIF2 gene,

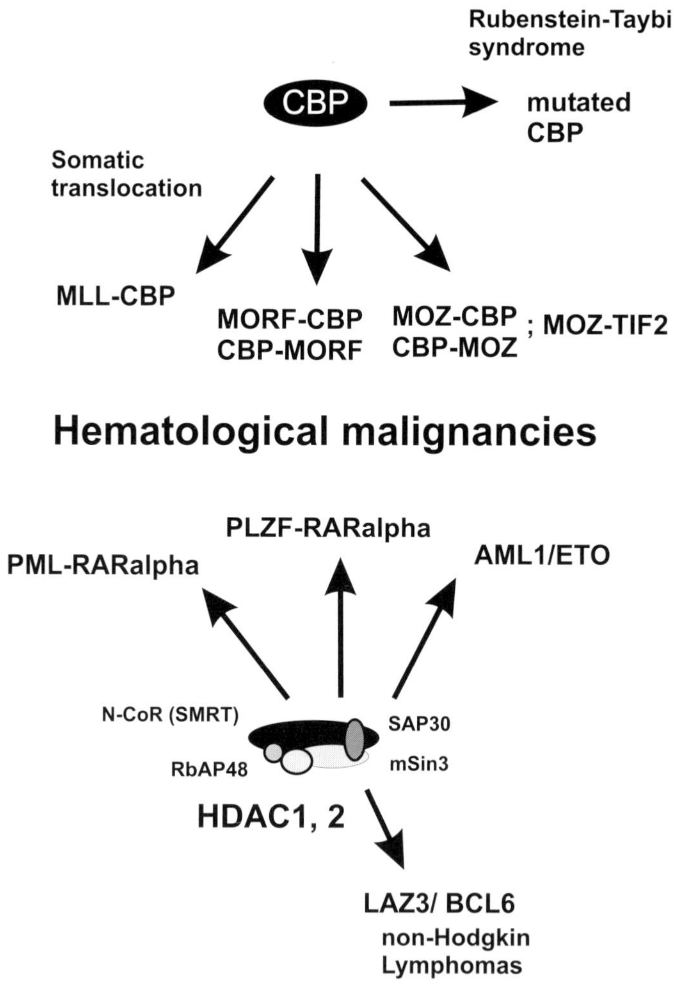

Figure 3. Somatic translocations and mis-targeting of HATs and HDACs. Somatic translocations and mutations involving the CBP gene can interfere with the normal function and targeting of CBP, a potent histone acetyltransferase and coactivator, resulting in a variety of hematological malignancies. Mis-targeting of HDAC complexes, which deacetylate histones and transcription factors and are corepressors, by fusion proteins arising from somatic translocations, can interfere with genetic programs, resulting in a variety of hematological malignancies.

generating a chimeric protein. This MOZ-TIF2 protein can bind CBP and mimic the MOZ-CBP fusion function that has been demonstrated to be necessary for transformation and leukemogenesis *in vitro* and *in vivo* [26].

In many cases, the development of leukemias can arise during recession after primary treatment. These disorders are not only limited to AML, they are also found in chronic myeloid leukemia (CLL) and myelodysplastic syndrome

(MDS). Rearrangements and consequently formation of fusion partners of mixed lineage leukemia (MLL) gene with CBP, p300 and MOZ have been associated with development of therapy-related disease due to defects in chromatin structure modulation and the inability of HAT fusion proteins to regulate differentiation and cell cycle control [21, 27–29]. The ability of the *MLL* gene to generate fusion partners with many HATs suggests that multiple avenues of cellular homeostasis can be affected, disturbed and deregulated.

Aberrant HATs associated with other defects

Many studies have demonstrated the presence of somatic mutations in solid tumors and carcinomas generating truncated, missense or nonfunctional HAT proteins [23]. Biallelic inactivating somatic point mutations of p300, and more rarely of CBP, have been observed in neoplasms such as gastric, colon, ovarian, breast and pancreatic cancers, which, depending on the mutation site and the domains that are affected, can yield a nonfunctional HAT or a HAT with decreased activities [20, 21, 23]. The HAT, amplified in breast cancer 1 (AIB1), is found commonly overexpressed in breast and ovarian cancer due to an amplification of chromosomal region containing the gene [27]. Moreover, loss of heterozygosity in 80% glioblastomas of p300 locus in chromosome 22 and of CBP locus in hepatocellular carcinomas has also been detected [21].

In patients affected by the autosomal dominant disorder Rubinstein-Taybi syndrome, the *crebbp* gene, which encodes for CBP, has a germline allele mutation that renders the HAT nonfunctional. Since CBP is a potent and multipurpose acetyltransferase involved in many cellular processes, affected individuals display a wide range of severity and clinical presentations from facial and limb abnormalities to physiological anomalies, and are consequently predisposed to developing other malignancies [30]. Surprisingly, although the close homolog p300 remains intact in patients affected by this syndrome, it does not replace CBP function, which further displays that each HAT has specific and nonredundant roles [21].

Since HATs have key roles in many cellular processes particularly in chromatin remodeling and transcriptional activation, the consequences of these mutations and formation of fusion proteins in the development of malignancies are manifested far beyond the molecular level. These fusion proteins, depending on their specific orientation and on the domains they retain from their parent proteins, can dictate the pattern of pathogenicity and phenotypic presentation of disease.

Chromatin consequences of aberrant HATs

HATs are critical players in normal cellular function and work in concert with deacetylases. They have roles in bridging the basal transcriptional machinery

promyelocytic leukemia patients having the PML-RARα translocation can achieve remission successfully through treatment with retinoic acid, PLZF-RARα translocations do not respond well to this therapy [47]. HDAC-containing co-repressor complexes are also implicated in other leukemias including AML. The AML1 gene product acts to upregulate genes related to hematopoiesis [41]. The fusion protein AML1-ETO, produced as the result of a translocation event in a significant percentage of AMLs, is able to bind the HDAC co-repressor complexes N-CoR and SMRT, thereby providing a mechanism by which HDAC can be aberrantly targeted to alter chromatin structure and transcription status [48]. Chromosomal rearrangements of TEL, a transcriptional repressor, result in common acute lymphoblastic leukemia [41]. The protein encoded by TEL represses transcription by working in complex with mSin3A, SMRT and HDAC3 [49, 50]. The evidence indicates that abnormal transcriptional repression of genes necessary for proper cellular differentiation by HDACs is an important factor in the progression of hematological malignancies [41].

HDAC inhibitors as therapeutic targets for cancer

Experiments performed on cultured cells and with animal models have shown that treatment of transformed cells with HDAC inhibitors leads to growth arrest, differentiation, and apoptosis [37, 51]. Profiling of cultured cells treated with HDAC inhibitors has determined that the treated cells exhibit an altered expression profile for a small percentage of genes [52–55]. Alterations in expression profiles include many increases such as $p21^{WAF1}$ and metallothionein, while decreases were observed for ErbB2, vascular endothelial growth factor and others [37].

Expression of $p21^{WAF1}$ is induced by several HDAC inhibitors, such TSA, sodium butyrate, phenyl butyrate, SAHA, FK-228, MS-275 and oxoflatins [37]. Higher levels of $p21^{WAF1}$ are also observed in embryonic cells deficient in HDAC1 displaying a reduced rate of proliferation [56]. Studies have demonstrated an increase in the acetylation of histones in the $p21^{WAF1}$ promoter region after HDAC inhibitor treatment [57, 58]. Thus events leading to the change in expression of genes following HDAC inhibitor treatment may in some cases be a direct result of an increase in acetylation levels of histones associated with the affected genes [37].

Treatment with HDAC inhibitors also alters the acetylation status of several non-histone proteins including p53, MyoD, GATA-1, α-tubulin and Hsp90 [37]. An increase in acetylation of chaperone protein Hsp90 results in decreased binding to other proteins and the subsequent degradation of those proteins [37, 59]. Thus, HDAC inhibitors may modulate their effects by influencing both gene expression and protein stability [37].

Studies have shown that both normal and tumor cells accumulate acetylated histones when treated with HDAC inhibitors [60–63]. Yet the growth arrest of tumors occurred without excessive toxicity in animal models [37]. This result may be explained by the finding that tumor cells have a tenfold higher sensitivity to HDAC inhibitors than do normal cells [37]. Several HDAC inhibitors are currently in clinical trials and provide a promising new strategy to treat cancer [64].

Alterations to chromatin structure by activation of the Ras-MAPK pathway

Signals from growth factors, stresses and cytokines are relayed by the mitogen-activated protein kinase (MAPK) pathway. Activation of either the Ras-MAPK or p38 MAPK pathway leads to the activation of transcription at immediate-early genes (Fig. 2). In 1999, Brown et al. [65] identified genes whose transcription was activated within 15 min of stimulating human fibroblasts with serum. A programmed transcriptional response occurred: genes such as c-*fos*, Jun B and MAP kinase phosphatase-1 were activated within 15 min of serum addition, followed later by the transcription of genes encoding proteins associated with wound repair [65].

The Ras-MAPK pathway is activated by growth factors and phorbol esters (12-O-tetradecanoylphorbol-13-acetate, TPA), while the p38 MAPK pathway is activated by stressors such as UV irradiation (Fig. 2). TPA works through PKC and/or RasGRP, which is expressed in a cell type-specific manner, to activate the Ras-MAPK [66–68]. Stimulation of these pathways activates a series of protein kinases (see [69] for a review of this process), leading to the phosphorylation of histone H3 and HMGN1 and modification of chromatin structure. Stimulation of the MAPK induces phosphorylation of H3 on S10 and S28 [70–72], and phosphorylation of HMGN1 on S6 [73]. As evidenced by MSK1 and MSK2 knockout mice, impaired phosphorylation severely limits transcription of immediate-early genes [70]. The location of H3 phosphorylated on S28 after activation of the MAPK pathway has not been determined, but it is known that phosphorylation on S10 occurs at immediate-early genes [74–77]. H3 phosphorylated at S10 in TPA-treated, serum-starved mouse 10 T½ fibroblasts appears in numerous small foci throughout the interphase nuclei [74]. These sites may represent the nuclear locations of immediate-early genes targeted for transcriptional activation. After stimulation of the MAPK pathway, H3 phosphorylated on S10 is associated with the c-*fos* promoter and is found at various regions of the c-*jun* gene [75, 77].

The MAPK pathway, histone phosphorylation and cancer

Several cancerous tissues and cell lines display constitutive activation of the MAPK pathway. An estimated 30% of human malignancies contain a mutation to one of the *ras* oncogenes that renders the protein constitutively active [78].

Kirsten-*ras* (Ki-*ras*) is the most commonly mutated, and defects in Harvey-*ras* (Ha-*ras*) and Neuroblastoma-*ras* (N-*ras*) are also detected [78–80]. Other members of the MAPK pathway have also been implicated in malignant transformation. An upregulation of *ras*-oncogene-related p21-rac1 and MAPK p38α and a downregulation of genes associated with apoptosis have been observed in low-grade dysplastic adenomas of the colon [81]. The study also found upregulation of the rho GDP dissociation factor [81]. An upregulation in members of the epidermal growth factor family has been observed in several cancers, and various tumor cell lines exhibit high levels of activated Erk1/2 [82, 83]. Mutations to the BRAF gene encoding a MAPK kinase kinase are frequently seen in cutaneous melanoma [84]. BRAF mutation can lead to constitutive activation of the MEK/ERK pathway independent of Ras and increases in B-RAF(V599E) activity, which could contribute to anchorage-independent growth [84].

Alterations in chromatin and nuclear structure of oncogene-transformed cells

Cancer can be diagnosed by changes to nuclear morphology. Fibroblasts transformed with oncogenes such as v-Mos, v-Fes, v-Raf, v-Src, and H-Ras display abnormal nuclear morphology such that the nucleus is more rigid and spherical [85]. The degree of changes in nuclear morphology correlate with metastatic potential [85]. Oncogene-transformed cells also exhibit abnormalities in the nuclear matrix [86]. Highly metastatic lines transformed with *ras* or kinase oncogenes show similar nuclear matrix profiles [86], suggesting that components of the nuclear matrix change as malignancy progresses [87].

Changes to chromatin structure have also been observed. *Ras*- and *myc*-transformed cells display a relaxed chromatin conformation [15, 16]. Micrococcal nuclease digestion of chromatin from parental and ras-transformed cells determined that the bulk chromatin, and chromatin at the immediate-early genes ornithine decarboxylase and c-*myc* was less condensed in the *ras*-transformed cells [16]. Experiments to determine the methylation status of DNA at the ornithine decarboxylase gene showed similar methylation levels in parental and transformed cells, thus it is not believed that hypomethylation plays a role in the formation of a relaxed chromatin structure in this case [88].

Phosphorylation of H3 on S10 following stimulation of the Ras-MAPK is known to occur at immediate-early genes [75–77]. Transcription requires that chromatin structure become less condensed. Parental and *ras*-transformed fibroblasts synchronized by serum starvation contain relatively condensed chromatin that becomes highly decondensed in late G_1 [16]. Both phosphorylation of H3 and transcription at immediate-early genes is impaired if the H3 kinases, MSK1 and MSK2, are knocked out [70]. This link between the MAPK pathway, histone modifications and gene expression strengthens the hypothesis that deregulation of the Ras-MAPK pathway leads to an abnormal chromatin conformation and aberrant gene expression in transformed cells [89].

Histone kinases and phosphatases and cancer

Phosphorylation and dephosphorylation of proteins play a major role in mechanisms controlling proper execution of cellular functions. The importance of histone phosphorylation in intracellular processes has long been recognized. Kinases and phosphatases responsible for the reversible process of histone phosphorylation have been recognized as crucial mediators of various cellular processes including mitosis, meiosis and transcriptional activation [90]. Further, abnormal expression of mitotic histone kinases have been implicated in some human cancers, and deregulated activities of histone kinases and phosphatases could underline some of the mechanisms leading to oncogenic transformation and malignancies. Histone kinases implicated in cell transformation and tumorigenesis are reviewed in the following sections.

Histone kinases and activated Ras-MAPK pathway in cancer

Histone phosphorylation is linked to relaxation of chromatin structure [91]. Phosphorylation of histone H3 plays a crucial role during mitosis and meiosis. Further, H3 phosphorylation at S10 and S28 has a role in transcriptional activation of immediate-early genes [92]. The kinases responsible for phosphorylating H3 during this process are mitogen- and stress-activated protein kinases (MSKs). Both MSK1 and MSK2 phosphorylate H3 at S10 and S28 [70]. To date, the expression and activities of MSKs in human cancers have not been analyzed. However, H3 phosphorylation at S10 is elevated in oncogene-transformed cells, and we recently reported that the increase in the steady-state level of phosphorylated H3 is due to an increase in MSK1 activity and not its expression [74, 93]. Further research needs to be conducted with respect to MSKs to elucidate possible roles these H3 kinases have in oncogenesis and tumorigenesis.

Histone H1 phosphorylation is also linked to chromatin relaxation, and phosphorylation of H1 is elevated in oncogene-transformed mouse fibroblasts [15]. The increased phosphorylation of H1 in the oncogene-transformed cells was shown to be due to elevated activity of H1 kinase, cyclin-dependent kinase 2 (Cdk2). The activity of the H1 phosphatase, PP1, is not altered in the oncogene-transformed cells [94]. Cdk2 overexpression has been described in colon carcinoma cell lines, and Cdk2 overexpression is postulated to be a prognostic indicator of oral cancer progression [95, 96]. However, no direct studies relating Cdk2 activity and chromatin remodeling in cancers have been conducted.

Aurora kinases and cancer

Human mitotic serine/threonine kinases termed Aurora kinases belong to the prototypic yeast *Ipl1* and *Drosophila Aurora* kinase family. There are three human Aurora kinases, Aurora A (Aurora-2, STK15, mouse STK6), Aurora B (AIM-1, Aurora-1, STK12) and Aurora C (STK13). These kinases are responsible for proper mitosis and meiosis in all eukaryotes [97]. Aurora kinases have been shown to associate with interphase chromosomes, mitotic spindle poles, mitotic microtubules and the spindle midbody, implicating these kinases in the tight regulation of chromosomal ploidy in cells. Aurora kinases have a highly conserved catalytic domain, a short C-terminal domain and an N-terminal domain of varying size [98]. All three Auroras are able to phosphorylate H3 *in vitro* [91], with only Aurora B being shown to phosphorylate H3 at S10 and S28 *in vivo*. This phosphorylation event catalyzed by Aurora B has been linked to proper mitotic chromosome condensation [99]. It has been shown that H3 S10 phosphorylation is absolutely critical for both chromosome condensation and segregation in *Tetrahymena* [100]. However, in mammals both S10 and S28 phosphorylation seem to be important for proper mitotic processes. Further, H3 phosphorylation events during mitosis are tightly governed not only by Aurora B but also by H3 phosphatase PP1 [99]. It is important to note that the H3 phosphorylation events are reversible, and both kinase and phosphatase activities play critical roles in this regulation. Moreover, overexpression of all three Aurora kinases has been observed in various human cancers. Aurora A, which is implicated in centrosome maturation and spindle assembly, is mapped to chromosomal 20q13 region often amplified in human cancers [101]. Overexpression and amplification of Aurora A has been detected in colon, bladder, ovarian, human breast and pancreatic cancers. Further, overexpression of Aurora A has been shown to correlate with induced aneuploidy, centrosomal anomalies and prognosis of naturally occurring tumors in animal model systems as well as with the induction of oncogenic transformation in cells [98]. The evidence implies the involvement of Aurora A in the cellular processes that are most likely deregulated in many human cancers. In addition, Aurora B, which is required for proper mitotic events and cytokinesis, has an abnormal expression profile detected in various human tumor cell lines, including colorectal cancer cell lines [98]. Elevated kinase activity and overexpression of Aurora B along with the overexpression of Aurora B interacting proteins, such as INCENP and Survivin [102], which target Aurora B to the centromere at metaphase for H3 phosphorylation, are likely to be responsible for anomalous effects observed in cancer cells [98]. Increased H3 phosphorylation due to Aurora B has been attributable to chromosome number instability, a feature often seen in many human cancer cells. Aurora B overexpression in Chinese hamster cells led to increased phosphorylation of H3 at S10, and Aurora B overexpressing cells were able to form aggressive tumors in nude mice [103]. The involvement of Aurora B in the generation of chromosomal instabilities in conjunction with increased H3 phosphorylation reinforces the role this kinase could be undertaking in carcinogenesis. Lastly, Aurora C overexpression has been detected in various cancer cell lines. There is also some correlative evidence that Aurora C could be involved in oncogenic signal transduction in somatic cells

[98]. Recently, a potent and selective inhibitor of all three human Aurora kinases, VX-680 has been shown to decrease H3 phosphorylation at S10 in the MCF-7 cell line as well as suppress tumor growth *in vivo* [104]. This finding implicates Aurora kinases in the processes leading to malignant transformation and carcinogenesis, and shows promise for a new approach for anticancer therapy since VX-680 was able to induce regression of a range of human tumor types. The VX-680 inhibitor is progressing into clinical development [104].

Histone methylation and chromatin

The four core histones are modified by methylation of lysines and arginines located in the N-terminal tail and histone-fold domains (Fig. 1). Histone methylation is catalyzed by histone methyltransferases, which are a large family of enzymes that have specificity for a histone, the modification site (lysine or arginine) and chromatin region (for review see [105]). In contrast to histone acetylation and phosphorylation, histone methylation is a stable modification. To date a histone demethylase has not been identified. However, histone exchange occurring during transcription is one mechanism by which the core histones are dislodged from the transcribed DNA and replaced by a histone that is not methylated [106].

Analyses of the distribution of methylated histones in nuclei of normal and tumor cell nuclei using an antibody recognizing methylated lysines independent of their lysine position in the histone revealed a differential distribution of methylated histones in these nuclei. In contrast to the homogeneous distribution of chromatin with methylated histones in normal G_0 lymphocytes, the leukemic T cell Jurkat cells had methylated histones located in numerous distinct clusters. Further, chromatin with lysine methylated histones was concentrated more peripherally in colon carcinoma compared to nuclei of normal colon epithelial cells [107].

H3 methylated at K4 and K79 is located in transcribed regions of the genome, while H3 methylated at K9 and H4 methylated at K20 are present in heterochromatin regions, the histones of which are hypoacetylated [13, 108, 109]. H3 methyl K9 avidly binds to the chromodomain of heterochromatin protein 1 (HP1), recruiting the protein to heterochromatic regions [105]. HP1 interacts with the H3 K9 methyltransferase SUVAR39H1. Thus, HP1 recruited by a nucleosome bearing an H3 methyl K9 will enable the HP1-bound H3 K9 methyltransferase to methylate neighboring nucleosomes, establishing a self-propagating mechanism for the spreading of heterochromatin. In addition to heterochromatic silencing, SUV39H1 H3 methyltransferase and HP1 are involved in repression of euchromatic genes. Downregulation of $HP1^{Hs\alpha}$ is observed in metastatic breast cancer. The reduced expression of HP1 may result in the reorganization of chromatin and activation of genes involved in metastasis [110].

45 Kramer OH, Gottlicher M, Heinzel T (2001) Histone deacetylase as a therapeutic target. *Trends Endocrinol MeTab.* 12: 294–300
46 de The H (1996) Altered retinoic acid receptors. *FASEB J* 10: 955–960
47 Lin RJ, Egan DA, Evans RM (1999) Molecular genetics of acute promyelocytic leukemia. *Trends Genet* 15: 179–184
48 Lutterbach B, Westendorf JJ, Linggi B, Patten A, Moniwa M, Davie JR, Huynh KD, Bardwell VJ, Lavinsky RM, Rosenfeld MG et al. (1998) ETO, a target of t(8;21) in acute leukemia, interacts with the N-CoR and mSin3 corepressors. *Mol Cell Biol* 18: 7176–7184
49 Chakrabarti SR, Nucifora G (1999) The leukemia-associated gene TEL encodes a transcription repressor which associates with SMRT and mSin3A. *Biochem Biophys Res Commun* 264: 871–877
50 Wang L, Hiebert SW (2001) TEL contacts multiple co-repressors and specifically associates with histone deacetylase-3. *Oncogene* 20: 3716–3725
51 Marks P, Rifkind RA, Richon VM, Breslow R, Miller T, Kelly WK (2001) Histone deacetylases and cancer: causes and therapies. *Nat Rev Cancer* 1: 194–202
52 Suzuki H, Gabrielson E, Chen W, Anbazhagan R, van Engeland M, Weijenberg MP, Herman JG, Baylin SB (2002) A genomic screen for genes upregulated by demethylation and histone deacetylase inhibition in human colorectal cancer. *Nat Genet* 31: 141–149
53 Della RF, Criniti V, Della P, V, Borriello A, Oliva A, Indaco S, Yamamoto T, Zappia V (2001) Genes modulated by histone acetylation as new effectors of butyrate activity. *FEBS Lett* 499: 199–204
54 Van Lint C, Emiliani S, Verdin E (1996) The expression of a small fraction of cellular genes is changed in response to histone hyperacetylation. *Gene Expr* 5: 245–253
55 Butler LM, Zhou X, Xu WS, Scher HI, Rifkind RA, Marks PA, Richon VM (2002) The histone deacetylase inhibitor SAHA arrests cancer cell growth, up-regulates thioredoxin-binding protein-2, and down-regulates thioredoxin. *Proc Natl Acad Sci USA* 99: 11700–11705
56 Lagger G, O'Carroll D, Rembold M, Khier H, Tischler J, Weitzer G, Schuettengruber B, Hauser C, Brunmeir R, Jenuwein T, Seiser C (2002) Essential function of histone deacetylase 1 in proliferation control and CDK inhibitor repression. *EMBO J* 21: 2672–2681
57 Sambucetti LC, Fischer DD, Zabludoff S, Kwon PO, Chamberlin H, Trogani N, Xu H, Cohen D (1999) Histone deacetylase inhibition selectively alters the activity and expression of cell cycle proteins leading to specific chromatin acetylation and antiproliferative effects. *J Biol Chem* 274: 34940–34947
58 Richon VM, Sandhoff TW, Rifkind RA, Marks PA (2000) Histone deacetylase inhibitor selectively induces p21WAF1 expression and gene-associated histone acetylation. *Proc Natl Acad Sci USA* 97: 10014–10019
59 Yu X, Guo ZS, Marcu MG, Neckers L, Nguyen DM, Chen GA, Schrump DS (2002) Modulation of p53, ErbB1, ErbB2, and Raf-1 expression in lung cancer cells by depsipeptide FR901228. *J Natl Cancer Inst* 94: 504–513
60 Richon VM, Emiliani S, Verdin E, Webb Y, Breslow R, Rifkind RA, Marks PA (1998) A class of hybrid polar inducers of transformed cell differentiation inhibits histone deacetylases. *Proc Natl Acad Sci USA* 95: 3003–3007
61 Scott GK, Marden C, Xu F, Kirk L, Benz CC (2002) Transcriptional repression of ErbB2 by histone deacetylase inhibitors detected by a genomically integrated ErbB2 promoter-reporting cell screen. *Mol Cancer Ther* 1: 385–392
62 Kelly WK, Richon VM, O'Connor O, Curley T, MacGregor-Curtelli B, Tong W, Klang M, Schwartz L, Richardson S, Rosa E et al. (2003) Phase I clinical trial of histone deacetylase inhibitor: suberoylanilide hydroxamic acid administered intravenously. *Clin Cancer Res* 9: 3578–3588
63 Qiu L, Kelso MJ, Hansen C, West ML, Fairlie DP, Parsons PG (1999) Anti-tumour activity *in vitro* and *in vivo* of selective differentiating agents containing hydroxamate. *Br J Cancer* 80: 1252–1258
64 Zelent A, Waxman S, Carducci M, Wright J, Zweibel J, Gore SD (2004) State of the translational science: summary of Baltimore workshop on gene re-expression as a therapeutic target in cancer January 2003. *Clin Cancer Res* 10: 4622–4629
65 Iyer VR, Eisen MB, Ross DT, Schuler G, Moore T, Lee JCF, Trent JM, Staudt LM, Hudson J Jr, Boguski MS et al. (1999) The transcriptional program in the response of human fibroblasts to serum. *Science* 283: 83–87

66 Brose N, Rosenmund C (2002) Move over protein kinase C, you've got company: alternative cellular effectors of diacylglycerol and phorbol esters. *J Cell Sci* 115: 4399–4411
67 Kazanietz MG (2000) Eyes wide shut: protein kinase C isozymes are not the only receptors for the phorbol ester tumor promoters. *Mol Carcinog* 28: 5–11
68 Chang F, Steelman LS, Lee JT, Shelton JG, Navolanic PM, Blalock WL, Franklin RA, McCubrey JA (2003) Signal transduction mediated by the Ras/Raf/MEK/ERK pathway from cytokine receptors to transcription factors: potential targeting for therapeutic intervention. *Leukemia* 17: 1263–1293
69 Hazzalin CA, Mahadevan LC (2002) MAPK-regulated transcription: a continuously variable gene switch? *Nat Rev Mol Cell Biol* 3: 30–40
70 Soloaga A, Thomson S, Wiggin GR, Rampersaud N, Dyson MH, Hazzalin CA, Mahadevan LC, Arthur JS (2003) MSK2 and MSK1 mediate the mitogen- and stress-induced phosphorylation of histone H3 and HMG-14. *EMBO J* 22: 2788–2797
71 Clayton AL, Mahadevan LC (2003) MAP kinase-mediated phosphoacetylation of histone H3 and inducible gene regulation. *FEBS Lett* 546: 51–58
72 Zhong S, Zhang Y, Jansen C, Goto H, Inagaki M, Dong Z (2001) MAP kinases mediate UVB-induced phosphorylation of histone H3 at serine 28. *J Biol Chem* 276: 12932–12937
73 Barratt MJ, Hazzalin CA, Zhelev N, Mahadevan LC (1994) A mitogen- and anisomycin-stimulated kinase phosphorylates HMG-14 in its basic amino-terminal domain *in vivo* and on isolated mononucleosomes. *EMBO J* 13: 4524–4535
74 Chadee DN, Hendzel MJ, Tylipski CP, Allis CD, Bazett-Jones DP, Wright JA, Davie JR (1999) Increased Ser-10 phosphorylation of histone H3 in mitogen-stimulated and oncogene-transformed mouse fibroblasts. *J Biol Chem* 274: 24914–24920
75 Thomson S, Clayton AL, Mahadevan LC (2001) Independent dynamic regulation of histone phosphorylation and acetylation during immediate-early gene induction. *Mol Cell* 8: 1231–1241
76 Clayton AL, Rose S, Barratt MJ, Mahadevan LC (2000) Phosphoacetylation of histone H3 on c-fos- and c-jun-associated nucleosomes upon gene activation. *EMBO J* 19: 3714–3726
77 Cheung P, Tanner KG, Cheung WL, Sassone-Corsi P, Denu JM, Allis CD (2000) Synergistic coupling of histone H3 phosphorylation and acetylation in response to epidermal growth factor stimulation. *Mol Cell* 5: 905–915
78 Bos JL (1989) ras oncogenes in human cancer: a review. *Cancer Res* 49: 4682–4689
79 Barbacid M (1987) ras genes. *Annu Rev Biochem* 56: 779–827
80 Andreyev HJ, Norman AR, Cunningham D, Oates JR, Clarke PA (1998) Kirsten ras mutations in patients with colorectal cancer: the multicenter "RASCAL" study. *J Natl Cancer Inst* 90: 675–684
81 Lechner S, Muller-Ladner U, Renke B, Scholmerich J, Ruschoff J, Kullmann F (2003) Gene expression pattern of laser microdissected colonic crypts of adenomas with low grade dysplasia. *Gut* 52: 1148–1153
82 Hoshino R, Chatani Y, Yamori T, Tsuruo T, Oka H, Yoshida O, Shimada Y, Ari-i S, Wada H, Fujimoto J, Kohno M (1999) Constitutive activation of the 41-/43-kDa mitogen-activated protein kinase signaling pathway in human tumors. *Oncogene* 18: 813–822
83 Campbell PM, Der CJ (2004) Oncogenic Ras and its role in tumor cell invasion and metastasis. *Semin Cancer Biol* 14: 105–114
84 Calipel A, Lefevre G, Pouponnot C, Mouriaux F, Eychene A, Mascarelli F (2003) Mutation of B-Raf in human choroidal melanoma cells mediates cell proliferation and transformation through the MEK/ERK pathway. *J Biol Chem* 278: 42409–42418
85 Fischer AH, Chadee DN, Wright JA, Gansler TS, Davie JR (1998) Ras-associated nuclear structural change appears functionally significant and independent of the mitotic signaling pathway. *J Cell Biochem* 70: 130–140
86 Samuel SK, Minish TM, Davie JR (1997) Altered nuclear matrix protein profiles in oncogene transformed fibroblasts exhibiting high metastatic potential. *Cancer Res* 57: 147–151
87 Davie JR, Samuel SK, Spencer VA, Holth LT, Chadee DN, Peltier CP, Sun J-M, Chen HY, Wright JA (1999) Organization of chromatin in cancer cells: role of signalling pathways. *Biochem Cell Biol* 77: 265–275
88 Laitinen J, Saris P, Holtta E (1995) DNA methylation is not involved in the structural alterations of ornithine decarboxylase or total chromatin of c-Ha-rasVal 12 oncogene-transformed NIH-3 T3 fibroblasts. *J Cell Biochem* 57: 670–679
89 Davie JR, Chadee DN (1998) Regulation and regulatory parameters of histone modifications. *J Cell Biochem* 30/31: 203–213

90 Berger SL (2002) Histone modifications in transcriptional regulation. *Curr Opin Genet Dev* 12: 142–148
91 Crosio C, Fimia GM, Loury R, Kimura M, Okano Y, Zhou H, Sen S, Allis CD, Sassone-Corsi P (2002) Mitotic phosphorylation of histone H3: spatio-temporal regulation by mammalian Aurora kinases. *Mol Cell Biol* 22: 874–885
92 Davie JR (2003) MSK1 and MSK2 mediate mitogen- and stress-induced phosphorylation of histone H3: a controversy resolved. *Sci STKE* 2003: E33
93 Drobic B, Espino PS, Davie JR (2005) MSK1 activity and histone H3 phosphorylation in oncogene-transformed mouse fibroblasts. *Cancer Res* 64: 9076–9079
94 Chadee DN, Peltier CP, Davie JR (2002) Histone H1(S)-3 phosphorylation in Ha-ras oncogene-transformed mouse fibroblasts. *Oncogene* 21: 8397–8403
95 Yamamoto H, Monden T, Miyoshi H, Izawa H, Ikeda K, Tsujie M, Ohnishi T, Sekimoto M, Tomita N, Monden M (1998) Cdk2/cdc2 expression in colon carcinogenesis and effects of cdk2/cdc2 inhibitor in colon cancer cells. *Int J Oncol* 13: 233–239
96 Mihara M, Shintani S, Nakahara Y, Kiyota A, Ueyama Y, Matsumura T, Wong DT (2001) Overexpression of CDK2 is a prognostic indicator of oral cancer progression. *Jpn J Cancer Res* 92: 352–360
97 Meraldi P, Honda R, Nigg EA (2004) Aurora kinases link chromosome segregation and cell division to cancer susceptibility. *Curr Opin Genet Dev* 14: 29–36
98 Katayama H, Brinkley WR, Sen S (2003) The Aurora kinases: role in cell transformation and tumorigenesis. *Cancer Metastasis Rev* 22: 451–464
99 Goto H, Yasui Y, Nigg EA, Inagaki M (2002) Aurora-B phosphorylates histone H3 at serine28 with regard to the mitotic chromosome condensation. *Genes Cells* 7: 11–17
100 Wei Y, Yu L, Bowen J, Gorovsky MA, Allis CD (1999) Phosphorylation of histone H3 is required for proper chromosome condensation and segregation. *Cell* 97: 99–109
101 Dutertre S, Descamps S, Prigent C (2002) On the role of aurora-A in centrosome function. *Oncogene* 21: 6175–6183
102 Bolton MA, Lan W, Powers SE, McCleland ML, Kuang J, Stukenberg PT (2002) Aurora B kinase exists in a complex with survivin and INCENP and its kinase activity is stimulated by survivin binding and phosphorylation. *Mol Biol Cell* 13: 3064–3077
103 Ota T, Suto S, Katayama H, Han ZB, Suzuki F, Maeda M, Tanino M, Terada Y, Tatsuka M (2002) Increased mitotic phosphorylation of histone H3 attributable to AIM-1/Aurora-B overexpression contributes to chromosome number instability. *Cancer Res* 62: 5168–5177
104 Harrington EA, Bebbington D, Moore J, Rasmussen RK, Ajose-Adeogun AO, Nakayama T, Graham JA, Demur C, Hercend T, Diu-Hercend A et al. (2004) VX-680, a potent and selective small-molecule inhibitor of the Aurora kinases, suppresses tumor growth *in vivo*. *Nat Med* 10: 262–267
105 Davie JR (2004) Histone modifications. In: J Zlatanova, S Leuba (eds): *Chromatin structure and dynamics: state-of-the-art*. Elsevier, Amsterdam, 205–240
106 Workman JL, Abmayr SM (2004) Histone H3 variants and modifications on transcribed genes. *Proc Natl Acad Sci USA* 101: 1429–1430
107 Cremer M, Zinner R, Stein S, Albiez H, Wagler B, Cremer C, Cremer T (2004) Three dimensional analysis of histone methylation patterns in normal and tumor cell nuclei. *Eur J Histochem* 48: 15–28
108 Liang G, Lin JC, Wei V, Yoo C, Cheng JC, Nguyen CT, Weisenberger DJ, Egger G, Takai D, Gonzales FA, Jones PA (2004) Distinct localization of histone H3 acetylation and H3–K4 methylation to the transcription start sites in the human genome. *Proc Natl Acad Sci USA* 101: 7357–7362
109 Schotta G, Lachner M, Sarma K, Ebert A, Sengupta R, Reuter G, Reinberg D, Jenuwein T (2004) A silencing pathway to induce H3–K9 and H4–K20 trimethylation at constitutive heterochromatin. *Genes Dev* 18: 1251–1262
110 Li Y, Kirschmann DA, Wallrath LL (2002) Does heterochromatin protein 1 always follow code? *Proc Natl Acad Sci USA* 99 Suppl 4: 16462–16469
111 Nielsen SJ, Schneider R, Bauer UM, Bannister AJ, Morrison A, O'Carroll D, Firestein R, Cleary M, Jenuwein T, Herrera RE, Kouzarides T (2001) Rb targets histone H3 methylation and HP1 to promoters. *Nature* 412: 561–565
112 Cao R, Zhang Y (2004) The functions of E(Z)/EZH2-mediated methylation of lysine 27 in histone H3. *Curr Opin Genet Dev* 14: 155–164

113 Hamamoto R, Furukawa Y, Morita M, Iimura Y, Silva FP, Li M, Yagyu R, Nakamura Y (2004) SMYD3 encodes a histone methyltransferase involved in the proliferation of cancer cells. *Nat Cell Biol* 6: 731–740
114 Verma M, Srivastava S (2002) Epigenetics in cancer: implications for early detection and prevention. *Lancet Oncol* 3: 755–763
115 Brown R, Strathdee G (2002) Epigenomics and epigenetic therapy of cancer. *Trends Mol Med* 8: S43–S48
116 Kondo Y, Shen L, Issa JP (2003) Critical role of histone methylation in tumor suppressor gene silencing in colorectal cancer. *Mol Cell Biol* 23: 206–215
117 Espada J, Ballestar E, Fraga MF, Villar-Garea A, Juarranz A, Stockert JC, Robertson KD, Fuks F, Esteller M (2004) Human DNA methyltransferase 1 is required for maintenance of the histone H3 modification pattern. *J Biol Chem* 279: 37175–37184
118 Stirzaker C, Song JZ, Davidson B, Clark SJ (2004) Transcriptional gene silencing promotes DNA hypermethylation through a sequential change in chromatin modifications in cancer cells. *Cancer Res* 64: 3871–3877
119 Fahrner JA, Eguchi S, Herman JG, Baylin SB (2002) Dependence of histone modifications and gene expression on DNA hypermethylation in cancer. *Cancer Res* 62: 7213–7218
120 Moore SC, Jason L, Ausio J (2002) The elusive structural role of ubiquitinated histones. *Biochem Cell Biol* 80: 311–319
121 Davie JR, Murphy LC (1990) Level of ubiquitinated histone H2B in chromatin is coupled to ongoing transcription. *Biochemistry* 29: 4752–4757
122 Vassilev AP, Rasmussen HH, Christensen EI, Nielsen S, Celis JE (1995) The levels of ubiquitinated histone H2A are highly upregulated in transformed human cells: partial colocalization of uH2A clusters and PCNA/cyclin foci in a fraction of cells in S-phase. *J Cell Sci* 108: 1205–1215
123 Okawa Y, Takada K, Minami J, Aoki K, Shibayama H, Ohkawa K (2003) Purification of N-terminally truncated histone H2A-monoubiquitin conjugates from leukemic cell nuclei: probable proteolytic products of ubiquitinated H2A. *Int J Biochem Cell Biol* 35: 1588–1600
124 Parseghian MH, Hamkalo BA (2001) A compendium of the histone H1 family of somatic subtypes: an elusive cast of characters and their characteristics. *Biochem Cell Biol* 79: 289–304
125 Lennox RW, Cohen LH (1988) The production of tissue-specific histone complements during development. *Biochem Cell Biol* 66: 636–649
126 Tan KB, Borun TW, Charpentier R, Cristofalo VJ, Croce CM (1982) Normal and neoplastic human cells have different histone H1 compositions. *J Biol Chem* 257: 5337–5338
127 Giancotti V, Bandiera A, Ciani L, Santoro D, Crane-Robinson C, Goodwin GH, Boiocchi M, Dolcetti R, Casetta B (1993) High-mobility-group (HMG) proteins and histone H1 subtypes expression in normal and tumor tissues of mouse. *Eur J Biochem* 213: 825–832

Cancer: Cell Structures, Carcinogens and Genomic Instability
Edited by Leon P. Bignold

Aneuploidy, stem cells and cancer

Sen Pathak and Asha S. Multani

Department of Molecular Genetics, The University of Texas M. D. Anderson Cancer Center, 1515 Holcombe Boulevard, Houston, Texas 77030, USA

Summary. Telomeres which protect the individual chromosomes from disintegration, end-to-end fusion and maintain the genomic integrity during the somatic cell divisions play an important role in cellular aging. Aging and cancer development are linked with each other because cancer is considered a group of complex genetic diseases that develop in old cells and, in both, telomere attrition is involved. Numeric chromosome imbalance also known as aneuploidy is the hallmark of most solid tumors, whether spontaneous or induced by carcinogens. We provide evidence in support of the hypothesis that telomere attrition is the earliest genetic alteration responsible for the induction of aneuploidy. Dysfunctional telomeres are highly recombinogenic leading to the formation of dicentric chromosomes. During cell divisions, such complex chromosome alterations undergo breakage fusion bridge cycles and may lead to loss of heterozygosity (LOH) and gene amplification. Furthermore, we have provided evidence in support of the hypothesis that all types of cancer originate in the organ- or tissue-specific stem cells present in a particular organ. Cancer cells and stem cells share many characteristics, such as, self-renewal, migration, and differentiation. Metaphases with abnormal genetic constitution present in the lymphocytes of cancer patients and in some of their asymptomatic family members may have been derived from the organ-specific stem cells. In addition, evidence and discussion has been presented for the existence of cancer-specific stem cells. Successful treatment of cancer, therefore, should be directed towards these cancer stem cells.

Key words: Aneuploidy, dysfunctional telomeres, fluorescence *in situ* hybridization, genetic instability, stem cell.

Introduction

Cancer is not a single disease. It comprises a group of complex genetic diseases of uncontrolled cell division and is also one of the characteristics of aged cells. Aging and cancer development, therefore, are linked with each other. Most cancers are caused by chromosome and gene mutations that accumulate in a specific tissue or organ during the cellular aging. Genetic instability is exhibited by aging cells, both *in vitro* and *in vivo*, in the form of numerical (aneuploidy) and structural chromosomal alterations (translocation, deletion, amplification and inversion) [1]. More than 90% cancers are caused by exposure to environmental carcinogens. The insult inflicted by carcinogens are first faced by the termini of chromosomes, called telomeres, which are attached to the inner nuclear wall. Among their many functions, telomeres determine the domain and stability of individual chromosomes within the nucleus and serve as guardian of the genome [2]. Functional telomeres are essential for the normal segregation and maintenance of chromosomes during mitotic and meiotic divisions [3]. More recent information has shown that the maintenance of the telomere depends on interactions with an enzyme, telomerase, with telomeric proteins, and with some still undiscovered factors regulating the telomeric functions. Dysfunctional telomeres support the survival of aneuploid cells, a characteristic of many human and murine cancers.

The single unifying cellular mechanism that influences both aging and cancer development is the telomere dynamics [4, 5]. Cancer cells stabilize their telomere repeats either by a telomerase-dependent pathway or by the telomere-independent or alternate lengthening of telomere (ALT) pathway [6]. Unlike the murine somatic cells, human somatic cells lack or have diminished telomerase activity. This major difference in human and murine cells can easily explain why it is difficult to transform normal human cells, but easy to transform mouse cells, *in vitro*. Mouse and human somatic cells differ in many other respects, for example, in their responses to oxidative stress [7].

Most hematological neoplasms are known to arise from stem cells, whereas epithelial malignancies are generally considered to originate in differentiated organ- or tissue-specific somatic cells. Recently, a hypothesis was proposed that not only the hematological malignancies but also all solid tumors originate in organ- or tissue-specific stem cells or their immediate progeny (progenitor) cells [5, 8, 9]. This hypothesis is based mainly on two recent observations: the presence of stem cells in each and every human organ or tissue [10–12], and the presence of poorly differentiated cancer cells signaling a poor prognosis for patients. Since stem cells, especially embryonic stem (ES) cells, have the potential to differentiate into all three major tissue lineages, ectoderm, mesoderm and endoderm and their derivative organs [13–15], it is not unreasonable to propose that organ- or tissue-specific cancers originate in the organ- or tissue-specific stem cells [2, 8, 9]. Human ES cells have four unique characteristics: (1) self renewal, (2) differentiation into other cell types, (3) migration *in vivo*, and (4) cell death under unfavorable conditions [16, 17].

The purpose of this chapter is to discuss, in brief, the relationships between aneuploidy, stem cells and cancer development. That stem cells and aneuploidy play crucial roles in cancer development and metastasis will also be discussed in some detail under separate subtitles.

Aneuploidy and carcinogenesis

Mammalian species, in general, contain a diploid (2n) complement of chromosomes in the somatic cells of both sexes. However, cancer cells originating from the same diploid cells are mostly aneuploid, especially the solid tumors. Aneuploid constitutions are generally due to random chromosome losses/gains from non-disjunctions and multipolar mitoses, mostly originating from tetraploid (4n) cells [18–21]. Tetraploid cells are formed either by fusion of two diploid (2n) cells or due to the endomitosis of a diploid cell. In addition, segmental chromosomal losses or gains are due to structural chromosome alterations, including translocations, amplifications or deletions of certain segments. Because of these inherent characteristics, most cancer cells have genomic heterogeneity. In other words, each cell of a given tumor has its own chromosomal features except for certain specific common marker chromosomes. Inherent chromosomal instability, which can be due to the telomere erosion, plays a major role in causation of most cancers [22–24].

As early as 1890, von Hansemann [25] first suggested that cancer originates in an alteration in the genetic content of a cell. Later, Theodore Boveri [18, 26], while working on chromosomes of *Ascaris* and *Paracentrotus* sea urchin eggs, proposed his famous theory of malignancy. According to Boveri's theory, the neoplastic properties of a cancer are the consequence of chromosomal aberrations, and a malignant transformation results from the clonal expansion of a single genetically altered somatic cell. In our earlier publications we asked [9], among several questions, 'Could this somatic cell undergoing neoplastic transformation be an organ- or tissue-specific stem cell?'. Boveri, who first introduced the term centrosome [27], postulated that cancer cells are formed due to abnormal chromosome distribution originating as the consequence of multipolar mitosis, caused by the formation of multiple centrosomes. Recent molecular studies have provided strong evidence in support of Boveri's theory of malignancy [28–37]. The centrosome is an important actor of the cell division machinery. Its malfunction may cause abnormal chromosome segregation or no segregation at all, resulting in aneuploidy, the hallmark of cancers [37–46].

The original definition of aneuploidy was deviation of one or more chromosomes from the haploid (1n) state [47], which in the human is 23 chromosomes, in mouse 20 chromosomes, in rat 21 chromosomes, in cat 19 chromosomes, in Syrian hamster 22 chromosomes and in Chinese hamster 11 chromosomes. Presence of an extra copy (trisomy) or the absence of a chromosome (monosomy) is generally considered an example of aneuploidy. However, recently, the term aneuploidy has been used even for the presence of an extra segment or the deletion of a segment from a chromosome without a gain or loss in the total chromosome/centromere numbers. Currently, this term is used ambiguously to encompass all kinds of structural and numerical chromosome instabilities. Is aneuploidy that is caused by a dysfunctional centrosome, an early genetic change that initiates cancer formation? Some researchers, including the present authors, favor the opinion that aneuploidy indeed is the first causal step in tumor development. According to the strict classical definition of aneuploidy, primary leukemia and lymphomas, which do not show numerical anomalies but are characterized by their specific structural alterations, including reciprocal translocations and inversions, should not be considered aneuploid [47]. Presence of t(9;22) in chronic myelogenous leukemia, and t(8;14) and t(14;18) in different lymphomas are typical examples of human cancers. Only in the blast phase or at advance stage of the disease have numerical (aneuploidy) and additional structural anomalies been reported. Practically all cancer types, hematological and solid, contain structural anomalies in their genomes [2, 5, 45, 46, 48, 49]. That structural chromosome anomalies precede aneuploidy has even been reported in immortal fibroblast cultures of Li-Fraumani syndromes [50]. The important question is: which comes first, the numerical alterations (aneuploidy) or the structural modifications in cancer initiation? The obvious reply is: structural alteration due to telomere erosion comes first in the multistage carcinogenic process (Fig. 1).

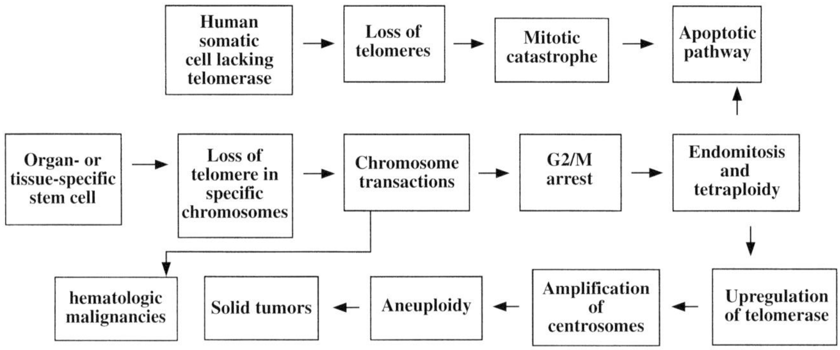

Figure 1. Pathways for apoptosis, aneuploidy and cancer initiation. Reprinted with permission and modified after [9].

Telomeres and chromosomal instability

Chromosomal anomalies have long been associated with, and are considered causal for, congenital birth defects and cancer initiation. The only difference is that children born with birth defects have minimal chromosomal alteration, whereas cancer cells and especially epithelial malignancies, have numerous defects. These abnormalities include numerical as well as structural alterations of different chromosomes in various neoplasias. This finding was possible due to the discovery of various chromosome banding techniques in early 1970s and the fluorescence *in situ* hybridization (FISH), comparative genomic hybridization (CGH) and spectral karyotyping (SKY) procedures later on. These molecular cytogenetic techniques have helped in dissecting break points of translocations, inversions, duplication and deletions, thus providing the pathological consequences of specific chromosome defects.

As mentioned earlier, the discovery of the Philadelphia (Ph) chromosome, t(9;22), in chronic myelogenous leukemia (CML) was the first reported cancer-specific cytogenetic defect [51, 52]. Since then, a number of cancer-specific chromosomal lesions (locations of oncogenes) have been identified [45, 46]. Chromosomal breakage and fusion (translocation) have been observed in all cancer cells. But, why do only certain chromosomes break and rejoin in a given cancer type? There is no satisfactory and definite answer available to such a question. We have, recently, hypothesized that only those chromosomes that have partially dysfunctional telomeres undergo such genetic changes. In other words, telomeres serve as the guardian of individual chromosomes and protect them from cellular challenges [2, 5]. It is also reported that not all chromosomes or both arms of the same chromosome have a similar number of telomeric repeats [53].

Telomeres are special DNA-protein structures present at the ends of linear eukaryotic chromosomes. Since the pioneer research of Muller and McClintock, the telomere has been recognized as protecting the chromosomal ends from degradation and fusion to other broken ends [54–57]. The telomeric DNA consists of G-rich sequences, for example, T_2AG_3 repeated many times in all vertebrate species (reviewed in [58]). In humans, all chromosome ends have approximately 5 kb of telomeric DNA [53]. Telomerase, a ribonucleoprotein reverse transcriptase, is responsible for stabilization of telomeres in cancer and germ cells. This enzyme is composed of an RNA subunit, hTERC, and a catalytic protein subunit, hTERT, in humans.

Chromosome instability and replication senescence

In most mammalian somatic cells, telomeres shorten with each round of cell division. In normal human somatic cells, the telomere length progressively decreases owing to an end replication problem, and the cell population undergoes either senescence or neoplastic transformation depending on the telomerase status. Upregulation/activation of the telomerase in human aged somatic cells may help stabilize the telomeres or cap them as functional. Under such conditions, these cells with genetic instability may get transformed and initiate cancer development. On the other hand, in the absence of telomerase activity, aged cells with continued telomere attrition may lead to apoptosis.

As shown diagrammatically in Figure 2, organ- or tissue-specific stem/progenitor cells, when insulted by clastogens, may either readily undergo altruistic apoptosis or may undergo chromosome rearrangements, gene amplification and aneuploidy, finally resulting in cancer formation, especially solid tumors. In the case of hematological malignancies, only specific translocations or inversions are sufficient to activate proto-oncogene(s) and the emergence of neoplastic cells. Aneuploidy in the strict classical sense may not be the primary cause of initiation of hematological malignancies. It may, however, be necessary at late stage during the blast phase of the disease. In solid tumors, the classical aneuploidy definitely plays an important role in initiation/promotion of the disease in which loss of heterozygosity (LOH) of tumor suppressor gene(s) is required [59, 60]. The presence of a high degree of aneuploidy is not uncommon among cancer cells [61]. Approximately 88% of all colon cancers are characterized by chromosome instabilities [62].

In the absence of telomerase, continued somatic cell division is accompanied by progressive shortening of telomeres. When the telomere lengths reach a critical stage, cells stop dividing and enter senescence. It was Hayflick who

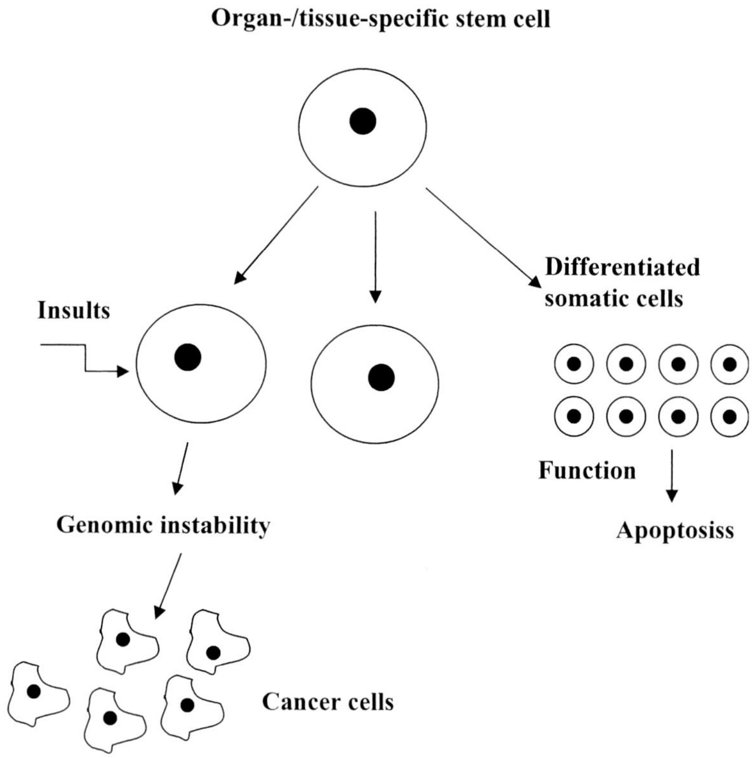

Figure 2. Diagram showing cancer formation from organ-specific stem cells.

first suggested that diploid human fibroblasts have a limited number of replication in culture, after which they stop dividing [63], and reach mortality stage 1 (M1) [64]. However, viral oncogenes, e.g., AgTsv40, may transform such cells even with critically shortened telomeres. Such cells may enter another mortality stage, stage 2 (M2), and in the senescent stage may remain metabolically active but without undergoing DNA synthesis. These cells with drastically changed morphology become large and express β-galactosidase activity [65]. Replicative senescence is also a genetically dominant phenotype [66].

The question here is: What factor(s) are responsible for driving primary somatic cells to enter replicative senescence? One of the answers lies in telomere erosion and the absence of telomerase. In mTERC$^{-/-}$ mutant mice, telomere attrition has been shown to cause genomic instability, progressive infertility and even the induction of epithelial malignancies in late generation animals [67–69]. Primary murine embryonic fibroblasts (MEFs) from the mTERC$^{-/-}$ mouse were used to study the mechanism of dysfunctional telomeres and a number of telomere-associated proteins (Tab. 1). These mice are telomerase null because they lack the gene that encodes for telomerase RNA. MEFs derived from such mutant mice, have reduced ability to immortalize sponta-

Table 1. Localization of telomerase gene and telomere-associated proteins on human chromosomes*

Name	Interaction	Functions at telomeres	Chromosome localization	References
Telomerase hTERT hTERC	With T_2AG_3 overhang, Telomerase	Telomere elongation RNA subunit	5p15 3q26	[70] [71]
Specific proteins				
Pot1	With G-rich strand	Length maintenance	7	[72]
TRF1	T loops	Negative length regulator (dependent on telomerase)	8q13	[73]
TRF2	T loops	Negative length regulator (independent of telomerase)	16q22	[74]
TANK1/2	With telomere	Positive length regulator	8p23/10q23	[75, 76]
TIN2	With TRF1	Positive length regulator	14q11	[77, 78]
RAP1	With TRF2	Length regulator	16	[79]
PINX1	With TRF1/Pin2	Telomerase inhibitor	8p23	[80]
Nonspecific proteins				
Ku70/Ku86	With telomeric repeats	Negative length regulator	2q35/22q11	[81, 82]
DNA-PKCA	With telomeric DNA	Capping of telomere	8q11	[83]
Rad50 NSB/ MRE11	With TRF2	T-loop sterilization	5q31	[84]
Rad51	In ALT cells	Recombination in ALT cells	15q15	[85]
WRN/BLM	With 3' overhang	Telomere structure maintenance	8p12/15q26	[86, 87]
p53	With single-strand T-loop	Telomere structure	17p13	[88, 89]
ATM	With TRF1	Telomere chromatin structure	11q22	[90, 91]

* Reprinted with permission and modified after [92].

neously in culture. These findings indicate that telomere attrition limits the replicative potential of MEF *in vitro*. A typical metaphase spread from a fifth generation (G5) mTERC$^{-/-}$ mouse fibroblast revealing a chromosome fusion product and 41 chromosomal arms is shown in Figure 3.

Stem cell characteristics of cancer and metastatic cells

As listed in Table 2, stem cells and cancer cells have many characteristics in common. Organ-specific stem cells are known to participate in organ or tissue homeostasis by constantly replacing differentiated somatic cells lost as a result

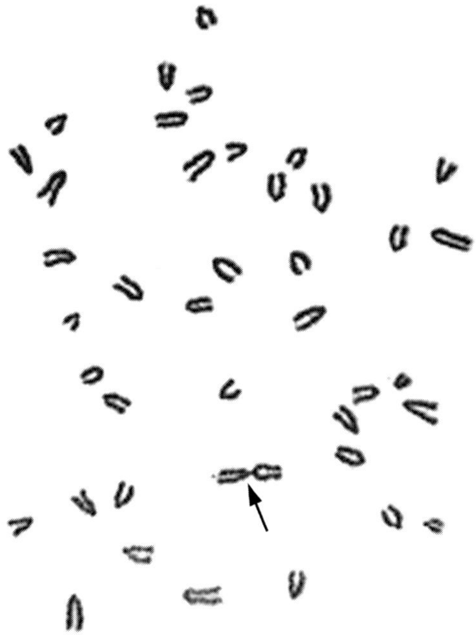

Figure 3. A typical metaphase spread from a fifth generation (G5) mTERC$^{-/-}$ mouse fibroblast showing a fusion product (arrow) and 41 chromosomal arms, an example of aneuploidy.

Table 2. Common characteristics of stem cells and cancer cells*

	Stem cells	Cancer cells
1	Proliferate indefinitely	Proliferate indefinitely
2	Self renewal by similar signals	Self-renewal by similar signals
3	Are heterogeneous, with different phenotypes	Are heterogeneous, with different phenotypes
4	Migrate	May metastasize (migrate)
5	Express telomerase	Express telomerase
6	Have extended telomere repeats	Metastatic cells have extended telomeric repeats
7	Differentiate	Differentiate
8	Can be tissue-specific	Can be tissue-specific
9	Undergo organogenesis	Undergo limited organogenesis
10	Undergo apoptosis	Undergo apoptosis

* Reprinted with permission and modified after [9].

of cellular aging, injury, or disease. A cancer is essentially an aberrant organ or tissue that acquires the characteristics for unregulated proliferation through the accumulation of genetic mutations. Like stem cells, cancer cells are able to proliferate indefinitely and in some cases may lead to metastasis. Cancer metastasis, which is analogous to species migration [93], is achieved by a series of genetic mutations and amplification of telomereic DNA in the cancer cells [94]. Some of these sequential steps are: dissociation from the primary tumor mass, extensive vascularization, invasion, detachment, embolization, extravasation into the organ parenchyma, and, finally, vascularization and proliferation within the organ site [95]. Almost all of these steps involve cell migration. That stem cells can migrate *in vivo* explains most of these steps, once we consider that cancers originate in the stem cells and that the poorly differentiated cancer cells still retain some characteristics of the stem/progenitor cells.

Angiogenesis or the formation of new blood vessels is a pre-requisite for cancer cell growth and metastasis to distant organs [96]. Cancer cells have the potential to form not only their own blood channel [97, 98] and mosaic-blood vessels [99], but also a variety of blood cell types [10]. It is therefore clear that stem cells are pluripotent and can differentiate into various cell types, including lymphocytes. That solid tumors arise from organ- or tissue-specific stem cells has profound implications for cancer treatment. The target for successful cancer treatment and chemoprevention must be the cancer-specific stem cells.

Do cancer cells have their own stem cells?

Stem cells, especially the ES cells, have infinite proliferative and developmental potentials. Self renewal, migration and differentiating characteristics of stem cells make them suitable as a source for organ and tissue regeneration. We have previously proposed that all cancers originate in organ- and tissue-specific stem cells [5, 8]. Stem cells, sometimes also called master cells, are known to participate in organ or tissue homeostasis by replacing aged somatic cells lost as a result of aging, injury or disease. It has been shown that mouse neural stem cells can differentiate into brain cells (astrocytes, oligodendrocytes and neurons) and can form a variety of blood cell types, including hematopoietic cells [10–12]. Neural stem cells from adult mouse brain are capable of forming chimeric chick and mouse embryos, and give rise to all germ layers and cell types [11]. These observations taken together suggest that adult neural stem cells have a pluripotent phenotype, and may have potential to produce a variety of cell and organ types for transplantation. Although the molecular mechanism of the neural stem cell proliferation and differentiation is not fully resolved, the phosphatase and tensin gene (PTEN) homolog, also known to be mutated in multiple advanced cancers (MMACI), from human chromosome 10q22-24 is a candidate tumor suppressor gene implicated in such cellular phenotypes [100].

The burning question is: Do cancer cells have their own stem cells? As early as in 1956, a famous cytogeneticist, Sajiro Makino from Japan, proposed that cancer cells may have their own stem cells [101]. This idea was has received further support [102, 103]. Recently, Reya and associates [104] and Kondo et al. [105] have brought this hypothesis into the limelight by presenting molecular evidence in support of this concept. The later group has isolated "cancer stem cells", as a small side population (SP), even from the long-term established tumor cell lines including C6 glioma, MCF-7 breast cancer, B104 neuroblastoma and HeLa cell lines [105].

Evidence that cancer develops from stem cells, not differentiated somatic cells

There is plenty of evidence to support the statement that dividing (cycling) cells are more susceptible than the quiescent (non-dividing) cells to accumulate mutations when challenged by the environmental mutagens. Most differentiated somatic cells perform their functions in an organ- and tissue-specific environment and are then replaced by proliferation of specialized stem or progenitor cells [106]. During wound healing and in disease conditions, organ- or tissue-specific stem cells or their progenitors divide, migrate and help in repair. The stem or progenitor cells form one cell that remains a stem cell and another cell that differentiates into the specialized function-oriented mature cell. Fully differentiated somatic cells perform their functions and then undergo apoptosis (Fig. 2). They are replaced by the newly differentiated somatic cells in the organ. Since cancer cells originate by the interactions of environmental carcinogens and the genetic make-up of the person, it is worth considering that stem cells, being poorly differentiated somatic cells with the phenotypes of their progenitors, also having telomerase activity, are the target of neoplastic transformation [104, 105, 107–109]. In a recent book chapter, Sell [110] speculated that most carcinomas and adenocarcinomas originate in the organ-specific stem cells. Of course, most hematological malignancies have their origin in stem cells. In fact, cancer originates from an imbalance between the rate of cell division and the rate of differentiation and cell death. Also, maturation arrest of stem cell differentiation has been considered a common pathway for the origin of teratocarcinoma and epithelial malignancies [111]. Inherently, the stem cells are characterized by substantial longevity, replication potential and telomerase activity. In addition, they are also known to have much longer mean telomere length, which is a survival factor for the cells [2].

There is mounting evidence to suggest that practically every organ has its own stem cell reservoir. Although these specialized cells are present in a small pocket in the organ-specific environment, their potential to replace damaged cells is controlled by the cell requirement. This is analogous to the "Seed and Soil" hypothesis proposed by Paget [112] for metastatic breast cancer cells. The presence of organ-specific stem cells has been reported in lung, mammary gland, liver, pancreas, kidney, heart, retina, muscle, skin and brain of mammalian species including human. Even during angiogenesis, angioblasts, which are the precursors of endothelial cells, act as progenitor cells with several stem cell characteristics.

Are circulating abnormal metaphases derived from organ- or tissue-specific stem cells of cancer patients?

The initial (primary) genetic (chromosomal) alterations associated with cancer development do not necessarily occur in every somatic cell [113]. It has been proposed that predisposed individuals inherit susceptibility traits that makes their specific chromosomes prone to breaking at a particular loci [49]. The chromosomes of a cancer-predisposed individual may undergo specific alterations at relatively low frequency in all tissues, including peripheral blood lymphocytes (PBL). Could this trait be the attrition of telomere in those chromosomes? Clastogens that are able to induce chromosome-specific aberrations have been described previously in a separate report [2].

Specific cytogenetic alterations were first identified in PBL as being associated with chromosome 13 in retinoblastoma, with chromosome 11 in Wilms' tumor, with chromosome 3 in renal cell carcinoma, with chromosomes 2, 5 and 11 in colorectal cancer, with chromosomes 1, 6 and 9 in melanoma, with chromosome 1 in endometrial cancer, and chromosomes 5, 7, 8, 10, and 16 in prostate cancer (see reviews [2, 49]). Subsequent reports have shown specific chromosomal changes in a small percentage (1–3%) of phytohemagglutinin-stimulated lymphocytic metaphases of various epithelial malignancies, including breast, lung, prostate and other adenocarcinomas [114, 115]. These circulating aberrant metaphases are not cancer cells because: (1) they have mainly chromatid breaks, simple translocations or deletions; (2) the patients whose tumor cells and PBL were both analyzed cytogenetically had mainly chromatid breaks in the PBL but stable marker chromosomes involving the same chromosomes in their tumor cells [116]; and (3) such types of chromosomal alterations are present even in the PBLs of some asymptomatic family members [117]. Could these rare abnormal metaphases be coming from the tissue- or organ-specific stem cells? Undoubtedly, more research in future is required to substantiate this hypothesis.

Conclusion

In conclusion, we provide evidence for the proposal that telomere attrition (dysfunction) is the earliest genetic alteration in the organ- or tissue-specific stem cells, which then is responsible for aneuploidy and is the cause of all types of cancer. Metaphases with abnormal genetic constitutions present in the PBL of cancer patients and some of their asymptomatic family members might be derived from the organ-specific stem cells. Successful treatment of cancer by different modalities and chemoprevention strategies should be directed towards these cancer stem cells. It would be most rewarding to develop isolation procedures for these organ-specific stem cells for their further biological characterization and to elucidate the mechanism of proliferation, progression and metastasis of cancer.

Acknowledgements
We would like to thank Amanda Weppler and Gunjan P. Multani for preparing the figures and Burr and Cynthia Furlong for their editorial comments. This work was supported in part by the institutional T.C. Hsu Molecular Cytogenetic Core established in the Department of Molecular Genetics. Our apology to all those investigators whose work could not be cited here due to constrain of space.

References

1 Sohn SH, Multani AS, Gugnani PK, Pathak S (2002) Telomere erosion-induced mitotic catastrophe in continuously grown Chinese hamster Don cells. *Exptl Cell Res* 279: 271–276
2 Pathak S (2001) Telomeres in human cancer research. 10th All India Congress of Cytology and Genetics Award Lecture. *Perspect Cytol Genet* 10: 13–22
3 Dynek JN, Smith S (2004) Resolution of sister telomere association is required for progression through mitosis. *Science* 304: 97–100
4 Multani AS, Narayan S, Jaiswal AS, Zhao Y-J, Barkley RA, Furlong CL, Pathak S (2002) Telomere dynamics, aging, and cancer: study of human syndromes characteristic of premature aging. *J Anti-Aging Med* 5: 271–281
5 Pathak S (2003) Cancer biology and genetic heterogeneity. *Proc Indian Natl Sci Acad* B69: 11–22
6 Henson J D, Neumann A A, Yeager T R, Reddel R R (2002) Alternative lengthening of telomeres in mammalian cells. *Oncogene* 21:598–610
7 Parrinello S, Samper E, Krtolica A, Goldstein J, Melov S, Campisi J (2003) Oxygen sensitivity severely limits the replicative lifespan of murine fibroblasts *Nature Cell Biol* 5: 741–747
8 Pathak S (2002) Organ- and tissue-specific stem cells and carcinogenesis. *Anticancer Res* 22: 1353–1356
9 Pathak S, Multani AS, Furlong CL, Sohn SH (2002) Telomere dynamics, aneuploidy, stem cells, and cancer (review). *Int J Oncol* 20: 637–641
10 Bjornson C R R, Rietze R L, Reynolds B A, Bagli M C, Vescovi A L (1999) Turning brain into blood: a hematopoietic fate adopted by adult neural stem cells *in vivo*. *Science* 283: 534–537
11 Clarke D L, Johansson C B, Wolbertz J, Veress B, Nilsson E, Karlström H, Lendahl U, Frisen J (2000) Generalized potential of adult neural stem cells. *Science* 288: 1660–1663
12 Temple S (2001) The development of neural stem cells. *Nature* 414: 112–117
13 Lagasse E, Connors H, Al-Dhalimy M, Reitsma M, Dohse M, Osborne L, Wang X, Finegold M, Weissna IL, Grompe M (2000) Purified hematopoietic stem cells can differentiate to hepatocytes *in vivo*. *Nature Med* 6: 1229–1234
14 Krause DS, Theise ND, Collector MI, Henegariu O, Hwang S, Gardner R, Neutzel S, Sharkis SJ (2001) Multi-organ, multi-lineage engraftment by a single bone marrow derived stem cell. *Cell* 105: 369–377
15 McCulloch EA (2003) Normal and leukemia hematopoietic stem cells and lineages. In: S Sell (ed.): *Stem Cells Handbook*. Humana Press, Totowa, NJ, 119–132
16 Brüstle O, Jones KN, Learish RD, Karrarn K, Choudhury K, Wiestler OD, Duncan ID, Mckay DG (1999) Embryonic stem cell-derived glial precursors: a source of myelinating transplants. *Science* 285: 754–756
17 Pittenger MF, Mackay AM, Beck SC, Jaiswal RK, Douglas R, Mosca JD, Moorman MA, Simonetti DW, Craig S, Marshak DR (1999) Multilineage potential of adult human mesenchyma stem cells. *Science* 284: 143–147
18 Boveri T (1902, 1964) On multipolar mitoses as a means of analysis of the cell nucleus. In: B Willer, J Oppenheimer (eds): *Foundations of Experimental Embryology*. Englewood Cliffs, NJ, Prentice Hall, 74–97
19 Hsu TC, Moorhead P (1956) Chromosome anomalies in human neoplasms with special reference

to the mechanisms of polyploidization and aeuploidization in the HeLa strain. *Ann N Y Acad Sci* 63: 1083–1094
20 Levine DS, Sanchez CA, Rabinovitch PS, Reid BJ (1991) Formation of the tetraploid intermediate is associated with the development of cells with more than four centrioles in the elastase-simian-virus 40 tumor antigen transgenic mouse model of pancreatic cancer. *Proc Natl Acad Sci USA* 88: 6427–6431
21 Rasnick D (2002) Aneuploidy theory explains tumor formation, the absence of immune surveillance, and the failure of chemotherapy. *Cancer Genet Cytogenet* 136: 66–72
22 Pathak S, Risin S, Brown NM, Berry K (1994) Telomeric association of chromosomes is an early manifestation of programmed cell death. *Int J Oncol* 4: 323–328
23 Pathak S, Dave BJ, Gagos SH (1994) Chromosome alterations in cancer development and apoptosis. *In Vivo* 8: 843–850
24 Feldser DM, Hackett JA, Greider CW (2003) Telomere dysfunction and the initiation of genomic instability. *Nat Rev Cancer* 3: 623–627
25 von Hansemann D (1890) Ueber asymmetrische Zelltheilung in Epithelkrebsen und deren biologische Bedeutung. *Virchows Arch [A]* 119: 229–326
26 Boveri T (1929) *The Origin of Malignant Tumors* (translated by Marcella Boveri). Williams and Wilkins, Baltimore
27 Boveri T (1901) *Zellenstudien IV*. Uever die Nature der Centrosomen. Fisher, Jena
28 Zhou H, Kuang J, Zhong L, Kuo WL, Gray JW, Sakin A, Brinkley BR, Sen S (1998) Tumor amplified kinase STK15/BTAK induces centrosome amplification, aneuploidy and transformation. *Nat Genet* 20: 189–193
29 Pihan GA, Doxsey SJ (1999) The mitotic machinery as a source of genetic instability in cancer. *Semin Cancer Biol* 9: 289–303
30 Lingle WL, Barrett SL, Negron VC, D'Assoro C, Salisbury JL (2002) Centrosome amplification drives chromosomal instability in breast tumor development. *Proc Natl Acad Sci USA* 99: 1978–1983
31 Brinkley BR, Goepfert TM (1998) Supernumerary centrosomes and cancer: Boveri's hypothesis resurrected. *Cell Motil Cytoskeleton* 41: 281–288
32 Duensing S, Duensing A, Crum CP, Munger K (2001) Human papilloma virus type 16E7 oncoprotein-induced abnormal centrosome synthesis is an early event in the evolving malignant phenotype. *Cancer Res* 61: 2356–2360
33 Max J (2001) Do centrosome abnormalities lead to cancer? *Science* 291: 426–429
34 Fukasawa K, Choi T, Kuriyana R, Rulong S, Vande Woude GF (1996) Abnormal centrosome amplification in the absence of p53. *Science* 271: 1744–1747
35 Duesberg P (1999) Are centrosomes or aneuploidy the key to cancer? *Science* 284: 2091–2092
36 Duesberg P, Rasnick D (2000) Aneuploidy, the somatic mutation that makes cancer a species of its own. *Cell Motil Cytoskeleton* 47: 81–107
37 Duesberg P, Fabarius A, Hehlmann R (2004) Aneuploidy, the primary cause of the multilateral genomic instability of neoplastic and preneoplastic cells. *IUBMB Life* 56: 65–81
38 Li R, Yerganian G, Duesberg P, Kraemer A, Willer A, Rausch C, Hehlmann R (1997) Aneuploidy correlates 100% with chemical transformation of Chinese hamster cells. *Proc Natl Acad Sci USA* 94: 14506–14511
39 Li R, Sonik A, Stindl R, Rasnick D, Duesberg P (2000) Aneuploidy *versus* gene mutation hypothesis of cancer: recent study claims mutation but is found to support aneuploidy. *Proc Natl Acad Sci USA* 97: 3236–3241
40 Rasnick D, Duesberg P (1999) How aneuploidy affects metabolic control and causes cancer? *Biochem J* 340: 621–630
41 Deusberg PH (2003) Are cancers dependent on Oncogenes or on aneuploidy? *Cancer Genet Cytogentics* 143: 89–91
42 Deusberg P, Rausch C, Rasnick D, Hehlmann R (1998) Genetic instability of cancer cells is proportional to their degree of aneuploidy. *Proc Natl Acad Sci USA* 95: 13692–13697
43 Duesberg P, Li R, Rasnick D, Rausch C, Willer A, Kraemer A, Yerganian G, Hehlmann R (2000) Aneuploidy precedes and segregates with chemical carcinogenesis. *Cancer Genet Cytogenet* 119: 83–93
44 Ghadimi BM, Sackett DL, Difilippantonio MJ, Schrock E, Neumann T, Jauho A, Auer G, Ried T (2000) Centrosome amplification and instability occurs exclusively in aneuploid, but not in diploid colorectal cancer cell lines, and correlates with numerical chromosomal aberrations.

Genes Chromosomes Cancer 27: 183–190
45 Sandberg AA (1990) *The Chromosomes in Human Cancer and Leukemia*. Elsevier, New York
46 Heim S, Mitelman F (1995) *Cancer Cytogenetics: Chromosomal and Molecular Genetic Aberrations of Tumor Cells*. John Wiley and Sons, Inc., New York
47 Rieger R, Michaelio A, Green MM (1976) *Glossary and Genetics and Cytogenetics: Classical and Molecular.* Springer-Verlag, Berlin, 30–31
48 Pathak S (1986) Cytogenetics of solid tumors: renal cell carcinoma, malignant melanoma, retinoblastoma and Wilms' tumor. In: AA Luderer, HH Weetal (eds): *The Human Oncogenic Viruses Molecular Analysis and Diagnosis*. The Humana Press, Clifton, NJ, 43–87
49 Pathak S (1990) Cytogenetic abnormalities in cancer: with special emphasis on tumor heterogeneity. *Cancer Metastasis Rev* 8: 299–318
50 Bischoff FZ, Sun OY, Pathak S, Grant G, Siciliano MJ, Giovanella BC, Strong LC, Tainsky MA (1990) Spontaneous abnormalities in normal fibroblasts from patients with Li-Fraumeni cancer syndrome: aneuploidy and immortalization. *Cancer Res* 50: 7979–7984
51 Nowell PC, Hungerford DA (1960) A minute chromosome in human chronic granulocytic leukemia. *Science* 132: 1497
52 Rowley JD (1973) A new consistent chromosomal abnormality in chronic myelogenous leukemia identified by quinacrine fluorescence and Giemsa staining. *Nature* 243: 290–293
53 Martens UM, Zijlmans JM, Poon SS, Dragowska W, Yui J, Chavez EA, Ward RK, Lansdorp PM (1998) Short telomeres on human chromosome 17p. *Nat Genet* 18: 76–80
54 Muller HJ (1938) The remaking of chromosomes. *Collecting Net* 13: 183–195
55 McClintock B (1938) The fusion of broken ends of sister half-chromatids following chromatid breakage at meiotic anaphases. *Missouri Agricultural Exptl Station Res Bull* 290: 1–48
56 McClintock B (1941) Spontaneous alterations in chromosome size and form in *Zea mays. Cold Spring Harbor Symp Quant Biol* 9: 72–81
57 McClintock B (1951) Chromosome organization of genic expression. *Cold Spring Harbor Sym Quant Biol* 16: 13–47
58 Greider CW (1996) Telomere length regulation. *Annu Rev Biochem* 65: 337–365
59 Knudson AG (1971) Mutation and cancer: statistical study of retinoblastoma. *Proc Natl Acad Sci USA* 68: 820–823
60 Weinberg RA (1989) Oncogenes, antioncogenes, and the molecular bases of multistep carcinogenesis. *Cancer Res* 49: 3713–3721
61 Sen S (2000) Aneuploidy and cancer. *Curr Opin Oncol* 12: 82–88
62 Lengauer C, Kinzler KW, Vogelstein B (1998) Genetic instabilities in human cancers. *Nature* 396: 643–649
63 Hayflick L (1965) The limited *in vitro* lifetime of human diploid cell strains. *Exp Cell Res* 37: 614–636
64 Wright WE, Pereira-Smith OM, Shay JW (1989) Reversible cellular senescence: implications for immortalization of normal human diploid fibroblasts. *Mol Cell Biol* 9: 3088–3092
65 Dimri GP, Lee X, Basile G, Acosta M, Scott G, Roskelley C et al. (1995) A biomarker that identifies senescent human cells in culture and in aging skin *in vivo*. *Proc Natl Acad Sci USA* 92: 9363–9367
66 Pereira-Smith OM, Smith JR (1988) Genetic analysis of indefinite division in human cells: identification of four complementation groups. *Proc Natl Acad Sci USA* 85: 6042–6046
67 Hande MP, Samper E, Lansdorp P, Blasco MA (1999) Telomere length dynamics and chromosomal instability in cells derived from telomerase null mice. *J Cell Biol* 144: 589–601
68 Artandi SE, Chang S, Lee S-L, Alson S, Gottlieb GJ, Chin L, DePinho RA (2000) Telomere dysfunction promotes non-reciprocal translocations and epithelial cancers in mice. *Nature* 406: 641–645
69 Blasco MA, Lee HW, Hande MP, Samper E, Lansdorp PM, DePinho RA, Greider CW (1997) Telomere shortening and tumor formation by mouse cells lacking telomerase RNA. *Cell* 91: 25–34
70 Meyerson M, Counter CM, Eaton EN, Ellisen LW, Steiner P, Caddle SD, Ziaugra L, Beijersbergen RL, Davidoff MJ, Liu Q et al. (1997) hEST2, the putative human telomerase catalytic subunit gene, is up-regulated in tumor cells and during immortalization. *Cell* 90: 785–795
71 Soder AI, Hoare SF, Muir S, Going JJ, Parkinson EK, Keith WN (1997) Amplification, increased dosage and *in situ* expression of the telomerase RNA gene in human cancer. *Oncogene* 14: 1013–1021
72 Baumann P, Cech TR (2001) Pot 1, the putative telomere end-binding protein in fission yeast and

humans. *Science* 292: 1171–1175
73 Broccoli D, Chong L, Oelmann S, Fernald AA, Marziliano N, van Steensel B, Kipling D, Le Beau MM, de Lange T et al. (1997) Comparison of the human and mouse genes encoding the telomeric protein, TRF 1: chromosomal localization, expression and conserved protein domains. *Hum Mol Genet* 6: 69–76
74 Sakaguchi AY, Padalecki SS, Mattern V, Rodriguez A, Leach RJ, McGill JR, Chavez M, Giambernardi TA (1998) Chromosomal sublocalization of the transcribed human telomere repeat binding factor 2 gene and comparative mapping in the mouse. *Somat Cell Mol Genet* 24: 157–163
75 Cook BD, Dynek JN, Chang W, Shostak G, Smith S (2002) Role for the related poly (ADP-Ribose) polymerases tankyrase 1 and 2 at human telomeres. *Mol Cell Biol* 22: 332–342
76 Kaminker PG, Kim SH, Taylor RD, Zebarjadian Y, Funk WD, Morin GB, Yaswen P, Campisi J (2001) TANK2, anew TRF1-associated poly (ADP-ribose) polymerase, causes rapid induction of cell death upon overexpression. *J Biol Chem* 276: 35891–35899
77 Kim SH, Kaminker P, Campisi J (1999) TIN2, a new regulator of telomere length in human cells. *Nat Genet* 23: 405–412
78 Simonsson T (2001) The human TINF2 gene organization and chromosomal localization. *Biochimie* 83: 433–435
79 Li B, Oestreich S, de Lange T (2000) Identification of human Rap1: implications for telomere evolution. *Cell* 101: 471–483
80 Zhou XZ, Lu KP (2001) The Pin2/TRF1- interacting protein PinX1 is a potent telomerase inhibitor. *Cell* 107: 347–359
81 Chen DJ, Marrone BL, Nguyen T, Stackhouse M, Zhao Y, Siciliano MJ (1994) Regional assignment of a human DNA repair gene (XRCC5) to 2q35 by X-ray hybrid mapping. *Genomics* 21: 423–427
82 Hsu HL, Gilley D, Blackburn EH, Chen DJ (1999) Ku is associated with the telomere in mammals. *Proc Natl Acad Sci USA* 96: 12454–12458
83 Sipley JD, Menninger JC, Hartley KO, Ward DC, Jackson SP, Anderson CW (1995) Gene for the catalytic subunit of the human DNA-activated protein kinase maps to the site of the XRCC7 gene on chromosome 8. *Proc Natl Acad Sc USA* 92: 7515–7519
84 Zhu XD, Kuster B, Mann M, Petrini JH, de Lange T (2000) Cell-cycle-regulated association of RAD50/MRE11/NBS1 with TRF2 and human telomeres. *Nat Genet* 25: 347–352
85 Takahashi E, Matsuda Y, Hori T, Yasuda N, Tsuji S, Mori M, Yoshimura Y, Yamamoto A, Morita T, Matsushiro A (1994) Chromosome mapping of the human (RECA) and mouse (Reca) homologs of the yeast RAD51 and *Escherichia coli* recA genes to human (15q15.1) and mouse (2F1) chromosomes by direct R-banding fluorescence *in situ* hybridization. *Genomics* 19: 376–378
86 Schellenberg GD, Martin GM, Wijsman EM, Nakura J, Miki T, Ogihara T (1992) Homozygosity mapping and Werner's syndrome. *Lancet* 339: 1002
87 German JL, Roe AM, Leppert MF, Ellis NA (1994) Bloom syndrome: an analysis of consanguineousfamilies assigns the locus mutated to chromosome band 15q26.1. *Proc Natl Acad Sci USA* 91: 6669–6673
88 Benchimol S, Lamb P, Crawford LV, Sheer D, Shows TB, Bruns GA, Peacock J (1985) Transformation associated p53 protein is encoded by a gene on human chromosome 17. *Somat Cell Mol Genet* 11: 505–510
89 Stansel R M, Subramanian D, Griffith JD (2002) p53 binds telomeric single strand overhangs and t-loop junctions *in vitro*. *J Biol Chem* 277: 11625–11628
90 Gatti RA, Berkel I, Boder E, Braedt G, Charmley P, Concannon P, Ersoy F, Foroud T, Jaspers NG, Lange K et al. (1988) Localization of an ataxia-telangiectasia gene to chromosome 11q22-23. *Nature* 336: 577–580
91 Pandita TK (2002) ATM function and telomere stability. *Oncogene* 21: 611–618
92 Mathieu N, Pirzio L, Freulet-Marriere M-A, Desmaze C, Sabatier L (2004) Telomeres and chromosomal instability. *CMLS Cellular and Molecular Life Sciences* 61: 641–656
93 Pathak S (1990) Chromosome alterations in speciation and neoplastic transformation: a parallelism. In: T Sharma (ed.): *Trends in Chromosome Research*. Springer-Verlag, Narosa Publishing House, New Delhi, 204–220
94 Multani AS, Ozen M, Sen S, Mandal AK, Price JE, Fan D, Radinsky R, Ali-Osman F, von Eschenbach AC, Fidler IJ, Pathak S (1999) Amplification of telomeric DNA directly correlates with metastatic potential of human and murine cancers of various histologic origin. *Int J Oncol* 15: 423–429

95 Fidler IJ (1996) Critical determinants of melanoma metastasis. *J Invest Dermatol Symp Proc* 1: 203–208
96 Folkman J (1996) Clinical applications of research on angiogenesis. Seminar in medicine of the Beth Israel Hospital, Boston. *New Engl J Med* 333: 1757–1763
97 Maniotis AJ, Folberg R, Hess A, Seftor EA, Garder LMG, Peer J, Trent JM, Meltzer PS, Hendrix MJC (1999) Vascular channel formation by human melanoma cells *in vivo* and *in vitro*: vasculogenic mimicry. *Am J Pathol* 155: 739–752
98 McDonald DM, Munn L, Jain RK (2000) Vasculogenic mimicry: how convincing, how novel, and how significant? *Am J Pathol* 156: 383–388
99 Chang YS, di Tomaso E, McDonald DM, Jones R, Jain RK, Munn LL (2000) Mosaic blood vessels in tumors: Frequency of cancer cells in contact with flowing blood. *Proc Natl Acad Sci USA* 97: 14608–14613
100 Groszer M, Erickson R, Scripture-Adams DD, Lesche R, Trumpp A, Zack JA, Kornblurn HI, Liu X, Wu H (2001) Negative regulation of neural stem/progenitor cell proliferation by the Pten tumor suppressor gene *in vivo*. *Science* 294: 2186–2189
101 Makino S (1956) Further evidence favoring the concept of the stem cell in ascites tumors of rats. *Ann N Y Acad Sci* 63: 818–830
102 Sell S (ed.): (2004) *Stem Cells Handbook*. Humana Press Inc., Totowa
103 Turksen K (ed.): (2004) *Adult Stem Cells*. Humana Press Inc., Totowa
104 Reya T, Morrison SJ, Clarke MF, Weissman IL (2001) Stem cells, cancer, and cancer stem cells. *Nature* 414: 105–111
105 Kondo T, Setoguchi T, Taga T (2004) Persistance of a small subpopulation of cancer stem-like cells in the C6 glioma cell line. *Proc Natl Acad Sci USA* 101: 781–786
106 van den Brink GR, de Santa Barbara P, Roberts DJ (2001) Epithelial cell differentiation – a matter of choice. *Science* 294: 2115–2116
107 Ueda M, Ouhtit A, Bito T, Nakazawa K, Liibbe J, Ichihashi M, Yamasaki H, Nakazawa H (1997) Evidence for UV-associated activation of telomerase in human skin. *Cancer Res* 57: 370–374
108 Bickenbach JR, Vormwald-Dogan V, Bachor C, Blemel K, Schnapp G, Boukamp P (1998) Telomerase is not an epidermal stem cell marker and is down regulated by calcium. *J Invest Dermatol* 111: 1045–1052
109 Sun W, Kang K-S, Morita I, Trosko JE, Chang C-C (1999) High susceptibility of a human breast epithelial cell type with stem cell characteristics to telomerase activation and immortalization. *Cancer Res* 59: 6118–6123
110 Sell S (2004) Stem cells What are they? Where do they come from? Why are they here? When do they go wrong? Where are they going? In: S Sell (ed.): *Stem Cell Handbook*. Humana Press Inc., Totowa, 1–18
111 Sell S, Pierce GB (1994) Maturation arrest of stem cell differentiation is a common pathway for the cellular origin of teratocarcinoma and epithelial cancers. *Lab Invest* 70: 6–22
112 Paget S (1889) The distribution of secondary growth in cancer of the breast. *Lancet* 1: 571–573
113 Pathak S, Dhaliwal MK, Hopwood VL (1989) Genetic susceptibility, somatic mosaicism, and predisposition to human cancer. *Anticancer Res* 9: 17–20
114 Pathak S (1992) Cytogenetics of epithelial malignancies. *Cancer* 70: 1660–1670
115 Pathak S, Berry KK, Hopwood VL, Burke TW, Baker VV (1995) Identification of primary chromosome abnormalities in a patient with endometrial carcinoma: analyses of tumor biopsy and lymphocyte cultures. *Int J Oncol* 7: 765–772
116 Ghayee HK, Dinney CP, Pathak S (1997) Do lymphocytes contain chromosomal lesions that are also stable markers in cancer cells? Lymphocytes and tumor cell karyotyping in a melanoma patient. *Int J Oncol* 11: 681–684
117 Pathak S, Hopwood VL, Hughes JI, Jackson GL (1991) Identification of colon cancer-predisposed individuals: a cytogenetic analysis. *Am J Gastroenterol* 86: 679–668

Cancer: Cell Structures, Carcinogens and Genomic Instability
Edited by Leon P. Bignold

The mode of action of organic carcinogens on cellular structures

Andreas Luch

Massachusetts Institute of Technology, Center for Cancer Research, 77 Massachusetts Avenue, E17-132, Cambridge, MA 02139, USA

Abstract. Most genotoxic organic carcinogens require metabolic activation to exert their detrimental effects. The present review summarizes the mechanisms of how organic carcinogens are bioactivated into DNA-reactive descendants. Beginning with the history of discovery of some important human organic carcinogens, the text guides through the development of the knowledge on their molecular mode of action that has grown over the past decades. Some of the most important molecular mechanisms in chemical carcinogenesis, the role of the enzymes involved in bioactivation, the target gene structures of some ultimate carcinogenic metabolites, and implications for human cancer risk assessment are discussed.

Key words: Aromatic amines, carcinogen-DNA adducts, cytochrome P450, metabolism, mutation profiles, phase II enzyme activation, polycyclic aromatic hydrocarbons.

Introduction

Epidemiological evidence based on geographic and temporal variations in cancer incidences and studies of migrant populations suggest that ‘environmental exposures’ have a substantial impact on the causation of human cancer [1]. These studies led to the conclusion that the majority of cancer deaths in Western industrial countries are attributable to exogenous factors such as tobacco, diet, infections, and occupational exposures. The notion that the environment has the principal role in the causation of sporadic cancer is also supported by analyses of cancer cases in cohorts of twins [2] and by analyses of family-cancer databases [3]. In either case there is strong evidence that the influence of ‘nonshared environmental factors’ predominates.

There is no doubt that a wide range of organic chemicals (as pure compounds or present in mixtures) are carcinogenic in humans [4]. Examples from the list of known or suspected human chemical carcinogens are aromatic and heterocyclic amines or amides (e.g., 2-naphthylamine, 2-NA; 2-acetylaminofluorene, 2-AAF; 2-amino-3-methylimidazo[4,5-*f*]quinoline, IQ), halogenated and unsubstituted olefines (e.g., tetra- or perchloroethylene, PER), halogenated paraffins (e.g., 1,2-dibromoethane, 1,2-DBE), *N*-nitroso compounds (e.g., *N*-nitrosodimethylamine, NDMA; 4-(methylnitrosamino)-1-(3-pyridyl)-1-butanone, NNK), benzene and polycyclic aromatic compounds (e.g., benzo[*a*]pyrene, B[*a*]P; 2,3,7,8-tetrachlorodibenzo-*p*-dioxin, TCDD), estrogen-receptor agonists or antagonists with residual agonistic effects (e.g., diethylstilbestrol, DES; tamoxifen), wood dust (e.g., from oak and beech), natural compounds (e.g., aflatoxins, ochratoxin A, OTA; pyrrolizidine alkaloids), and so forth. Some selected structures are shown in Figure 1.

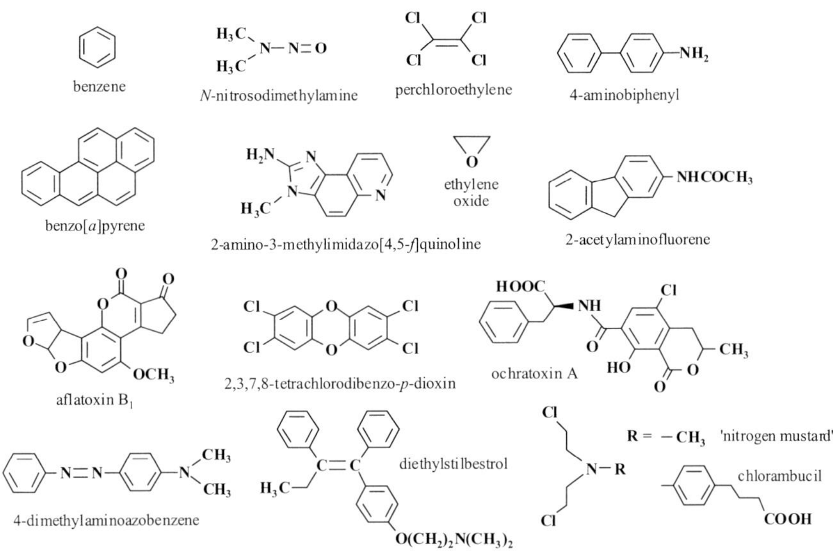

Figure 1. Some examples of organic chemicals that are known or reasonably anticipated to be carcinogenic in humans.

Discovery of organic carcinogens

In the 18th century, the two English physicians John Hill (1716–1775) and Percivall Pott (1714–1788) described the occurrence of cancer in the nose and at the skin of the scrotum, and associated it with local long-term exposure to snuff [5] and local contamination by soot at a young age [6], respectively. The interest of the latter observation lay in the first clear relationship between an occupation (chimney sweeping) and one particular form of squamous cell carcinoma; and in the possibility to prevent this disease by personal hygiene.

In the following years it became evident that certain exposures to pure organic chemicals or mixtures are associated with human carcinogenesis. Richard von Volkmann (1830–1889) and Joseph Bell (1837–1911) confirmed the early observation from Pott by describing several cases of skin affections and scrotal skin tumors among workers in the paraffin industry [7, 8]. In 1895, the surgeon Ludwig Wilhelm Carl Rehn (1849–1930) had recognized the appearance of tumors of the bladder of men employed in the German aniline dyestuff industry to produce red magenta ('Fuchsin') [9]. It was then noticed that this particular occupation was associated with increased rates of cancer in the urinary bladder, subsequently referred to as 'aniline cancers'. In 1898, the German internist Otto Michael Ludwig Leichtenstern (1845–1900) considered 2-NA most likely to be involved in human bladder tumorigenesis [10].

Until the dawn of the 20th century, physicians were only able to collect the unexpected outcome of an undesigned and undesirable grand scale (occupational exposure) experiment based on the rise of industrialization. By the year 1907 it was officially recognized by the 'Workmen's Compensation Act' of Great Britain that epidermal cancer can be caused by pitch, tar or tarry compounds [11]. The imperative next step was that of experimental reproduction of cancer. After many failures to reproduce the known human outcome in laboratory animals, the Japanese pathologist Katsusaburo Yamagiwa (1863–1930) and his assistant Ichikawa successfully produced malignant tumors through application of a chemical mixture (coal tar) to the ear of rabbits [12]. While epithelial proliferation could be chemically induced already some years earlier [13, 14], this experiment produced undoubted malignant epithelial cancer for the first time. It was then experimentally and epidemiologically confirmed that tar and soot is carcinogenic in the skin of mice and in humans exposed at work place, respectively [15–17]. After chemical synthesis routes for pure higher molecular polycyclic aromatic hydrocarbons (PAHs) had been first described, Sir Ernest Laurence Kennaway (1881–1958) and his colleagues at the Royal Cancer Hospital in London successfully proved that single PAHs such as 1,2:5,6-dibenzanthracene (dibenz[*a,h*]anthracene, DB[*a,h*]A) and others are tumorigenic [18, 19]. As an indicator assay, they applied the mouse skin bioassay, which had been introduced to chemical cancer research some years earlier [20]. In 1933, Cook, Hewett and Hieger from the Cancer Hospital were successful in isolating the 'carcinogenic principle' out of coal tar pitch [21]. It turned out to be another PAH, the pentacyclic 3,4-benzpyrene (B[*a*]P), which nowadays is one of the most investigated carcinogens ever and a standard compound used in many cancer experiments as a 'positive control' (Fig. 1).

Studies with aromatic amines, aminoazo dyes and related compounds supplemented the large volume of experimental data on the carcinogenicity of industrial chemicals that had been released in high amounts during this time. The incriminated aniline itself failed to produce tumors in the urinary bladder of rabbits, as did other aromatic amines such as 2-NA [22]. Later, bladder papillomas and carcinomas were successfully induced in dogs by gavage or dermal application of 2-NA [23] – an experiment that supported the early prediction from Leichtenstern. In the meantime aminoazo dyes, such as *o*-aminoazotoluene (reduction product of 'Scarlet Red') [24] and 4-dimethylaminoazobenzene (DAB, 'butter yellow'; Fig. 1), were shown to be tumorigenic in rat liver [25]. Wilson et al. [26] reported on the tumorigenicity of 2-AAF in bladder, liver, and various other organs of rats. 2-AAF is an arylamide which was intended to be used as a pesticide, but has been later introduced as a model compound in experimental liver cancer research (Fig. 1).

With the beginning of the 1940s, a huge amount of experimental data on the bioactivity of pure and structurally defined organic compounds present in the industrial environment had been collected.

Metabolic activation of organic carcinogens – Achilles' heel of defense?

Crucial hint: DNA adducts

Based on the assumption that cancer derives from clonal expansion of a single cell [27], it seems axiomatic that chemical carcinogens must induce tumorigenesis through interaction with subcellular components that are intimately involved in mediating the underlying heritable loss of growth control. In principle, these interactions may be noncovalent and reversible or covalent, but reversible only in case repair mechanisms would restore the naïve structure. In the 'pre-Watson and Crick era', binding of carcinogens such as (brightly colored) DAB [28] or (fluorescent) B[*a*]P [29] to proteins at target tissue sites *in vivo* had been reported, and considered to be associated with the initiation of the disease [30]. Later, when the sensitivity was sufficiently high, due to the availability of radioactively labeled chemicals, binding of carcinogens, such as *N*-methyl-bis(2-chloroethyl)amine ('nitrogen mustard') [31], bis(β-chloroethyl)sulfide ('mustard gas') [32], *N*-nitrosamines (e.g., NDMA) [33], PAHs (e.g., B[*a*]P) [34], aromatic amines/amides (e.g., 2-AAF) [35], and aminoazo dyes (e.g., DAB) [36], to DNA *in vivo* was discovered (see Fig. 1). By the end of the 1960s, accumulating evidence for a correlation between the level of DNA binding of a particular carcinogen and its biological potency was overwhelming. In addition, all of these compounds were soon regarded as being mutagens as well [37]. To react with cellular macromolecules, however, most carcinogens require enzymatic activation. The parent compounds are, therefore, considered as precarcinogens or indirect carcinogens to be bioactivated into their ultimate carcinogenic forms.

The role of biotransformation

First evidence for metabolic conversion of organic carcinogens *in vivo* had been obtained through detection of hydroxylated derivatives and their conjugates in the urine of animals treated with pure compounds [38–40]. In 1948, James A. Miller (1915–2000) and co-workers were the first to demonstrate the oxidative metabolism of DAB by rat liver microsomes [41]. Later, the activating enzymes present in microsomes were characterized as mixed-function oxidases [42], cytochrome P450 had been discovered [43], and cytochrome P450-dependent monooxygenases (CYPs) were shown to work in concert with an NADPH-dependent reductase [44] (Fig. 2).

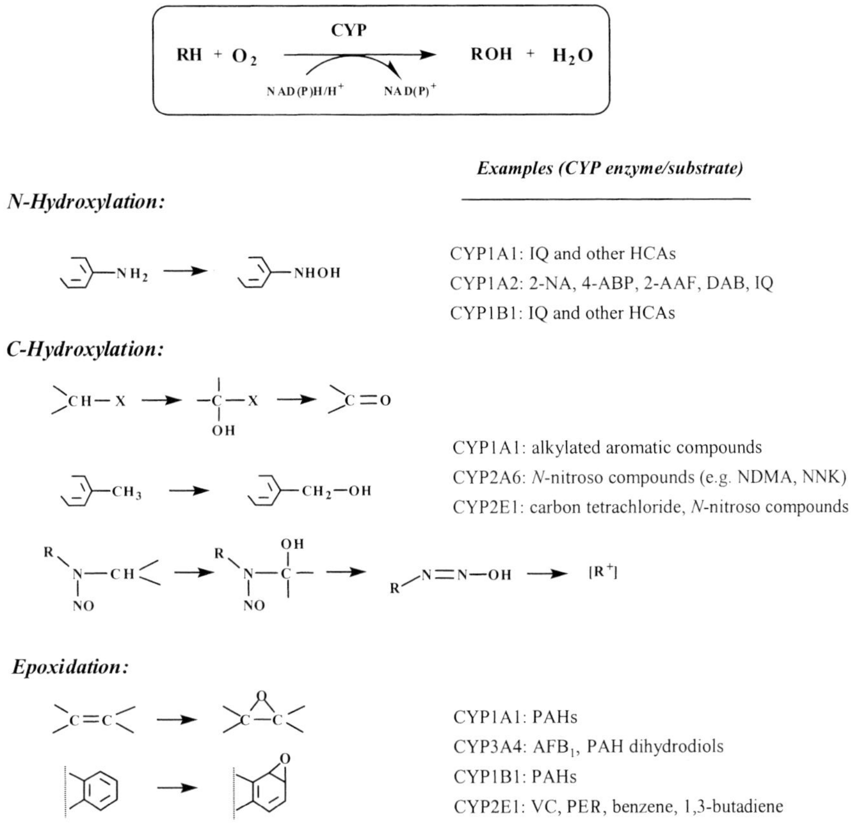

Figure 2. Cytochrome P450-dependent biotransformation (monooxygenation) of organic chemicals (cf. Fig. 1) [46, 47]. The principal reaction is shown at the top (scheme). 2-AAF, 2-acetylaminofluorene; 4-ABP, 4-aminobiphenyl; AFB_1, aflatoxin B_1; CYP, cytochrome P450-dependent monooxygenase; DAB, 4-dimethylaminoazobenzene; HCAs, heterocyclic amines, also known as 'cooked food mutagens' due to their generation from amino acids during cooking processes; IQ, 2-amino-3-methylimidazo[4,5-*f*]quinoline; 2-NA, 2-naphthylamine; NDMA, *N*-nitrosodimethylamine; NNK, 4-(methylnitrosamino)-1-(3-pyridyl)-1-butanone; PAH, polycyclic aromatic hydrocarbons; R = alkyl residue, e.g., methyl, ethyl; PER, tetra- or perchloroethylene; VC, vinyl chloride. See text for further explanations.

In principle, enzymatic biotransformation of xenobiotic compounds is aimed at ensuring detoxification and subsequent elimination via excretion pathways rather than leading to descendants of enhanced biological activity. Conversion of organic and mostly aromatic compounds into hydrophilic and excretable derivatives is catalyzed not only by CYP enzymes, from which at present 57 genes are identified in humans [45]. Rather, a great variety of additional 'xenobiotic metabolizing enzymes' (XMEs) contribute to this process which, in general, proceeds through activated intermediates that are capable of undergoing subsequent conjugation reactions with hydrophilic functional groups or molecules. Formation of activated intermediates via CYP-mediated monooxygenation is considered as 'phase-I' in biotransformation (Fig. 2). Modification catalyzed by hydrolases, dehydrogenases, peroxidases, and reductases may also contribute to this phase. The resulting derivatives then may or may not enter 'phase-II', in which transferases catalyze conjugation to polar molecules, such as glucuronic acid (UDP-glucuronosyltransferases, UGTs) and glutathione (glutathione *S*-transferases, GSTs), or to small residues such as sulfate (sulfotransferases, SULTs) or acetic acid (*N*-acetyltransferases, NATs). On the other hand, bioactivation towards electrophilic intermediates may also lead to covalent interaction with cellular proteins or DNA. Therefore, this approach poses an inherent risk to the physiological integrity of living cells, and may thus be regarded as the Achilles' heel in the cellular defense against biohazards.

Monooxygenation of xenobiotics in mammalian species including humans is mainly catalyzed by members of the CYP families 1–3 (i.e., CYP1A1, 1A2, 1B1, 2A6, 2E1, 3A4) [46, 47]. These forms are considered as the mainstay in biotransformation of a large number of carcinogenic chemicals. For instance, CYP1A1 is the major form in human lung, CYP1A2, 2A6, 3A4, or 2E1 are mainly expressed in liver, while CYP1B1 is the main extrahepatic and extrapulmonary form with highest levels in prostate and uterus [48]. Figure 2 summarizes the major CYP-mediated bioactivation reactions of organic carcinogens.

Under certain circumstances, phase-II enzymes may also contribute to the activation of precarcinogens *in vivo*. SULT and NAT enzymes catalyze sulfonation and acetylation of nucleophilic metabolites such as *N*-hydroxylamines, phenols, benzylic alcohols and others. *N*-Hydroxylation as a bioactivation step was discovered in studies on liver carcinogenesis induced by 2-AAF [49] (Figs 2 and 3). Since *N*-hydroxy-2-AAF was found to be a more potent carcinogen than 2-AAF itself [50], it is regarded as the proximate carcinogenic metabolite. With the beginning of the 1970s, SULT-catalyzed *O*-sulfonation of *N*-hydroxy-2-AAF had been discovered [51, 52]. With time, it was then established that *O*-acetylation (by NATs, cofactor: acetyl-CoA) as well as *O*-sulfonation (by SULTs, cofactor: 5'-phosphoadenosine-3'-phosphosulfate) of *N*-hydroxy derivatives of arylamines (e.g., 2-NA), arylamides (e.g., 2-AAF), aminoazo dyes (e.g., DAB), or heterocyclic amines (HCAs, e.g., 'cooked food mutagens' such as IQ) yield highly reactive ester intermediates *in vivo* that bind to C8 or exocyclic amino groups in purine bases via intermediate release of arylnitrenium ions (Fig. 3) [53, 54].

The discovery of mercapturic acids (*N*-acetylcysteinyl conjugates) dates back to the 19th century, yet GST-mediated transfer of electrophilic metabolites onto glutathione (GSH) was not described before 1961 [55], shortly after the relationship between GSH conjugates and their excretable follow-up products had been uncovered [56]. GST enzymes interact with electrophiles via the reactive cysteinyl residue in their cofactor (Fig. 4). While GST-mediated

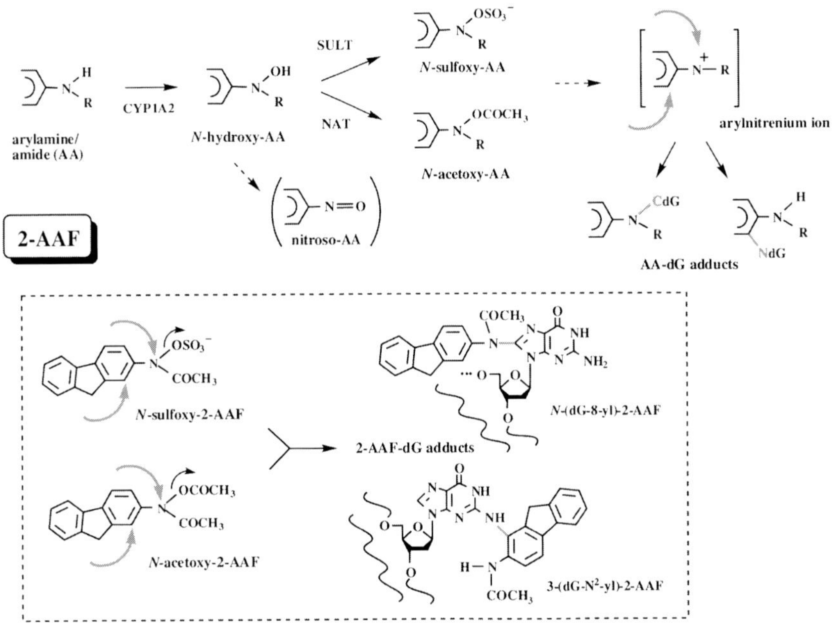

Figure 3. Bioactivation of aromatic amines or amides (AA) towards ultimate DNA-reactive sulfate or acetoxy esters. The ultimate electrophiles and major DNA binding products are exemplified in the case of 2-AAF. 2-AAF, 2-acetylaminofluorene; CYP, cytochrome P450-dependent monooxygenase; dG, 2'-deoxyguanosine; SULT, sulfotransferase; NAT, *N*-acetyltransferase; The arylnitrenium ion is a putative reactive intermediate. The grey arrows point to the position of a nucleophile attack of DNA, protein, or GSH. See text and Figure 2 for further explanations.

detoxification of activated metabolites such as epoxides or dihydrodiol epoxides (e.g., from PAHs) has been widely acknowledged since then, more recent evidence points to an additional but detrimental role of GST enzymes in the activation of certain industrial chemicals from the classes of haloalkanes (e.g., 1,2-DBE) and haloalkenes (e.g., PER; see Fig. 1). These compounds are likely to be human carcinogens due to similarities between susceptible animals and humans in bioactivation [4]. In the case of PER, there is also some limited evidence from cohort studies of laundry and dry-cleaning workers, among whom a higher than normal occurrence of non-Hodgkin's lymphoma, esophageal and cervical cancer was found [4]. Both groups of organic carcinogens are bioactivated into genotoxic GSH conjugates. While dihaloalkanes may undergo GST-catalyzed conversion into DNA-binding 'GSH half mustard' and GSH episulfonium electrophiles (Fig. 4) [57], haloalkene-GSH conjugates can be further converted via a kidney-specific cysteine conjugate β-lyase-dependent pathway. After release of the terminal amino acids γ-Glu and Gly, lysis of the β-bond in the remaining Cys adduct may eventually lead to the generation of electrophilic and toxic thioketenes, which are capable of binding to macromolecules in this tissue [58].

Molecular specificity of genotoxic carcinogens?

Carcinogen-induced DNA damage and mutations disrupt the expression of genes involved in the surveillance of cellular growth, proliferation, and death (Fig. 5). What is known about the molecular specificity in organic carcinogens' modes of action?

The impact of DNA binding level

In their landmark paper from 1964, Brookes and Lawley worked on the tissue binding levels of six PAHs, i.e., naphthalene, dibenz[*a,c*]anthracene, DB[*a,h*]A, B[*a*]P, 3-methylcholanthrene, and DMBA [34]. Based on their experiments, the authors concluded that there is a 'significant positive correlation' between the binding to DNA and the carcinogenic potency of these compounds according to Iball's index (that is, ratio between tumor incidence and average tumor latency period [106]). In contrast, this correlation was not found for the binding to proteins or RNA [34]. In more recent years it became clear that vicinal dihydrodiol epoxides of PAHs are the actual DNA-binding metabolites that mediate the biological effects associated with their parent structures (cf. above; and see Fig. 6). By direct application of these reactive descendants it could be shown that their mutagenic potency correlates nicely to the level of PAH-DNA adduct formation [107]. Given these data, it therefore seems plausible that the quantity of DNA damage rather than the individual (qualitative) adduct structure is the main factor determining the mutagenic activity of a particular genotoxic species. In addition, the tumor-inducing potencies of different PAHs in the lungs of strain A/J mice, another experimental tumor model, were found to correlate to the time-integrated DNA adduct levels (TIDAL) calculated as area under the curves of total dihydrodiol epoxide-DNA adduct levels during a time course of 30 days after injection [108]. This parameter represents the total effective molecular dose delivered to target lung DNA, and it linearly correlated to the PAH doses administered. The intimate relationships between DNA binding level and mutagenicity and between the TIDAL and carcinogenicity observed in mice *in vivo* support the notion that the DNA binding level can serve as an important (bio)indicator of the tumor threat that may result from certain exposures of humans to carcinogenic PAHs [109], and other genotoxic carcinogens as well [110].

The impact of stereochemistry

The DNA adduct level itself at a given time point is an integrated product of compound's (structure-related) toxicokinetic and toxicodynamic behavior, including metabolic activation and detoxification prior to covalent binding, as well as the effectiveness of the repair of those lesions that have been formed.

XMEs usually operate with high regio- and stereoselectivity. For instance, CYP-mediated toxification of aflatoxin B_1 (AFB_1), a natural carcinogen produced by *Aspergillus* mould and linked to human liver carcinogenesis [111], occurs at its 8,9-position (Fig. 7). Based on the isolation of the main DNA adduct (at N7 in guanine bases), formation of an 8,9-epoxide intermediate had been proposed and subsequently confirmed [112]. CYP3A4, the most important enzyme involved in AFB_1 activation in human liver, exclusively forms the *exo* isomer (Fig. 7). In contrast, another CYP form, CYP1A2, may add some small amounts of the diastereomeric *endo* epoxide [113]. However, the predominant AFB_1 *exo*-8,9-oxide is about 1000-fold more genotoxic than its *endo* diastereomer due to the spatial configuration of the epoxide moiety within the AFB_1 *exo*-8,9-oxide-DNA intercalation complex. Intercalation of the furanocoumarine residue between DNA bases directs the *exo* epoxide ring in a favorable position for an S_N2 attack by the N7 atom of guanine [114, 115]. Follow-up products of the main N7-DNA adduct of AFB_1 then result from depurination or ring opening of the purine base. These products, along with the detoxification products produced by GST- or mEH-mediated conversions, are depicted in Figure 7.

Similarly, metabolic activation of PAHs is highly selective [80]. As demonstrated for a wide range of carcinogenic PAHs the initial epoxidation → hydrolysis sequence produces a dihydrodiol with *R,R*-configuration in high enantiomeric excess (cf. above). Subsequent (diastereo)selective epoxidation at the vicinal double bond then predominantly generates the *R,S*-dihydrodiol *S,R*-epoxide with the epoxide moiety *trans* to the benzylic hydroxy group. In the case of B[*a*]P, all four possible stereoisomeric 7,8-dihydrodiol 9,10-epoxides are depicted in Figure 6, with (+)-*anti*-B[*a*]PDE as the major species formed during bioactivation. Depending on the activation system, some small amounts of the other isomers, the (–)-*syn*-(*R,S,R,S*)-, (+)-*syn*-(*S,R,S,R*) and (–)-*anti*-(*S,R,R,S*) dihydrodiol epoxides, may also be generated – the latter two through monooxygenation of the *S,S*-dihydrodiol (Fig. 6). However, B[*a*]P-induced DNA damage *in vitro* or *in vivo* predominantly results from covalent interaction of (+)-*anti*-B[*a*]PDE, most of which is trapped by 2'-deoxyguanosine (dG) residues via *trans* opening of the epoxide moiety [80, 116] (Fig. 7). If not repaired properly, the product of this reaction is likely to cause nucleotide misincorporation at the opposite DNA strand during the next round of DNA replication, and, therefore, has the potential of inducing mutations (i.e., dG base substitutions or frame shifts). DNA lesions such as the (+)-*trans*-*anti*-B[*a*]PDE-N^2-dG adduct are subject to nucleotide excision repair (NER). Induction of NER activity requires both the disruption of normal base pairing and the presence of a chemical modification ('bipartite damage recognition') [117]. It could be demonstrated that (+)-*anti*-B[*a*]PDE induces a considerably different degree of NER activity in a certain DNA base context depending on the way of epoxide ring opening during adduct formation [118]. While the *cis*-opened adduct, i.e., the (+)-*cis*-*anti*-B[*a*]PDE-N^2-dG, adopts an intercalative, internal adduct conformation with the benzo[*a*]pyrenyl moiety *anti*-B[*a*]PDE-N^2-dG adducts were almost exclusively formed at methylated CpG dinucleotides of the nontranscribed DNA strand, possibly as a result of the slow repair of this strand as compared to its transcribed counterpart [134]. On the other hand, comparison of *TP53* mutational hotspots in hepatocellular carcinoma with AFB_1-induced DNA damage in human liver HepG2 cells rather questions the relationship between compound and disease [135]. This analysis is, however, contrasted by several lines of evidence: (i) a wealth of epidemiological data point to a correlation between AFB_1 exposure and specific mutations in codon 249 of *TP53* [136, 137]; (ii) new insights on the biological activity and repair resistance of a major AFB_1-DNA adduct (the AFB_1-FAPY adduct, cf. Fig. 7) support the notion of its contribution to the occurrence of G $\rightarrow$ T transversions at the third position in this codon [138]; and (iii) the presence of AFB_1-DNA adducts in hepatocellular tumors is accompanied by a higher frequency of methylated CpG sites in the promotors of the tumor suppressors *p16* and *RASSF1A*, two genes epigenetically silenced in a high percentage (47% and 85%) of liver cancer cases from Taiwan [139].

Risk assessment in chemical carcinogenesis – implications

The toxicological approach

Studies in animal tumor models or in cells *in vitro* lend support to the notion that the level of DNA binding along with the intrinsic (structural) quality of the DNA adducts induced and their qualitative and quantitative distribution both at the organismic level (i.e., within different kinds of organs and tissues) and at the cellular level (i.e., within the cell's genome) are the main determinants for the biological activity of a particular genotoxic carcinogen. Because of this intimate correlation between DNA damage and carcinogenic potency it can be anticipated that differences in the activity of enzymes that produce DNA-reactive intermediates, detoxify these species, or repair the DNA lesions formed, may play a major role in mediating the individual tumor susceptibility in human populations (see below). However, there are essentially three major issues that limit the predictive value of the knowledge obtained in experimental tumor models with regard to human cancer:

1. All XMEs (including polymorphic variants) can contribute to both activation and detoxification of organic carcinogens (e.g., GST enzymes: AFB_1 *exo*-8,9-oxide *versus* haloalkanes/alkenes; see Figs 4 and 7). Occasionally, toxication and detoxication routes of the very same carcinogen are served by the very same enzyme (e.g., mEH: formation of PAH dihydrodiols, which then may be subject to either conjugation $\rightarrow$ excretion or further activation towards dihydrodiol epoxides; see Figs 6 and 7). An interesting example of these interactions is the protective effect of CYP1A1 and 1A2

in vivo against DNA adduct formation and toxicity of B[*a*]P [140] or 4-ABP [91]. Both enzymes are involved in activation of B[*a*]P (CYP1A1) or 4-ABP (CYP1A2; see Figs 2, 3 and 6), yet animals that lack the corresponding genes have been found to suffer from higher DNA adduct levels in internal organs and increased toxicity. It seems likely that this effect is due to the well-balanced expression levels of CYP1A1 and 1A2 along with detoxifying phase-II enzymes in wild-type animals, but also a result of the overcompensatory induction of other oxidative enzymes, such as CYP1B1 or flavin-dependent monooxygenases, which would substitute for the absence of CYP1A1 and 1A2 [141].

2. Humans are mostly exposed to complex mixtures of compounds rather than to single carcinogens. For instance, due to the manner of their generation (incomplete combustion), more than 100 different PAHs can be detected in airborne particulates [142]. Cigarette smoke entails the risk of being exposed to about 60 known carcinogens from a variety of chemical classes including 4-ABP, B[*a*]P, and traces of metal ions [143]. The interactions of those individual compounds may result in synergistic effects within the biological system. At the level of DNA repair, for example, co-exposure of cells to Ni ions [144], or As and its methylated metabolites [145], was found to enhance B[*a*]P-mediated mutagenesis through inhibition of NER-catalyzed removal of *anti*-B[*a*]PDE-N^2-dG adducts. Even the presence of structurally and stereochemically different DNA lesions in the same genome may cause repair inhibition through sequestration of critical NER subunits by those modifications that are more repair resistant (so-called 'decoy adducts') [146]. Despite being refractory to excision, 'decoy adducts' immobilize NER factors, and may therefore contribute to synergistic interactions between multiple genotoxic agents present in complex environmental mixtures. Thus, the results obtained from single compound experiments would not allow direct extrapolation to the corresponding effects expected to be exerted from mixtures.
3. In experimental tumor models, animals or cells are treated with single compounds in very high doses as compared to the levels of human background exposures. Therefore, regulatory toxicologists have to extrapolate the dose-response relationships found in animals into the low-dose exposure ranges of humans in order to assess the accompanying risk. Depending on the kind of approach applied, the results may differ tremendously, and are often subject to believe or disbelieve [147, 148].

The role of molecular epidemiology

Many studies conducted in recent years tried to explore the influence of metabolic gene polymorphisms in the human population and their interactions on the levels of biomarkers of carcinogen exposure such as urinary metabolites, DNA adducts, chromosomal aberrations and others [149]. The goal of these studies is to use this information and to identify subpopulations of humans that are likely to be more susceptible ('to be at high risk') of developing tumors in response to carcinogen exposure. Due to their low frequency in Caucasian populations, however, significant contributions of gene polymorphisms related to carcinogen activation/inactivation are usually hard to detect [150–152] (Tab. 1). One exception is the lack of GSTM1 activity (*GSTM1* null genotype), which is most common (50% of Caucasians) and hence entails some marked effects on cancer susceptibility (Tab. 1).

Given the chemical complexity of most environmental matrices, it seems difficult, if not, impossible to uncover any causative connections between certain

Table 1. Genetic polymorphisms of metabolic enzymes related to chemical carcinogenesis and their role in tumor susceptibility (selection)[a]

Genetic polymorphisms →	Biological effects/tumor susceptibility (selection)
Cytochrome P450-dependent monooxygenases (CYP, EC 1.14.14.1)[b]	
CYP1A1	*CYP1A1*2A* (*Msp*I, 3' untranslated region) and **2B* ($Ile^{462}Val$, exon 7) are associated with increased catalytic activity, higher levels of PAH-DNA adducts and *TP53* mutations in smokers. *CYP1A1 Msp*I combined with *GSTM1 0/0* leads to an 'at risk' genotype for tobacco-accociated DNA damage and lung cancer.
CYP1A2	Limited evidence for higher risk of bladder cancer in smokers with *CYP1A2*1F* ($^{163}C \rightarrow A$: SNP intron 1) entailing increased inducibility. Combination effect with slow NAT2 acetylator phenotype.
CYP1B1	Effects of known polymorphisms are only marginal *in vitro*. However, *CYP1B1*3* ($Val^{432}Leu$) entails a higher frequency of smoking-induced *TP53* mutations [166]. Increased OR with combined *GSTM1*2* or *GSTT1*2*. Limited evidence.
CYP2A6	*CYP2A6*2* ($Leu^{160}His$) and **3* (exon deletion): impaired activity, yet no *in vivo* evidence.
CYP2E1	*CYP2E1*6* ($^{7632}T \rightarrow A$: *Dra*I RFLP intron 6): N7-alkyl levels in lung samples elevated. Inadequate evidence for increased lung and breast cancer risk.
Microsomal epoxide hydrolase (mEH, EC 3.3.2.3)	
mEH	Substrates: arene oxides (e.g., AFB_1, B[*a*]P). Limited evidence for higher AFB_1 protein adduct levels and liver cancer susceptibility in subjects with *mEH* $Tyr^{113}His$ (exon 3 polymorphism). Limited evidence for reduced lung cancer risk in smokers [167].
Sulfotransferases (SULT, EC 2.8.2.1)	
SULT1A1	Substrates: toxication of aromatic *N*-hydroxylamines (e.g., 2-AAF), benzylic alcohols and phenols (e.g., PAHs, benzene); detoxication of hydroxy derivatives (e.g., PAHs). Large interindividual variation in activity. Various alleles, e.g., *SULT1A1*2* ($Arg^{213}His$) with reduced activity. Limited evidence for a protective role of *SULT1A1*2* in colon, but higher risk in lung.
SULT1A2	Several alleles, e.g., *SULT1A2*2* ($Asn^{235}Thr$), cause a strong decrease in activity. Functional role *in vivo* unclear.
SULT2A1	Main form in liver. Several rare alleles with moderate effects on enzyme activity (<2).

(Continued on next page)

Table 1. (Continued)

Genetic polymorphisms → Biological effects/tumor susceptibility (selection)	
N-Acetyltransferases (NAT, EC 2.3.1.5)	
NAT1	Predominantly extrahepatic. Substrates: *N*-acetylation (detoxication), and *O*-acetylation or *N,O*-transacetylation (toxication) of arylamines/amides and some HCAs. *NAT1*10* allele (elevated activity) entails higher risk for colon and bladder cancer (high rate of *O*-acetylation of HCAs in colon and arylamines in bladder).
NAT2	Predominantly hepatic. Substrates and reactions as NAT1, main form for HCAs, but no *N,O*-transacetylation. Several NAT2 alleles (e.g. *NAT2*5B*) are associated with a slow acetylator phenotype. Slow acetylators have higher levels of 4-ABP-hemoglobin adducts and an increased risk of bladder cancer (impaired hepatic *N*-acetylation of arylamine); yet they are at lower risk for colon cancer.
Glutathione *S*-transferases (GST, EC 2.5.1.18)	
GSTM1	Substrates: epoxides of PAHs and olefines (AFB_1, 1,3-butadiene, etc.), arylamine esters. *GSTM1 0/0* genotype (loss of activity): although mainly expressed in liver, higher PAH-DNA adduct levels, cytogenetic damage, and *TP53* mutations in lungs of smokers; effect pronounced with combined *CYP1A1*2A* genotype. Higher 4-ABP-hemoglobin levels in smokers and non-smokers; effect pronounced with combined slow NAT2 acetylator genotype. *GSTM1 0/0* alone is only a weak modifier of lung (OR 1.41) and bladder cancer (OR 1.44). For lung cancer, the effect significantly increases in the presence of an active *CYP1A1 Msp*I genotype (OR 3-10).
GSTT1	Substrates: epoxides of PAHs, 1,3-butadiene, ethylene oxide (detoxication), dihaloalkanes/-methanes (toxication). Accordingly, *GSTT1 0/0* polymorphism entails a higher or lower risk for genomic damage *in vitro*, depending on the substrate. At present, no evidence for modulation of human cancer risk.
GSTP1	Equivocal results *in vitro*. Limited evidence for association between *GSTP1* $Ile^{105}Val$ polymorphism and increased bladder cancer susceptibility. Inadequate evidence for a role in human lung cancer.
Arylhydrocarbon receptor (AHR) pathway	
AHR	Several polymorphisms in humans reported. Combination of $Lys^{554}Leu$ and $Val^{570}Ile$ impairs TCDD-mediated CYP1A1 induction *in vitro*. Strong evidence for a correlation of human lung, laryngeal, and oral cavity cancer with AHR phenotype.
ARNT	Polymorphisms known, but functional roles as yet unknown [169].
DNA repair proteins	
XRCC1/3	Involved in base excision and DNA double strand repair. *XRCC1* $Arg^{399}Gln$ (exon 10) and *XRCC3* $Thr^{241}Met$ (exon 7) are associated with higher 'bulky' DNA adduct levels in non-smokers and both non- and ex-smokers, respectively. Combined effects of multiple gene variants on DNA damage levels observed. As yet only insufficient evidence for an important role in human (lung) cancer susceptibility.
XPD	Helicase involved in NER. *XPD* $Asp^{312}Asn$ (exon 10) and $Lys^{751}Gln$ (exon 23) correlate to significant elevated DNA adduct levels in non-smokers. Combination effects observed. Limited evidence for role in cancer susceptibility: OR (lung cancer) of 1.06-3.2.

[a]Adapted from [159, 166–176]
[b]cf. Figure 2.
Additional websites: *http://www.imm.ki.se/CYPalleles/* (genetic polymorphisms of CYPs); *http://www.louisville.edu/medschool/pharmacology/NAT.html* (genetic polymorphisms of NATs); HCAs, heterocyclic amines; OR, odds ratio; RFLP, restriction fragment length polymorphism.

forms of human cancer and the exposure to particular carcinogenic compounds. In addition to any epidemiological hints, collective evidence obtained from molecular toxicology and molecular epidemiology together may nevertheless be able to uncover the role of individual compounds (or single classes of compounds), and to extract their contribution from the overall human biological response to environmental mixtures. Although highly debated, one of the most well worked-out examples in this regard is the crucial role of carcinogenic PAHs (i.e., B[*a*]P) in the etiology of human lung cancer based on their presence in cigarette smoke. Their importance in tumor initiation is strongly supported by several lines of evidence that can be summarized as follows:

1. As early as 1973, it was reported that the extent of inducibility of 'AHH' (now known as CYP1A1 and 1B1) was increased in lung cancer patients as compared to controls [153]. CYP1A1 and 1B1 are the two major monooxygenases involved in activation of B[*a*]P and other carcinogenic PAHs (Fig. 6), and it appears that CYP1A1 inducibility is related to lung cancer risk in smokers [154].
2. The activity of pulmonary CYP1A1 correlates to 'bulky' *anti*-B[*a*]PDE-DNA adduct levels in human lung tissue from cancer patients [154, 155]. The high variability in lung PAH-DNA adduct levels can be rationalized by large interindividual differences in pulmonary CYP1A1 expression [154], which in turn result from polymorphisms in the regulation of the corresponding gene [156].
3. The levels of *anti*-B[*a*]PDE-DNA adducts in lung tissue *in vivo* are generally higher in smokers as compared to ex- or non-smokers [157].
4. The levels of *anti*-B[*a*]PDE-DNA adducts in lung tissue from individuals carrying a *CYP1A1 Msp*I-*GSTM1*-null genotype are increased as compared to wild-type allele carriers [158] (Tab. 1). This combined 'at risk' genotype correlates to a higher frequency of *TP53* exon 5–8 mutations in lung tumor tissue [159], and is related to high susceptibility for developing lung cancer [152]. Hence, germline polymorphisms of two important genes involved in PAH metabolism (i.e., *CYP1A1*, *GSTM1*) and associated with increased levels of PAH-DNA adducts in lung target tissue, are also related to smoking-associated *TP53* mutations in this tissue.
5. Exposure of human bronchial epithelial cells to *anti*-B[*a*]PDE leads to the formation of *anti*-B[*a*]PDE-N^2-dG adduct hotspots within *K-Ras* (codon 12) and *TP53* (codons 157, 248, 273) at positions identical to mutational hotspots detected in lung tumor tissue from smokers (see above). Both genes are frequently mutated in cigarette smoke-induced lung cancer [126, 160]. In accordance with the preferential formation of *anti*-B[*a*]PDE adducts at N^2-dG sites (Fig. 7), G $\rightarrow$ T transversions were detected as the principal mutations at all of these mutational hotspots.

The coincidence of mutational hotspots in *K-Ras* or *TP53* and *anti*-B[*a*]PDE-N^2-dG adduct hotspots suggests that B[*a*]P is involved in transformation of human lung tissue in smokers. There is no doubt that cigarette smoke contains a highly heterogenous mixture of hundreds of compounds including tobacco-specific *N*-nitrosamines such as NNK, which constitute another major class of strong lung carcinogens present in this matrix (Fig. 1) [161]. However, *N*-nitrosamines predominantly induce G → A and C → T transitions in *K-Ras* and *TP53* of animal models [162]. Although some of these (alkylating) compounds are principally capable of producing G → T transversions, *N*-nitrosamines are activated by CYP2E1, CYP2A6 and some other isoforms, but not by CYP1A1 [46, 163] (Fig. 2), and they barely contribute to 'bulky' DNA adducts induced in the lungs of smokers. The presence of G → A transitions in the *TP53* gene of human lung cancer tissue [164], and the modulating effects of *CYP2E1* gene polymorphisms in human lung cancer risk [165] (Tab. 1), however, suggest that synergistic effects may have a particular importance in chemical-induced lung cancer in man.

Prospects

Background levels of carcinogen-DNA adducts in human tissue samples from unexposed individuals were found to be in the range of 1 per 10^5 (oxidative lesions/human placenta), 1 per 10^7 (B[*a*]P/human placenta), 0.3–3.9 per 10^8 (4-ABP/human bladder), or 0.2 per 10^8 nucleotides (tobacco-specific *N*-nitrosamines/peripheral lung) [177]. The improvement of our knowledge on the efficiency of the enzymatic repair of such lesions and on the various biological effects (modes of action) exerted by genotoxic carcinogens (Fig. 5) is necessary to determine molecularly defined no-adverse-effect levels as the basis for setting 'practicable' thresholds in the human environment [178]. Further, any risk assessment that would not consider the interindividual variability within the human population would be prone to severely underestimate the risks to those subjects who are most vulnerable to carcinogen-induced DNA damage. It therefore becomes important to identify individuals 'at risk' by means of toxicogenetics and molecular epidemiology [179, 180]. This implies that we learn more about the role and interplay of additional, as yet unknown, susceptibility and resistance genes targeted by human carcinogens or involved in modulating human responses to carcinogenic compounds. Beyond the range of known polymorphic enzymes in carcinogen metabolism and repair (Tab. 1), additional genetic variants are likely to contribute to the development of sporadic cancer [181, 182]. The discovery of these variants and the characterization of their interactions with environmental exposures are clearly among the major topics in chemical-related cancer research in the years to come.

Acknowledgments
I am very grateful to my colleague and friend Dr. Gregory P. Tochtrop for his critical reading of the manuscript. The work of the author was supported by the German Research Foundation (DFG: LU 841/2-1).

References

1 Doll R, Peto R (1981) The causes of cancer: quantitative estimates of avoidable risks of cancer in the United States today. *J Natl Cancer Inst* 66: 1191–1308
2 Lichtenstein P, Holm, NV, Verkasalo PK, Iliadou A, Kaprio J, Koskenvuo M, Pukkala E, Skytthe A, Hemminki K (2000) Environmental and heritable factors in the causation of cancer. *N Engl J Med* 343: 78–85
3 Czene K, Lichtenstein P, Hemminki K (2002) Environmental and heritable causes of cancer among 9.6 million individuals in the Swedish family-cancer database. *Int J Cancer* 99: 260–266
4 National Toxicology Program (2004) *10th Report on Carcinogens*. U.S. Department of Health and Human Services, Public Health Service (website: *http://ehp.niehs.nih.gov/docs/allpubs.html*, accessed October 2004) Research Triangle Park, NC, USA
5 Hill J (1761) *Cautions Against the immoderate Use of Snuff. Founded on the known Qualities of the Tobacco Plant; And the Effects it must produce when this Way taken into the Body: And Enforced by Instances of Persons who have perished miserably of Diseases, occasioned, or rendered incurable by its Use.* R. Baldwin and J. Jackso, London
6 Pott P (1775) Cancer Scroti. In: P Pott (ed.): *The Chirurgical Works, Vol. 5, Chirurgical Observations Relative to the Cataract, The Polypus of the Nose, The Cancer of the Scrotum, The Different Kinds of Ruptures, and The Mortification of the Toes and Feet*, Hawes, W. Clarke, and R. Collins, London pp 60–68
7 Volkmann R (1874) Ueber Theer- und Russkrebs. *Berl Klin Wochenschr* 11: 218
8 Bell J (1876) Paraffin epithelioma of the scrotum. *Edinb Med J* 22: 135–137
9 Rehn L (1895) Blasengeschwülste bei Fuchsin-Arbeitern. *Arch Klin Chir* 50: 588–600
10 Leichtenstern O (1898) Ueber Harnblasenentzündung und Harnblasengeschwülste bei Arbeitern in Farbfabriken. *Dtsch Med Wochenschr* 24: 709–713
11 Henry SA (1947) Occupational cutaneous cancer attributable to certain chemicals in industry. *Br Med Bull* 4: 389–401
12 Yamagiwa K, Ichikawa K (1915) Experimentelle Studie über die Pathogenese der Epithelialgeschwülste. *Mitt Med Fak Kaiserl Univ Tokio* 15: 295–344 [English version: *J Cancer Res* 3: 1–29 (1918)]
13 Fischer B (1906) Die experimentelle Erzeugung atypischer Epithelwucherungen und die Entstehung bösartiger Geschwülste. *Münch Med Wochenschr* 53: 2041–2047
14 Bayon H (1912) Epithelial proliferation induced by the injection of gasworks tar. *Lancet* II: 1579
15 Bloch B, Dreifuss W (1921) Ueber die experimentelle Erzeugung von Carcinomen mit Lymphdrüsen- und Lungenmetastasen durch Teerbestandteile. *Schweiz Med Wochenschr* 51: 1033–1037
16 a) Scott A (1922) On the occupation cancer of the paraffin and oil workers of the Scottish shale oil industry. *Br Med J* II: 1108–1109; b) Legge TM (1922) Epitheliomatous ulceration in industry. *Br Med J* II: 1110–1111
17 Passey RD (1922) Experimental soot cancer. *Br Med J* II: 1112–1113
18 Kennaway E, Hieger I (1930) Carcinogenic substances and their fluorescence spectra. *Br Med J* 1: 1044–1046
19 Cook JW, Haslewood GAD, Hewett CL, Hieger I, Kennaway EL, Mayneord WV (1937) Chemical compounds as carcinogenic agents. *Am J Cancer* 29: 219–259
20 Tsutsui H (1918) Über das künstlich erzeugte Cancroid bei der Maus. *Gann* 12: 17–21
21 Cook JW, Hewett CL, Hieger I (1933) The isolation of a cancer-producing hydrocarbon from coal tar. *J Chem Soc* 395–405
22 Berenblum I, Bonser GM (1937) Experimental investigation of "aniline cancer". *J Ind Hyg Toxicol* 19: 86–92
23 Hueper WC, Wiley FH, Wolfe HD, Ranta KE, Leming MF, Blood FR (1938) Experimental production of bladder tumors in dogs by administration of beta-naphthylamine. *J Ind Hyg Toxicol* 20: 46–84
24 Yoshida T (1933) Über die serienweise Verfolgung der Veränderungen der Leber der experimentellen Hepatomerzeugung durch *o*-Aminoazotoluol. *Trans Jap Path Soc* 23: 636–638
25 a) Kinosita R (1936) Researches on the carcinogenesis of the various chemical substances. (In Japanese) *Gann* 30: 423–426; b) Kinosita R (1940) Studies on the cancerogenic azo and related compounds. *Yale J Biol Med* 12: 287–300
26 Wilson RH, DeEds F, Cox AJ Jr, (1941) The toxicity and carcinogenic activity of 2-acetamino-

fluorene. *Cancer Res* 1: 595–608
27 Nowell PC (1976) The clonal evolution of tumor cell populations. *Science* 194: 23–28
28 Miller EC, Miller JA (1947) The presence and significance of bound amino azodyes in the livers of rats fed *p*-dimethylaminoazobenzene. *Cancer Res* 7: 468–480
29 Miller EC (1951) Studies on the formation of protein-bound derivatives of 3,4-benzpyrene in the epidermal fraction of mouse skin. *Cancer Res* 11: 100–108
30 Miller EC, Miller JA (1952) *In vivo* combinations between carcinogens and tissue constituents and their possible role in carcinogenesis. *Cancer Res* 12: 547–556
31 Wheeler GP, Skipper HE (1957) Studies with mustards. III. *In vivo* fixation of C^{14} from nitrogen mustard-$C^{14}H_3$ in nucleic acid fractions of animal tissues. *Arch Biochem Biophys* 72: 465–475
32 Brookes P, Lawley PD (1960) The reaction of mustard gas with nucleic acids *in vitro* and *in vivo*. *Biochem J* 77: 478–484
33 Magee PN, Farber E (1962) Toxic liver injury and carcinogenesis. Methylation of rat-liver nucleic acids by dimethylnitrosamine *in vivo*. *Biochem J* 83: 114–124
34 Brookes P, Lawley PD (1964) Evidence for the binding of polynuclear aromatic hydrocarbons to the nucleic acid of mouse skin: relation between carcinogenic power of hydrocarbons and their binding to deoxyribonucleic acid. *Nature* 202: 781–784
35 Sporn MB, Dingman CW (1966) 2-Acetamidofluorene and 3-methylcholanthrene: differences in binding to rat liver deoxyribonucleic acid *in vivo*. *Nature* 210: 531–532
36 Dingman CW, Sporn MB (1967) The binding of metabolites of aminoazo dyes to rat liver DNA *in vivo*. *Cancer Res* 27: 938–944
37 Ames BN, Durston WE, Yamasaki E, Lee FD (1973) Carcinogens are mutagens: a simple test system combining liver homogenates for activation and bacteria for detection. *Proc Natl Acad Sci USA* 70: 2281–2285
38 Wiley FH (1938) The metabolism of β-naphthylamine *J Biol Chem* 124: 627–630
39 Boyland E, Levi AA, Mawson EH, Roe E (1941) Metabolism of polycyclic compounds. 4. Production of a dihydroxy-1:2:5:6-dibenzanthracene from 1:2:5:6-dibenzanthracene. *Biochem J* 35: 184–191
40 Stevenson ES, Dobriner K, Rhoads CP (1942) The metabolism of dimethylaminoazobenzene (Butter Yellow) in rats. *Cancer Res* 2: 160–167
41 Mueller GC, Miller JA (1948) The metabolism of 4-dimethylaminoazobenzene by rat liver homogenates. *J Biol Chem* 176: 535–544
42 Brodie BB, Axelrod J, Cooper JR, Gaudette L, La Du BN, Mitoma C, Udenfried S (1955) Detoxication of drugs and other foreign compounds by liver microsomes. *Science* 121: 603–604
43 Omura T, Sato R (1962) A new cytochrome in liver microsomes. *J Biol Chem* 237: 1375–1376
44 Lu AYH, Coon MJ (1968) Role of hemoprotein P-450 in fatty acid ω-hydroxylation in a soluble enzyme system from liver microsomes. *J Biol Chem* 243: 1331–1332
45 Nelson DR (2004) Cytochrome P450 gene superfamily: *http://drnelson.utmem.edu/cytochrome P450.html* (accessed October 2004)
46 Guengerich FP, Shimada T (1991) Oxidation of toxic and carcinogenic chemicals by human cytochrome P-450 enzymes. *Chem Res Toxicol* 4: 391–407
47 Shimada T, Oda Y, Gillam EMJ, Guengerich FP, Inoue K (2001) Metabolic activation of polycyclic aromatic hydrocarbons and other precarcinogens by cytochromes P450 1A1 and P450 1B1 allelic variants and other human cytochromes P450 in *Salmonella typhimurium* NM2009. *Drug Metab Disp* 29: 1176–1182
48 Nishimura M, Yaguti H, Yoshitsugu H, Naito S, Satoh T (2003) Tissue distribution of mRNA expression of human cytochrome P450 isoforms assessed by high-sensitivity real-time reverse transcription PCR. *Yakugaku Zasshi* 123: 369–375
49 Cramer JW, Miller JA, Miller JC (1960) *N*-Hydroxylation: a new metabolic reaction observed in the rat with the carcinogen 2-acetylaminofluorene. *J Biol Chem* 235: 885–888
50 Miller EC, Miller JA, Hartmann HA (1961) *N*-Hydroxy-2-acetylaminofluorene: a metabolite of 2-acetylaminofluorene with increased carcinogenic activity in the rat. *Cancer Res* 21: 815–824
51 DeBraun JR, Smith JYR, Miller EC, Miller JA (1970) Reactivity *in vivo* of the carcinogen *N*-hydroxy-2-acetylaminofluorene: increased by sulfate ion. *Science* 167: 184–186
52 Weisburger JH, Yamamoto RS, Williams GM, Grantham PH, Matsushima T, Weisburger EK (1972) On the sulfate ester of *N*-hydroxy-*N*-2-fluorenylacetamide as a key ultimate hepatocarcinogen. *Cancer Res* 32: 491–500
53 Dipple A (1995) DNA adducts of chemical carcinogens. *Carcinogenesis* 16: 437–441

54 Guengerich FP (2002) N-Hydroxyarylamines. *Drug Metab Rev* 34: 607–623
55 Booth J, Boyland E, Sims P (1961) An enzyme from rat liver catalysing conjugations with glutathione. *Biochem J* 79: 516–524
56 a) Barnes MM, James SP, Wood PB (1959) The formation of mercapturic acids. 1. Formation of mercapturic acid and the levels of glutathione in tissues. *Biochem J* 71: 680–690; b) Bray HG, Franklin TJ, James SP (1959) The formation of mercapturic acids. 2. The possible role of glutathionase. *Biochem J* 71: 690–696; c) Bray HG, Franklin TJ, James SP (1959) The formation of mercapturic acids. 3. *N*-Acetylation of *S*-substituted cysteines in the rabbit, rat and guinea pig. *Biochem J* 73: 465–473
57 Guengerich FP (2003) Activations of dihaloalkanes by thiol-dependent mechanisms. *J Biochem Mol Biol* 36: 20–27
58 Anders MW, Dekant W (1998) Glutathione-dependent bioactivation of haloalkenes. *Annu Rev Pharmacol Toxicol* 38: 501–537
59 McGregor DB, Partensky C, Wilbourn J, Rice JM (1998) An IARC evaluation of polychlorinated dibenzo-*p*-dioxins and polychlorinated dibenzofurans as risk factors in human carcinogenesis. *Environ Health Perspect* 106, Suppl 2: 755–760
60 Huff J, Lucier G, Tritscher A (1994) Carcinogenicity of TCDD: experimental, mechanistic, and epidemiologic evidence. *Annu Rev Pharmacol Toxicol* 34: 343–372
61 Piskorska-Pliszczynska J, Keys B, Safe S, Newman MS (1986) The cytosolic receptor binding affinities and AHH induction potencies of 29 polynuclear aromatic hydrocarbons. *Toxicol Lett* 34: 67–74
62 Conney AH, Miller EC, Miller JA (1957) Substrate-induced synthesis and other properties of benzpyrene hydroxylase in rat liver. *J Biol Chem* 228: 753–766
63 Nebert DW, Gelboin HV (1968) Substrate-inducible microsomal aryl hydroxylase in mammalian cell culture. I. Assay and properties of induced enzyme and II. Cellular responses during enzyme induction. *J Biol Chem* 243: 6242–6261
64 Poland A, Glover E, Kende AS (1976) Stereospecific, high affinity binding of 2,3,7,8-tetrachlorodibenzo-*p*-dioxin by hepatic cytosol. Evidence that the binding species is receptor for induction of aryl hydrocarbon hydroxylase. *J Biol Chem* 251: 4936–4946
65 Gu YZ, Hogenesch JB, Bradfield CA (2000) The PAS superfamily: sensors of environmental and developmental signals. *Annu Rev Pharmacol Toxicol* 40: 519–561
66 Nebert DW, Puga A, Vasiliou V (1993) Role of the Ah receptor and the dioxin-inducible [*Ah*] gene battery in toxicity, cancer, and signal transduction. *Ann NY Acad Sci* 685: 624–640
67 Gonzalez FJ, Fernandez-Salguero P (1998) The arylhydrocarbon receptor. Studies using the AHR-null mice. *Drug Metab Dispos* 26: 1194–1198
68 Fernandez-Salguero PM, Hilbert DM, Rudikoff S, Ward JM, Gonzalez FJ (1996) Aryl-hydrocarbon receptor-deficient mice are resistant to 2,3,7,8-tetradibenzo-*p*-dioxin-induced toxicity. *Toxicol Appl Pharmacol* 140: 173–179
69 Andersson P, McGuire J, Rubio C, Gradin K, Whitelaw ML, Pettersson S, Hanberg A, Poellinger L (2002) A constitutively active dioxin/aryl hydrocarbon receptor induces stomach tumors. *Proc Natl Acad Sci USA* 99: 9990–9995
70 Moennikes O, Loeppen S, Buchmann A, Andersson P, Ittrich C, Poellinger L, Schwarz M (2004) A constitutively active dioxin/aryl hydrocarbon receptor promotes hepatocarcinogenesis in mice. *Cancer Res* 64: 4707–4710
71 Mimura J, Fujii-Kuriyama Y (2003) Functional role of AHR in the expression of toxic effects by TCDD. *Biochim Biophys Acta* 1619: 263–268
72 Puga A, Maier A, Medvedovic M (2000) The transcriptional signature of dioxin in human hepatoma HepG2 cells. *Biochem Pharmacol* 60: 1129–1142
73 Frueh FW, Hayashibara KC, Brown PO, Whitlock JP Jr, (2001) Use of cDNA microarrays to analyze dioxin-induced changes in human liver gene expression. *Toxicol Lett* 122: 189–203
74 Carlson DB, Perdew GH (2002) A dynamic role for the Ah receptor in cell signaling? Insights from a diverse group of Ah receptor interacting proteins. *J Biochem Mol Toxicol* 16: 317–325
75 Enan E, Matsumura F (1996) Identification of c-Src as the integral component of the cytosolic Ah receptor complex, transducing the signal of 2,3,7,8-tetrachlorodibenzo-*p*-dioxin (TCDD) through the protein phosphorylation pathway. *Biochem Pharmacol* 52: 1599–1612
76 Johnson CD, Balagurunathan Y, Tadesse MG, Falahatpisheh MH, Brun M, Walker MK, Dougherty ER, Ramos KS (2004) Unraveling gene-gene interactions regulated by ligands of the aryl hydrocarbon receptor. *Environ Health Perspect* 112: 403–412

77 Boyland E, Levi AA (1935) Metabolism of polycyclic compounds. I. Production of dihydroxydihydroanthracene from anthracene. *Biochem J* 29: 2679–2683
78 a) Boyland E, Sims P (1960) Metabolism of polycyclic compounds. 16. The metabolism of 1:2-dihydronaphthalene and 1:2-epoxy-1:2:3:4-tetrahydronaphthalene. *Biochem J* 77: 175–181; b) Booth J, Boyland E, Sato T, Sims P (1960) Metabolism of polycyclic compounds. 17. The reaction of 1:2-dihydronaphthalene and 1:2-epoxy-1:2:3:4-tetrahydronaphthalene with glutathione catalysed by tissue preparations. *Biochem J* 77: 182–186
79 Sims P, Grover PL, Swaisland A, Pal K, Hewer A (1974) Metabolic activation of benzo[*a*]pyrene proceeds by a diol-epoxide. *Nature* 252: 326–328
80 Luch A, Baird WM (2005) Metabolic activation of polycyclic aromatic hydrocarbons. In: A Luch (ed.): *The Carcinogenic Effects of Polycyclic Aromatic Hydrocarbons*. Imperial College Press, London, 19–96
81 Shimizu Y, Nakatsuru Y, Ichinose M, Takahashi Y, Kume H, Mimura J, Fujii-Kuriyama Y, Ishikawa T (2000) Benzo[*a*]pyrene carcinogenicity is lost in mice lacking the aryl hydrocarbon receptor. *Proc Natl Acad Sci USA* 97: 779–782
82 Gonzalez FJ (2001) The use of gene knockout mice to unravel the mechanisms of toxicity and chemical carcinogenesis. *Toxicol Lett* 120: 199–208
83 Nakatsuru Y, Wakabayashi K, Fujii-Kuriyama Y, Ishikawa T, Kusama K, Ide F (2004) Dibenzo[*a,l*]pyrene-induced genotoxic and carcinogenic responses are dramatically suppressed in aryl hydrocarbon receptor-deficient mice. *Int J Cancer* 112: 179–183
84 Berenblum I, Shubik P (1947) A new, quantitative, approach to the study of the stages of chemical carcinogenesis in the mouse's skin. *Br J Cancer* 1: 383–391
85 Luch A (2005) Polycyclic aromatic hydrocarbon-induced carcinogenesis – an introduction. In: A Luch (ed.): *The Carcinogenic Effects of Polycyclic Aromatic Hydrocarbons*. Imperial College Press, London, 1–18
86 Friedewald WF, Rous P (1944) The initiating and promoting elements in tumor production. An analysis of the effects of tar, benzpyrene, and methylcholanthrene on rabbit skin. *J Exp Med* 80: 101–126
87 Balmain A, Pragnell IB (1983) Mouse skin carcinomas induced *in vivo* by chemical carcinogens have a transforming Harvey-*ras* oncogene. *Nature* 303: 72–74
88 Robles AI, Rodriquez-Puebla ML, Glick AB, Trempus C, Hansen L, Sicinski P, Tennant RW, Weinberg RA, Yuspa SH, Conti CJ (1998) Reduced skin tumor development in cyclin D1-deficient mice highlights the oncogenic *ras* pathway *in vivo*. *Genes Dev* 12: 2469–2474
89 Campbell SL, Khosravi-Far R, Rossman KL, Clark GJ, Der CJ (1998) Increasing complexity of Ras signaling. *Oncogene* 17: 1395–1413
90 Freudenthal RI, Stephens E, Anderson DP (1999) Determining the potential of aromatic amines to induce cancer in the urinary bladder. *Int J Toxicol* 18: 353–359
91 Tsuneoka Y, Dalton TP, Miller ML, Clay CD, Shertzer HG, Talaska G, Medvedovic M, Nebert DW (2003) 4-Aminobiphenyl-induced liver and urinary bladder DNA adduct formation in Cyp1a2(–/–) and Cyp1a2(+/+) mice. *J Natl Cancer Inst* 95: 1227–1237
92 Beland FA, Kadlubar FF (1985) Formation and persistance of arylamine DNA adducts *in vivo*. *Environ Health Perspect* 62: 19–33
93 Neumann HG, Ambs S, Bitsch A (1994) The role of nongenotoxic mechanisms in arylamine carcinogenesis. *Environ Health Perspect* 102, Suppl 6: 173–176
94 Klöhn PC, Soriano ME, Irwin W, Penzo D, Scorrano L, Bitsch A, Neumann HG, Bernardi P (2003) Early resistance to cell death and to onset of the mitochondrial permeability transition during hepatocarcinogenesis with 2-acetylaminofluorene. *Proc Natl Acad Sci USA* 100: 10014–10019
95 Van Delft JHM, van Agen E, van Breda SGJ, Herwijnen MH, Staal YCM, Kleinjans JCS (2004) Discrimination of genotoxic from non-genotoxic carcinogens by gene expression profiling. *Carcinogenesis* 25: 1265–1276
96 Skorski T (2002) Oncogenic tyrosine kinases and the DNA-damage response. *Nat Rev Cancer* 2: 351–360
97 Bombail V, Moggs JG, Orphanides G (2004) Perturbation of epigenetic status by toxicants. *Toxicol Lett* 149: 51–58
98 Newbold RR (2004) Lessons learned from perinatal exposure to diethylstilbestrol. *Toxicol Appl Pharmacol* 199: 142–150
99 Herbst AL, Ulfelder H, Poskanzer DC (1971) Adenocarcinoma of the vagina. Association of maternal stilbestrol therapy with tumor appearance in young woman. *N Engl J Med* 284: 878–881

146 Buterin T, Hess MT, Gunz D, Geacintov NE, Mullenders LH, Naegeli H (2002) Trapping of DNA nucleotide excision repair factors by nonrepairable carcinogen adducts. *Cancer Res* 62: 4229–4235
147 Lewtas J, DeMarini D, Favor J, Layton D, MacGregor J, Ashby J, Lohman P, Haynes R, Mendelsohn M (1994) Risk characterization strategies for genotoxic environmental agents. In: DJ Brusick (ed.): *Methods for Genetic Risk Assessment*. Lewis Publishers, Boca Raton, FL, 125–169
148 Calabrese EJ, Baldwin LA (2002) Applications of hormesis in toxicology, risk assessment and chemotherapeutics. *Trends Pharmacol Sci* 23: 331–337
149 Perera FP, Weinstein IB (2000) Molecular epidemiology: recent advances and future directions. *Carcinogenesis* 21: 517–524
150 Gonzalez FJ (1997) The role of carcinogen-metabolizing enzyme polymorphism in cancer susceptibility. *Reprod Toxicol* 11: 397–412
151 Wormhoudt LW, Commandeur NM, Vermeulen NPE (1999) Genetic polymorphisms of human *N*-acetyltransferase, cytochrome P450, glutathione-*S*-transferase, and epoxide hydrolase enzymes: relevance to xenobiotic metabolism and toxicity. *Crit Rev Toxicol* 29: 59–124
152 Bartsch H, Nair U, Risch A, Rojas M, Wikman H, Alexandrov K (2000) Genetic polymorphism of *CYP* genes, alone or in combination, as a risk modifier of tobacco-related cancers. *Cancer Epidemiol Biomarkers Prev* 9: 3–28
153 Kellermann G, Shaw CR, Luyten-Kellerman M (1973) Aryl hydrocarbon hydroxylase inducibility and bronchogenic carcinoma. *N Engl J Med* 289: 934–937
154 Bartsch H, Castegnaro M, Rojas M, Camus AM, Alexandrov K, Lang M (1992) Expression of pulmonary cytochrome P4501A1 and carcinogen DNA adduct formation in high risk subjects for tobacco-related lung cancer. *Toxicol Lett* 64/65: 477–483
155 Alexandrov K, Rojas M, Geneste O, Castegnaro M, Camus AM, Petruzzelli S, Giuntini C, Bartsch H (1992) An improved fluorometric assay for dosimetry of benzo[*a*]pyrene diol-epoxide-DNA adducts in smokers' lung: comparisons with total bulky adducts and aryl hydrocarbon hydroxylase activity. *Cancer Res* 52: 6248–6253
156 Smart J, Daly AK (2000) Variation in induced CYP1A1 levels: relationship to CYP1A1, Ah receptor and GSTM1 polymorphisms. *Pharmacogenetics* 10: 11–24
157 Phillips DH (2002) Smoking-related DNA and protein adducts in human tissues. *Carcinogenesis* 23: 1979–2004
158 Alexandrov K, Cascorbi I, Rojas M, Bouvier G, Kriek E, Bartsch H (2002) *CYP1A1* and *GSTM1* genotypes affect benzo[*a*]pyrene DNA adducts in smokers' lung: comparison with aromatic/hydrophobic adduct formation. *Carcinogenesis* 23: 1969–1977
159 Kawajiri K, Eguchi H, Nakachi K, Sekiya T, Yamamoto M (1996) Association of *CYP1A1* germ line polymorphisms with mutations of the *p53* gene in lung cancer. *Cancer Res* 56: 72–76
160 Greenblatt MS, Bennett WP, Hollstein M, Harris CC (1994) Mutations in the *p53* tumor suppressor gene: clues to cancer etiology and molecular pathogenesis. *Cancer Res* 54: 4855–4878
161 Hecht SS (2003) Tobacco carcinogens, their biomarkers and tobacco-induced cancer. *Nat Rev Cancer* 3: 733–744
162 Hecht SS (1998) Biochemistry, biology, and carcinogenicity of tobacco-specific *N*-nitrosamines. *Chem Res Toxicol* 11: 559–603
163 Goldman R, Shields PG (2003) Food mutagens. *J Nutr* 133, Suppl 3: 965–973
164 Gao WM, Mady HH, Yu GY, Siegfried JM, Luketich JD, Melhem MF, Keohavong P (2003) Comparison of p53 mutations between adenocarcinoma and squamous cell carcinoma of the lung: unique spectra involving G to A transitions and G to T transversions in both histologic types. *Lung Cancer* 40: 141–150
165 Wu X, Shi H, Hong J, Kemp B, Hong WK, Delclos GL, Spitz MR (1997) Association between cytochrome *P4502E1* genotype, mutagen sensitivity, cigarette smoking and susceptibility to lung cancer. *Carcinogenesis* 18: 967–973
166 Ko Y, Abel J, Harth V, Bröde P, Anthony C, Donat S, Fischer HP, Ortiz-Pallardo ME, Thier R, Sachinidis A et al. (2001) Association of CYP1B1 codon 432 mutant allele in head and neck squamous cell cancer is reflected by somatic mutations of p53 in tumor tissue. *Cancer Res* 61: 4398–4404
167 Lee WJ, Brennan P, Boffetta P, London SJ, Benhamou S, Rannug A, To-Figueras J, Ingelman-Sundberg M, Shields P, Gaspari L, Taioli E (2002) Microsomal epoxide hydrolase polymorphisms and lung cancer risk: a quantitative review. *Biomarkers* 7: 230–241

168 Nebert DW, McKinnon RA, Puga A (1996) Human drug-metabolizing enzyme polymorphisms: effects on risk of toxicity and cancer. *DNA Cell Biol* 15: 273–280
169 Scheel J, Hussong R, Schrenk D, Schmitz HJ (2002) Variability of the human aryl hydrocarbon receptor nuclear translocator (ARNT) gene. *J Hum Genet* 47: 217–224
170 Glatt H, Boeing H, Engelke CEH, Ma L, Kuhlow A, Pabel U, Pomplun D, Teubner W, Meinl W (2001) Human cytosolic sulphotransferases: genetics, characteristics, toxicological aspects. *Mutat Res* 482: 27–40
171 Matullo G, Palli D, Peluso M, Guarrera S, Carturan S, Celentano E, Krogh V, Munnia A, Tumino R, Polidoro S et al. (2001) *XRCC1, XRCC3, XPD* gene polymorphisms, smoking and ^{32}P-DNA adducts in a sample of healthy subjects. *Carcinogenesis* 22: 1437–1445
172 Spitz MR, Wei Q, Dong Q, Amos CI, Wu X (2003) Genetic susceptibility to lung cancer: the role of DNA damage and repair. *Cancer Epidemiol Biomarkers Prev* 12: 689–698
173 Kiyohara C, Otsu A, Shirakawa T, Fukuda S, Hopkin JM (2002) Genetic polymorphisms and lung cancer susceptibility. *Lung Cancer* 37: 241–256
174 Kiyohara C, Shirakawa T, Hopkin JM (2002) Genetic polymorphism of enzymes involved in xenobiotic metabolism and the risk of lung cancer. *Environ Health Prev Med* 7: 47–59
175 Vineis P (2004) Individual susceptibility to carcinogens. *Oncogene* 23: 6477–6483
176 Matullo G, Peluso M, Polidoro S, Guarrera S, Munnia A, Krogh V, Masala G, Berrino F, Panico S, Tumino R et al. (2003) Combination of DNA repair gene single nucleotide polymorphisms and increased levels of DNA adducts in a population-based study. *Cancer Epidemiol Biomarkers Prev* 12: 674–677
177 Farmer PB, Shuker DEG (1999) What is the significance of increases in background levels of carcinogen-derived protein and DNA adducts? Some considerations for incremental risk assessment. *Mutat Res* 424: 275–286
178 Wiltse JA, Dellarco VL (2000) U.S. Environmental Protection Agency's revised guidelines for carcinogen risk assessment: evaluating a postulated mode of carcinogen action in guiding dose-response extrapolation. *Mutat Res* 464: 105–115
179 Orphanides G, Kimber I (2003) Toxicogenetics: applications and opportunities. *Toxicol Sci* 75: 1–6
180 Mohrenweiser HW (2004) Genetic variation and exposure related risk estimation: will toxicology enter a new era? DNA repair and cancer as a paradigm. *Toxicol Pathol* 32, Suppl 1: 136–145
181 Balmain A, Gray J, Ponder B (2000) The genetics and genomics of cancer. *Nat Genet Suppl* 33: 238–244
182 Frank SA (2004) Genetic predisposition to cancer – insights from population genetics. *Nat Rev Genet* 5: 764–772

Cancer: Cell Structures, Carcinogens and Genomic Instability
Edited by Leon P. Bignold

Metal ions and carcinogenesis

Troy R. Durham and Elizabeth T. Snow

Centre for Cellular and Molecular Biology, School of Biological and Chemical Sciences, Deakin University, 221 Burwood Highway, Burwood, Victoria, Australia 3125

Abstract. Metals are essential for the normal functioning of living organisms. Their uses in biological systems are varied, but are frequently associated with sites of critical protein function, such as zinc finger motifs and electron or oxygen carriers. These functions only require essential metals in minute amounts, hence they are termed trace metals. Other metals are, however, less beneficial, owing to their ability to promote a wide variety of deleterious health effects, including cancer. Metals such as arsenic, for example, can produce a variety of diseases ranging from keratosis of the palms and feet to cancers in multiple target organs. The nature and type of metal-induced pathologies appear to be dependent on the concentration, speciation, and length of exposure. Unfortunately, human contact with metals is an inescapable consequence of human life, with exposures occurring from both occupational and environmental sources. A uniform mechanism of action for all harmful metals is unlikely, if not implausible, given the diverse chemical properties of each metal. In this chapter we will review the mechanisms of carcinogenesis of arsenic, cadmium, chromium, and nickel, the four known carcinogenic metals that are best understood. The key areas of speciation, bioavailability, and mechanisms of action are discussed with particular reference to the role of metals in alteration of gene expression and maintenance of genomic integrity.

Key words: Arsenic, cadmium, carcinogenesis, chromium, nickel, oxidative stress.

Introduction

The association of metal exposure with cancer is a well-documented phenomenon. Metals such as arsenic (As), cadmium (Cd), chromium (Cr), and nickel (Ni) are part of an ever growing list of environmental agents that have been formally classified by the International Agency for Cancer Research (IARC) as being known carcinogens [1–4]. For other metals such as iron, copper, beryllium, lead, and mercury there exists an ever increasing body of evidence to support their inclusion in the IARC listings [5–8]. Iron [8] and copper [7], in particular, are carcinogenic in excess, but are highly regulated and generally only produce cancer in animal models or in people with genetic diseases that prevent appropriate metabolic regulation. There is even less information on beryllium carcinogenesis, and no definitive studies that indicate the species, conditions or length of exposure by which lead and mercury metals cause cancer in humans. For these reasons, this review will focus on the known carcinogenic metals: As, Cd, Cr and Ni.

Despite increasing numbers of researchers in the field and the expanding role of metals in environmental health issues, the nature of cancer induction by metals remains a complex and poorly understood process. However, what is known is that metals can promote change in normal cellular functions, leading to aberrant cell growth and development [6]. All metals are now thought to promote cancer by a number of common mechanisms. These include the formation of free radicals, either actively as key players in redox reactions, or through less direct means such as biomethylation [5, 6, 9, 10]. Similarly, many metals can also influence cell control by altering gene regulation [7, 11–14]. In terms of direct damage to DNA, most metals are only weakly mutagenic; however, many are strong co-carcinogens, promoting a synergistic effect in the presence of other cancer-causing agents [5, 6, 15]. Hence, the ability of metals to promote cellular alterations may be far more dynamic than has been classically assumed. Thus, it is the purpose of this review to evaluate mechanisms that are central to the role of metals as carcinogenic agents. This review outlines current evidence related to the mechanisms of genotoxicity and gene expression, as well as other mechanisms unique to specific metals. The principle focus is on those metals for which the IARC has deemed there to be sufficient evidence to classify as carcinogens, and that there is the most information regarding genotoxic mechanisms. Because of the great amount of data now available on this topic, this chapter does not claim to be exhaustive, but will hopefully provide a useful survey of the field, with selected references focusing heavily on recent reviews.

Mechanisms of metal carcinogenesis: an overview

The process of carcinogenesis has classically been described as occurring in four stages: initiation, promotion, progression, and metastasis. Initiation is generally thought to be the result of genotoxicity leading to DNA mutation. Cancer initiation by metals most often involves the production of free radicals which can potentially damage DNA [5, 6, 16, 17]. This process can occur by multiple mechanisms, such as redox cycling, metabolism, and the induction of genes producing reactive species. The products of oxidative damage in DNA are frequently single base lesions, most notably the 7,8-dihydro-8-oxo-2'-deoxyguanosine base modification [18, 19]. Of all the metals, chromate salts produce the greatest genotoxic response in the shortest period of time [5, 13, 20, 21]. Chromium salts also exhibit the ability to form mutation-inducing crosslinks between DNA and protein [22]. Iron similarly is highly reactive and readily able to donate or accept electrons from a variety of sources. It is, however, highly regulated, such that it is unlikely that genotoxic affects or adduct formation would occur, except in circumstances where iron overload occurs [8].

Initiation of cancer, however, is not solely the result of point mutations. It may also be caused by DNA strand breaks, which can result in chromosomal rearrangements, or through alterations in DNA repair that reduce the capacity for repair of lesions not associated with the metal. Metal-dependent changes in transcriptional control of specific genes may also play a role in cancer initiation.

Altered gene expression through the induction of specific signal cascades is most often associated with cancer promotion. Metal salts can alter gene regulation through a number of mechanisms, most frequently by activation of transcription factors or through changes to gene methylation patterns [23–25]. Similarly, signal cascades may be modified by the interaction of metals with any of the steps in the pathway, often through direct binding to receptors or intermediate proteins [10, 13, 14].

Progression of tumors is akin to the point where aberrant control mechanisms in the cell begin to predominate, and is characterized by changes to the cell phenotype and metabolic processes. Metals have strong affects on cells, particularly with regard to redox status. Both chromium and arsenic have been shown to alter redox potential in mammalian cells *in vitro* [26–28]. Changes of this nature can facilitate an increased competitiveness of these types of cells as against their non-tumorigenic counterparts, with greater growth potential and rate.

The final phase of malignant tumorigenesis involves the migration of cancerous cells to other regions of the body. This process normally sees the formation of secondary sites of colonization by cells that have altered cell signaling cascades, phenotypic characteristics and proliferative capacity. Of the metals presented here, cadmium is the only one that has been shown to affect the extracellular matrix of cells [29] by interfering with cadherins, which link cells together, preventing their formation and subsequently enabling cells to move to other sites.

Speciation, uptake and health effects of specific metals

Arsenic

Arsenic has had a long and somewhat chequered history. Unlike many of the other carcinogenic metals, it has been used as a prophylactic agent for around 2500 years. It has been said that Hippocrates used arsenic sulfides to treat ulcers and that arsenic may have been used to treat the plague in the Middle Ages [30]. In the 1800s, arsenic formed the basis of the common medication, Fowler's reagent. This reagent was used for the better part of 200 years to treat anything from the common cold, to asthma and psoriasis [30]. Despite its wide medical use, arsenic was also a highly successful murder agent due to its relatively high toxicity at low doses. Clear, odorless and almost tasteless by nature, arsenic was easy to conceal in food or drink. It has been suggested that Napoleon was killed by an overdose of arsenic [31]. More recently, arsenic has been found to be an effective herbicide, as well as being useful in reducing discoloration in glass and as a preservative of wood. Medically, it has again found favor as an agent in the fight against acute promyelocytic leukemia in individuals who have become resistant to normal therapies [32]. Occupational exposure to arsenic is greatest in mining and metal smelting industries; however, it can also occur through glass manufacturing and as the result of coal burning for power production [33]. But the greatest extent of exposure is from arsenic-contaminated water sources.

Arsenic is normally found in close association with heavy metals such as gold, copper and silver. Mining of these heavy metals brings arsenic to the surface where it is concentrated through the refining processes [34]. Arsenic can also be brought to the surface when it is leached from the rock surrounding underground aquifers. It is in this circumstance that arsenic has had its most profound effects on human health. Since the 1980s countries such as Bangladesh, India, and China, where surface water is frequently contaminated with microbial pathogens, have invested heavily in alternate water sources that are now known to be heavily contaminated with arsenic [34, 35]. Concentrations of arsenic in these water sources vary wildly; however, in many regions they exceed by 10 to 15 times the current World Health Organization's (WHO) recommended level of 10 ppm [36]. Even at these levels, arsenic is not acutely toxic. However, as early as 1968, similar high levels of arsenic in the artesian well water in regions of southern Taiwan were recognized as a likely cause of carcinogenesis [37]. Increased cancer rates associated with arsenic-contaminated drinking water have now been recorded in many countries, including Taiwan, Argentina, Chile, and Mexico. The arsenic-associated cancer incidence in Bangladesh and West Bengal, India is expected to reach catastrophic levels over the next several decades [38, 39].

Arsenic in the environment can take a range of forms, both organic and inorganic. Inorganic arsenic has two possible valencies, arsenite, or As(III) and arsenate, As(V). Arsenite is the more toxic of the two species with cell viability assays indicating that concentrations anywhere from 1 to 10 μM and upwards are able to promote toxicity [5]. Arsenate is approximately three to fivefold less toxic than arsenite, presumably because As(V) requires reduction to As(III) to exert its toxicity. Organoarsenic species can also be formed by biometabolism. Many organoarsenic species are significantly less toxic than inorganic As(III). However, methyl As(III) species can be significantly more toxic than inorganic As(III) [40] and may contribute to arsenic carcinogenesis. The relative toxicity of the different forms of arsenic is predominantly the result of their different chemical properties, but may also relate to the relative efficiency of their uptake [41, 42], the duration of the exposure, and the time when the toxicity assay is performed [42, 43]. Arsenic excretion rates vary, but it is generally accepted that arsenic, unlike other carcinogenic metals, is rapidly excreted by the body, to the extent that more than 50% is removed within 2 days in acute poisoning cases [33].

Organic metabolites of arsenic that are of most interest are the monomethyl and dimethyl species of both arsenite and arsenate. These species are generated through biomethylation of inorganic arsenic, followed by reduction and subsequent methylation of monomethyl As(III) to produce dimethyl As(V)

(DMA). The intermediate methylated As(III) species are thought to be considerably more toxic than either methyl As(V) species or even inorganic As(III) species [40, 44]. However, the levels of available organic As(III) species in human tissues relative to other arsenic species are still largely undetermined.

Other organoarsenic species such as arsenobetaine and arsenosugars are commonly found in marine species. Although arsenobetaine is often found in high concentration in marine animals, it is largely excreted unmetabolized, and has very low toxicity [45]. Arsenosugars are frequently found in seaweed and in crustaceans [46]. Recent evidence suggests that these compounds may be metabolized to DMA, which can then be further metabolized to more toxic species [46]. This raises the possibility that consumption of seafoods can be a source of considerable arsenic intake, some of which may result in the retention of some relatively toxic arsenic species. This may have implications for Asian populations, particularly those which live on high fish diets, such as the Japanese [46].

Arsenic pathology is complex. When ingested at very high doses, in excess of 200 mg, it produces acute toxicity characterized by nausea, vomiting, sloughing off of epithelial tissues, internal bleeding, changes to blood pressure, and atrial fibrillation [33]. This can lead to heart attack, coma, and death. At sublethal doses arsenic ingestion can be treated with a range of metal chelators, which reduce its effects; however, few if any other treatments for arsenic ingestion exist. At very low doses, arsenic appears to have minimal short-term effect however, over longer periods a range of pathologies are seen [2]. Chronic low-dose exposure initially produces blotching of the skin, followed by hyperkeratosis of the palms and soles of the feet. If exposure continues, alterations to peripheral vasculature are seen along with the formation of skin lesions, which left untreated, can become cancerous [47, 48]. Arsenic is also associated with an increased risk for cancer of the lungs, liver and bladder [47].

Induction of cancer by arsenic is not thought to originate from a single exposure, but rather is the result of gradual changes to a variety of processes within the cell. Different arsenic species enter cells by different mechanisms. Arsenate is able to mimic phosphate, and hence is able to enter cells using phosphate transport proteins. Arsenite, however, is thought to enter through aquaglyceroporins [49]. Once in the blood stream, arsenic is taken to the liver where biometabolism occurs. This process involves the progressive methylation of arsenic, with As(III) converted to the less toxic methyl As(V) species. The ingested arsenic is excreted predominantly in the urine as inorganic As(III) and As(V), methyl As(V), and dimethyl As(V), with the proportions of these being variable and related to arsenic dose [50–52]. Some intermediate trivalent arsenic metabolites are also produced, and can be found in the urine [53]. Despite their greater toxicity, relative to either inorganic As or organic As(V) species, it is yet to be determined whether these methyl As(III) and dimethyl As(III) species play a significant role in carcinogenesis.

Cadmium

Unlike many other metals, cadmium is found in only one valence state, that of Cd(II). Exposure to cadmium has also been far less common than other carcinogenic metals. Of greatest note was the historical use of cadmium as a paint additive giving rise to the bright yellows seen in many paintings, such as those of Claude Monet [54]. Industrial use of cadmium is only a recent phenomenon, beginning in the 1940s. Cadmium is now most commonly encountered in cadmium-nickel battery production [10], although it continues to be used in paints, as well as in plastic production where it is an effective stabilizing agent. Like arsenic, occupational exposure to cadmium can occur through metal refining processes, where cadmium is often associated with copper and can be released into the atmosphere during heating [55]. The greatest exposure to cadmium, however, comes from cigarette smoke [10]. Particulate cadmium in cigarette smoke collects in the lungs where it can be transported into the bloodstream across the alveoli. Unlike arsenic, cadmium has a long biological half life, considered to be somewhere between 15 and 25 years [4, 56]. This means cadmium can accumulate to levels many times greater than an individual would be subjected to in a single exposure. Cadmium is only a weak mutagen, but is a strong co-mutagen [4, 57, 58]. This is of particular concern for cigarette smokers who simultaneously inhale cadmium and benzo[a]pyrene, as well as a range of other chemicals, including arsenic and other metals.

Health effects of cadmium are quite dissimilar to other metals. Non-toxic doses of cadmium produce a wide variety of effects, many of which are related to bone development and maintenance. Individuals exposed to cadmium can develop osteoporosis, anemia, eosinophilia, emphysema, and renal tubular damage [59]. Long-term cadmium toxicity can produce Itai-Itai disease, in which individuals suffer from bone fractures, severe pain, proteinuria and severe osteomalacia [59]. Acute high-level exposure to cadmium is also able to produce severe lung damage. However, like other metals, prolonged repeated exposures are required to induce carcinogenesis. Target organs for cadmium are varied however, lung cancers predominate [4]. Other tissues subject to malignant transformation by cadmium include the prostate, pancreas and kidney. The testes are also thought to be a site of cadmium carcinogenesis; however, this has only been shown in animal models. Like arsenic, cadmium is only a weak mutagen. This suggests that tumors result from either epigenetic or co-carcinogenic effects, particularly in cases of smoking-induced lung cancer [10].

Chromium

Chromium is widely available, complex in action, and used industrially in a myriad of applications including, pigment production, chrome plating, welding, production of ferrochrome metals, leather tanning and as a dietary supplement [3, 60]. Dietary supplementation is of particular interest because of the critical nature of Cr(III) for optimum insulin binding [61]. Occupational exposure to Cr(VI) is a well-established source of human carcinogenesis; however, occupational health initiatives have had a considerable impact in reducing incidence levels. Non-occupational sources of exposure are thought to originate from engine emissions, atmospheric particles released from smelting and refining industries, as well as through cigarette smoke [13]. Chromium speciation is complex, and chromium is often found in compounds with other metals. Environmental chromium is generally found in two principle valency states, the more toxic and carcinogenic Cr(VI) [60] and the essential Cr(III). Cr(VI) species are readily taken up into cells by phosphate/sulfate anion channels [62–64]. Cr(III), however, cannot move into cells by the same mechanism, and is required at considerably higher concentrations to produce toxicity in cells. It must be noted, however, that not all Cr(VI) species are of equal carcinogenic risk. Animal models have shown that the largely insoluble chromium compounds are far more carcinogenic than their soluble counterparts [3, 65]. It appears that particulate matter containing insoluble chromium is deposited on the epithelial surface of the lung where it accumulates to levels high enough to produce cancer [66].

The mechanism of chromium carcinogenesis is unclear; however, the complex intracellular redox cycling of chromium is thought to produce a range of reactive species as well as producing DNA-protein crosslinks. Generally, Cr(VI) on entry into cells is rapidly reduced by interaction with any of a number of low molecular weight thiols, from glutathione (GSH) to cysteine, as well as a range of other reductants such as ascorbic acid, hydrogen peroxide, cytochrome P450 reductase and NADPH [13]. Of these, GSH, ascorbic acid and cysteine residues appear to be the most critical. The reduction process itself is thought to occur either by sequential single electron transfers, progressively reducing Cr(VI) to Cr(V) and then Cr(III), or by a two electron transfer to Cr(IV) then by single electron transfer to Cr(III) [13, 16, 67]. These reactions can produce a variety of other reactive intermediates and provide the mechanism for crosslinking of DNA to proteins by means of a bifunctional Cr(III) intermediate. Both the oxidative DNA damage caused by redox reactive intermediates and, more importantly, the Cr(III)-mediated DNA-protein crosslinks [22] can cause mutations, thereby initiating the process of carcinogenesis. Similarly, it is possible that the interaction of reactive species may also alter cell signaling pathways causing alterations in gene regulation [13].

Although Cr(III)-DNA adducts generated by reduction of Cr(VI) are known to be mutagenic, it is often believed that Cr(III) compounds are non-toxic and, in fact, Cr(III) is promoted as a highly beneficial dietary additive. However, some forms of Cr(III) are known to be capable of producing DNA damage *in vitro* [68] and the possibility that excess Cr(III) supplementation might eventually lead to increased cancer risk is seldom acknowledged [69].

Unlike arsenic and cadmium, chromium is an essential trace element in its trivalent form. That said, Cr(VI) species can be highly toxic to humans [13].

to suppress tumor formation in a range of tissues [204, 206]. The effect of cadmium on cell adhesion was first characterized by a significant loss of tissue integrity that was not initially due to apoptosis [207–209]. Later studies, especially those of Prozialeck et al. [210] showed that cell adhesion and, in particular, the integrity of E-cadherin was an early target of cadmium toxicity. It was also shown that cadmium was able to exert its greatest effects on E-cadherins when calcium levels were low, suggesting that cadmium competes for calcium binding sites [211, 212] The loss of E-cadherins are thought to enhance tumor metastasis, promote toxicity, and promote changes to gene expression profiles through altered β-catenin signaling [204, 206].

Chromium

Complexes formed by chromium are considerably more varied than those of other metals discussed here. The binding of chromium to DNA does not occur with Cr(VI); however, the reduced metabolites of chromium, Cr(III, IV, and V) have all been shown to be reactive towards DNA [213, 214]. Although the structure and efficiency of formation of these chromium-DNA complexes is strongly affected by the reductant involved, such as GSH, ascorbate, or cysteine, most of the resulting adducts seem to be both genotoxic and mutagenic [22]. Binding of chromium species to DNA appears to be preferential for guanine nucleotides, and occurs largely with phosphates in the backbone [22, 215]. The formation of chromium-DNA adducts has a twofold effect, they both inhibit DNA replication and prevent DNA repair, thereby promoting mutagenesis.

Nickel

Nickel shows a strong affinity for histidines and, to a lesser extent, cysteines, and is able to form complexes with a wide variety of proteins [216, 217]. As a result, nickel is frequently used to extract and purify proteins that have been histidine tagged [218, 219]. Proteins that have been shown to bind nickel include: serum albumins, the neuroblastoma-associated tumor suppressor (DAN), and histones [220–222]. Like other metals that form protein complexes, it is thought that nickel interacts with proteins, altering their conformation in such a way that they are no longer able to perform normal cellular functions. Nickel has also been shown to crosslink DNA as the result of oxidation of DNA-associated proteins [12, 223].

Effects on gene regulation: direct and epigenetic changes

Metals have been shown to alter the expression of a great number of genes, too many to cover in detail here. These changes in gene expression are generally plement [3, 60]. Dietary supplementation is of particular interest because of the critical nature of Cr(III) for optimum insulin binding [61]. Occupational exposure to Cr(VI) is a well-established source of human carcinogenesis; however, occupational health initiatives have had a considerable impact in reducing incidence levels. Non-occupational sources of exposure are thought to originate from engine emissions, atmospheric particles released from smelting and refining industries, as well as through cigarette smoke [13]. Chromium speciation is complex, and chromium is often found in compounds with other metals. Environmental chromium is generally found in two principle valency states, the more toxic and carcinogenic Cr(VI) [60] and the essential Cr(III). Cr(VI) species are readily taken up into cells by phosphate/sulfate anion channels [62–64]. Cr(III), however, cannot move into cells by the same mechanism, and is required at considerably higher concentrations to produce toxicity in cells. It must be noted, however, that not all Cr(VI) species are of equal carcinogenic risk. Animal models have shown that the largely insoluble chromium compounds are far more carcinogenic than their soluble counterparts [3, 65]. It appears that particulate matter containing insoluble chromium is deposited on the epithelial surface of the lung where it accumulates to levels high enough to produce cancer [66].

The mechanism of chromium carcinogenesis is unclear; however, the complex intracellular redox cycling of chromium is thought to produce a range of reactive species as well as producing DNA-protein crosslinks. Generally, Cr(VI) on entry into cells is rapidly reduced by interaction with any of a number of low molecular weight thiols, from glutathione (GSH) to cysteine, as well as a range of other reductants such as ascorbic acid, hydrogen peroxide, cytochrome P450 reductase and NADPH [13]. Of these, GSH, ascorbic acid and cysteine residues appear to be the most critical. The reduction process itself is thought to occur either by sequential single electron transfers, progressively reducing Cr(VI) to Cr(V) and then Cr(III), or by a two electron transfer to Cr(IV) then by single electron transfer to Cr(III) [13, 16, 67]. These reactions can produce a variety of other reactive intermediates and provide the mechanism for crosslinking of DNA to proteins by means of a bifunctional Cr(III) intermediate. Both the oxidative DNA damage caused by redox reactive intermediates and, more importantly, the Cr(III)-mediated DNA-protein crosslinks [22] can cause mutations, thereby initiating the process of carcinogenesis. Similarly, it is possible that the interaction of reactive species may also alter cell signaling pathways causing alterations in gene regulation [13].

Although Cr(III)-DNA adducts generated by reduction of Cr(VI) are known to be mutagenic, it is often believed that Cr(III) compounds are non-toxic and, in fact, Cr(III) is promoted as a highly beneficial dietary additive. However, some forms of Cr(III) are known to be capable of producing DNA damage *in vitro* [68] and the possibility that excess Cr(III) supplementation might eventually lead to increased cancer risk is seldom acknowledged [69].

Unlike arsenic and cadmium, chromium is an essential trace element in its trivalent form. That said, Cr(VI) species can be highly toxic to humans [13].

Inhalation of particulate Cr(VI) can cause irritation to the nasal tissue, leading to nose bleeds, ulceration and formation of lesions in the nasal passage [60]. Damage to lung tissue is also not uncommon [70, 71]. Ingestion of Cr(VI) can cause nausea, vomiting, ulceration of the stomach, damage to the liver and kidney, and finally death [60]. Both species of chromium can cause contact hypersensitivity, leading to rashes, swelling and ulcerations. Cr(VI) is the most carcinogenic form of chromium, with insoluble particulate chromium compounds being the most persistent [66] and the most hazardous [72].

Nickel

Nickel has many common industrial uses, thanks largely to its unique chemical properties. Industrially, it is used in electroplating, electroforming, in circuitry, and in nickel-cadmium batteries. Nickel alloys, including stainless steel, are used in a wide variety of objects, from kitchen knives to building tools [73]. Nickel is also used in jewelry and medical implements. Metallic nickel is non-carcinogenic to humans; however, all other nickel compounds, such as nickel sulfides, oxides, and silicates, and other soluble salts, are known carcinogens [12]. Carcinogenic nickel exposure is greatest through the inhalation of nickel-containing particulates. The burning of fossil fuels, as well as the refining of metals such as copper, introduces considerable amounts of nickel into the atmosphere [12]. Like arsenic, nickel can also be leached from soils and rock, thereby contaminating water supplies. In lower organisms such as bacteria, nickel is an essential trace element found in up to seven different enzymes [74]. Higher organisms, however, have failed to show any definitive role for nickel in normal cellular function. That said, studies in the 1970s and 80s showed that the removal of nickel from the diet of rats had significant effects both physically and mentally, which, with continued exclusion of nickel from the diet, were more profound in the subsequent generations [75]. It may be that nickel is not required for normal cellular function in humans, but rather is essential for our intestinal microflora. Like both arsenic and chromium, nickel occurs in different oxidation states, ranging from I to IV, with Ni(II) being most common in biological systems.

As with chromium, particulate nickel is most harmful to humans, especially in the lung where crystalline nickel becomes lodged in the mucous prior to being phagocytized by both epithelial cells and macrophages [76]. Once inside the cells, the nickel compounds are gradually broken down releasing reactive nickel ions. The phagocytic nature of nickel uptake means considerable amounts of nickel are able to accumulate over time, damaging lung tissue and frequently causing latent effects in individuals who may have been exposed to nickel many years earlier [76].

Nickel is not overly toxic to individuals at low doses; however, nickel-containing jewelry can produce contact hypersensitivity in many people [73]. This normally results in rashes and inflammation of the region of contact. However, in more extreme reactions, individuals can suffer from asthma attacks. Individuals who inhale nickel fumes for prolonged periods of time frequently develop bronchitis and chronic lung infections. While ingestion of large quantities of nickel is not normally fatal, it can produce stomach aches, kidney pain and blood in the urine [73]. Nickel carcinogenesis is generally limited to the lung, because phagocytosis is necessary to bring the nickel ions to the DNA in the target tissue [12, 77].

Metals and oxidative stress

Most, if not all, of the carcinogenic metals, have the capacity to produce a variety of radical species that can damage cells. Arsenic, chromium, copper, iron, nickel and, to a lesser extent, cadmium, have all been shown to be able to participate in reactions resulting in the formation of reactive oxygen, sulfur or nitrogen species (for reviews see [6, 7, 11, 14, 27, 76–79]). In most cases these metals produce either radicals based on oxygen species or those based on nitrogen species; however, the formation of oxygen species appears to predominate. The formation of radical species can originate from a variety of sources, from redox cycling, through Fenton/Haber-Weiss chemistry, as products of biometabolism, as messengers in signal cascades, and as normal products of cellular metabolism [6, 80, 81]. Essential transition metals, such as iron and copper, are most likely to participate in redox cycling and Fenton/Haber-Weiss chemistry; however, these metals are highly regulated and are of less concern with regard to carcinogenesis. Nevertheless, other carcinogenic metals may also react in similar fashion and thereby produce reactive species that can cause DNA damage and mutations. Some of the key reactions responsible for the metal-related formation of reactive oxygen species (ROS) are described briefly below.

The production of reactive oxygen species

Superoxide ($O_2^{\cdot-}$) was first shown to be produced in phagocytic cells by membrane-bound NADPH oxidase [82, 83]. More recently, it has been observed that epithelial and endothelial cells also express NADPH oxidase [84, 85]. Phagocytic cells produce large concentrations of $O_2^{\cdot-}$ as a killing agent. However, in most non-phagocytic cells, superoxide is primarily formed as a byproduct of mitochondrial metabolism [86], although it is also used as an intracellular messenger in signal cascades [87, 88]. Recent data suggest that at least some of the ROS induced by both arsenic [85, 89, 90] and chromium [91] at low doses is due to activation of NADPH oxidase. This is in direct contrast to earlier theories that assumed that the majority of metal-induced ROS were the result of direct metal-catalyzed redox reactions. Because of its reactivity, the level of $O_2^{\cdot-}$ in cells is normally tightly regulated by superoxide dismutase

(SOD), thereby producing the less reactive, but more mobile, H_2O_2 [92]. Like $O_2^{\cdot -}$, H_2O_2, is also tightly regulated by a multiplicity of catalase and peroxidase enzymes.

Production of ROS by arsenic

The formation of ROS by arsenic is considered one of the most probable mechanisms of arsenic carcinogenesis [79, 93]. However, unlike iron and copper, arsenic does not actively participate in the generation of ROS by conventional processes, such as the Fenton reaction [94]. Importantly, methyl metabolites of arsenic can be more reactive and capable of producing ROS than inorganic arsenic [44, 95]. DMA(V), for example, can be reduced to form either the very reactive DMA(III) [96] or the highly toxic and reactive dimethylarsine gas [95]. Dimethylarsine can react with molecular oxygen to produce both superoxide and dimethylarsenic radicals, which in turn can interact with free transition metals, producing the highly damaging hydroxyl radical [95]. Because DMA(V) appears to specifically target the lung [95], the formation of oxidative DNA damage through the intermediary of DMA(V) and its metabolites may well correlate with the high incidence of lung cancers seen in chronically exposed individuals. The presence of methylated As(III) metabolites in the urine also correlates with increased levels of bladder cancer [96, 97].

Arsenite, through the upregulation of hepatic and renal heme oxygenase, has been shown to release free iron, carbon monoxide and biliverdin from heme, making them available for free radical-generating reactions [93, 98, 99]. The release of bound iron by this mechanism is dependent on the arsenic species, with dimethylarsenite, DMA(III), being the most effective [100]. Thus, the methylated metabolites of arsenic, which are produced almost exclusively in the liver [96, 101], are most capable of producing ROS, such as superoxide, hydroxyl radicals, singlet oxygen and H_2O_2. Although the skin, lungs, and bladder seem to be the primary targets for arsenic carcinogenesis, increased levels of liver cancer have been reported in chronically exposed populations, as well as in experimental animal models [25, 35, 102].

Inorganic arsenic species can also produce ROS in non-hepatic mammalian cells. A number of groups have reported the production of ROS using the DCF fluorescence assay [103–105]. Similarly, Shi et al. [106] have shown that arsenite is able to produce superoxide species in keratinocyte cells; however, significant amounts of superoxide were only detected at concentrations that promote apoptosis. Indirect evidence to support the formation of ROS has also been reported. Various studies have shown that arsenic promotes the upregulation of GSH and antioxidants [107–110]. Similarly, depletion of GSH results in an increase in the toxic and clastogenic effects of arsenic [111]. Biomarkers for oxidative stress, such as 8-oxo-dG, have also been shown to be increased after exposure to arsenic in mammalian cell culture and human tissues [44,

103, 109, 112, 113]. In myeloid leukemia (NB4) and epithelial cells, arsenic treatment at low doses has been shown to induce NADPH oxidase [85, 90]. Recent data show that arsenic can also activate NADPH oxidase in endothelial cells [85, 89].

In addition to ROS, nitrogen-based radicals, such as nitric oxide and peroxynitrite, have also been implicated in oxidative damage by arsenic. The formation of micronuclei and induction of poly(ADP-ribosylation) in Chinese hamster ovary (CHO) cells and bovine endothelial cells and the formation of oxidative DNA damage [measured by cleavage with formamidopyrimidine-DNA glycosylase (Fpg) enzyme] have all been shown to be effectively blocked by the addition of inhibitors of nitric oxide synthase, suggesting that these radicals may account for some of the damage seen in cells [114, 115]. In all, the formation of radical species by arsenic appears to be an important mechanism by which arsenic may promote its carcinogenic effects.

Chromium

Chromium, like arsenic, has been shown to produce oxidative stress in cells by multiple mechanisms; however, the extent to which these are able to produce cancer is still subject to debate. As mentioned above, Cr(VI) can undergo a series of reductions leading to the formation of Cr(III). Chromium(VI) is a strong oxidizing agent and, like copper and iron, can produce ROS directly through Fenton type chemistry, whereby Cr(VI), or one of its metabolites, is able to interact with H_2O_2 in the presence of a reductant to produce both superoxide and hydroxyl radicals [116–119]. However, it is not only the ROS produced by the reduction of chromium species that can produce oxidative damage in cells, there is a growing body of evidence to suggest that the genotoxicity of chromium can be caused in part by the reactive chromium species themselves, such as Cr(V) [120]. O'Brien et al. [13] have raised the possibility that these species may in fact be the direct cause of the oxidative stress response measured by DCFH and rhodamine 123. Even the use of ROS scavengers is not sufficient to rule out this possibility, since these scavanges can also react directly with Cr(V) to prevent DNA damage [121, 122]. It must be noted, however, that the formation of radicals by this mechanism has only been shown to occur when both chromium and H_2O_2 were present at concentrations that are unlikely to be physiologically achievable within cells.

Like most metals that have the capacity to undergo redox reactions, chromium has been shown to deplete intracellular GSH and alter the regulation of the redox enzymes such as catalase and SOD [123–125]. Glutathione has shown to be a critical factor in the reduction of Cr(VI) to Cr(III). The relationship between chromium-induced oxidative stress, DNA damage and repair processes, and apoptotic cell death are complex [13, 22]. Moreover, the relationship between these processes and the induction of cancer is far from well understood.

to suppress tumor formation in a range of tissues [204, 206]. The effect of cadmium on cell adhesion was first characterized by a significant loss of tissue integrity that was not initially due to apoptosis [207–209]. Later studies, especially those of Prozialeck et al. [210] showed that cell adhesion and, in particular, the integrity of E-cadherin was an early target of cadmium toxicity. It was also shown that cadmium was able to exert its greatest effects on E-cadherins when calcium levels were low, suggesting that cadmium competes for calcium binding sites [211, 212] The loss of E-cadherins are thought to enhance tumor metastasis, promote toxicity, and promote changes to gene expression profiles through altered β-catenin signaling [204, 206].

Chromium

Complexes formed by chromium are considerably more varied than those of other metals discussed here. The binding of chromium to DNA does not occur with Cr(VI); however, the reduced metabolites of chromium, Cr(III, IV, and V) have all been shown to be reactive towards DNA [213, 214]. Although the structure and efficiency of formation of these chromium-DNA complexes is strongly affected by the reductant involved, such as GSH, ascorbate, or cysteine, most of the resulting adducts seem to be both genotoxic and mutagenic [22]. Binding of chromium species to DNA appears to be preferential for guanine nucleotides, and occurs largely with phosphates in the backbone [22, 215]. The formation of chromium-DNA adducts has a twofold effect, they both inhibit DNA replication and prevent DNA repair, thereby promoting mutagenesis.

Nickel

Nickel shows a strong affinity for histidines and, to a lesser extent, cysteines, and is able to form complexes with a wide variety of proteins [216, 217]. As a result, nickel is frequently used to extract and purify proteins that have been histidine tagged [218, 219]. Proteins that have been shown to bind nickel include: serum albumins, the neuroblastoma-associated tumor suppressor (DAN), and histones [220–222]. Like other metals that form protein complexes, it is thought that nickel interacts with proteins, altering their conformation in such a way that they are no longer able to perform normal cellular functions. Nickel has also been shown to crosslink DNA as the result of oxidation of DNA-associated proteins [12, 223].

Effects on gene regulation: direct and epigenetic changes

Metals have been shown to alter the expression of a great number of genes, too many to cover in detail here. These changes in gene expression are generally transient, and can be produced or caused by a multitude of different factors. Accordingly, this section looks at a limited number of genes that best illustrate the effects of carcinogenic metals on gene expression. For more detailed information on gene expression, the following recent reviews cover each metal in detail [10, 12–14]. Changes in gene expression are often thought to be the indirect result of signal cascades, DNA methylation changes and ROS; however, metals may also be directly responsible for changes in transcription factor activity.

Epigenetic mechanisms are heritable changes that can impart effects on the regulation of genes without altering the genomic sequence itself. Hypermethylation generally causes genes to be downregulated or effectively switched off, while hypomethylation often results in increased levels of gene expression. A number of agents that induce carcinogenesis, such as X-rays, have been shown to affect cells in this manner [224]. Similarly, nickel, arsenic, and, to a lesser extent, cadmium and chromium, are able to produce extensive alterations in genomic methylation [10, 23, 25, 97, 225–228].

Arsenic

Arsenic can both induce and suppress gene expression, depending on its concentration and the length of exposure. Microarray analyses of gene expression in arsenic-treated cells have identified hundreds of genes, most of which fall within several categories: cellular stress response, cell cycle control, redox regulation, and DNA repair [25, 99, 229, 230]. Different cell types and different treatment conditions can produce different effects on gene expression. In some cases, for example, such as arsenic-induced Bowen's disease, p53 expression is upregulated compared to non-arsenic-related disease controls [231]. Low-dose arsenic also promotes upregulation of p53 in cultured fibroblasts after both acute and longer treatments [232, 233]. In contrast, microarray analysis of normal human keratinocytes exposed to between 0.005 and 5 μM As(III), showed a generalized downregulation of p53 [234]. Transcription factors such as AP-1 and NF-κB are also regulated by arsenic, presumably the result of the activation of signal transduction pathways and the formation of ROS [235–237]. Hu et al. [237] showed that acute low-dose treatments of human fibroblasts with arsenite produced upregulation of both AP-1 and NF-κB expression, while chronic exposures lead to a downregulation of AP-1 and NF-κB. The AP-1 transcription factor is important for the regulation of DNA repair, inflammatory responses and cell growth [238, 239]. The activation of NF-κB can increase the expression of cytokines and growth factors, which may be responsible for tumor promotion [240, 241]. In aortic endothelial cells, it has been shown that acute low-dose arsenic treatments promote nuclear accumulation of NF-κB, similar to results seen in rat lung slices [242, 243]. These changes in transcription factor expression and activation are likely to lead to the observed changes in gene expression, and, more importantly perhaps, the observed inhibition of BER by arsenic.

It has only recently been discovered that arsenic can also modify DNA methylation patterns [244]. Dose-dependent hypermethylation of gene promoters was first noticed in regions of the p53 gene following exposure of cultured cells to either As(III) or As(V) [245]. Similarly, the p53 promoter region was shown to be hypermethylated in basal cell carcinomas (BCCs) from arsenic-exposed individuals relative to BCCs from non exposed patients [246]. In contrast, Zhao et al. [244] showed that chronic treatment of rat liver cells with arsenic caused global hypomethylation of promoters and malignant transformation. This hypomethylation was thought to occur as the result of depletion of *S*-adenosyl-methionine [244]. Changes to methylation patterns induced by arsenic are persistent, destabilizing [247], and have the potential to promote aberrant expression of genes involved in cell development and regulation, leading to cancer induction [25, 97, 226].

Cadmium

Like most metals, cadmium is responsible for alterations in the expression of many genes, including the immediate early response genes, *c-fos, c-jun* and *c-myc*; stress response proteins, such as metallothionein and heat shock proteins; and transcription factors, such as NF-κB [248–251]. Zheng et al. [252] have also shown that the livers of mice treated with 10 μmol/kg $CdCl_2$ exhibit increased expression of c-jun and p53. All of these proteins are believed to be involved in tumor promotion. Immediate early response genes (IEGs) induce mitogenic growth signals causing increased proliferation, particularly of cells that already possess mutations in critical regulatory genes.

Activation of stress response genes in response to changes in the extracellular environment enables cells to both protect themselves against oxidative stress and maintain normal cellular function. Cadmium activates a variety of these genes, the most notable of which is metallothionein. Metallothionein is a cysteine-rich low molecular weight protein, which binds excess heavy metal ions preventing their toxic effects [253]. Differential tissue expression of metallothioneins is thought to be a major reason for the tissue specificity of cadmium carcinogenesis [10, 254]. Cadmium is readily able to induce metallothionein expression in the liver and kidneys, but not in the testes or prostate [253, 255, 256]. Reduced expression of metallothionein in the testes and prostate relative to the liver of rats, correlates with increased levels of tumors and toxicity in these tissues [253]. Similarly, the use of transgenic mice has demonstrated that metallothionein reduces cadmium-induced ROS formation and activation of other genes that protect against oxidative stress [257]. Several key antioxidant genes in cells, most notably SOD and catalase, show reduced levels of expression in response to cadmium treatment [10, 130, 131, 258]. Depression of these enzymes can facilitate an increased build up of ROS, which can cause significant damage to cells.

Cadmium-induced methylation changes are less well characterized than those produced by other metals. However, the effects of DNA methylation changes on cadmium toxicity have long been of interest, since it was shown that methylation of the metallothionein promoter results in decreased expression and increased cadmium toxicity [259]. Subsequently, it has been shown that demethylation of the silenced metallothionein promoter with 5-azacytidine (5-aza-CR) is able to induce cadmium resistance in cells that were previously cadmium sensitive [260]. More recently, it has been shown that cadmium, itself, can induce alterations in DNA methylation patterns [228, 261]. However, unlike arsenic, cadmium appears to interfere with methylation by direct interaction with the DNA binding domains of the DNA methyltransferases [228].

Chromium

Chromium can also induce changes in gene expression due to its ability to produce radical species and oxidative stress. For example, as with arsenic and cadmium, both NF-κB and AP-1 are modulated by chromium exposure, with NF-κB being up regulated, which in turn activates *c-myc* [16, 262, 263]. Microarray studies in various cell cultures and *in vitro* models exposed to low to medium doses of chromium show an increase in a variety of genes, including those of the oxidative stress response, particularly those involved in redox regulation [70, 262, 264, 265]. Chromium species, like nickel species, have also been shown to affect the expression of hypoxia-inducible factor-1 (HIF-1) proteins [266]. Unlike the other metals described here, there is very little evidence to suggest that chromium also produces epigenetic changes, with the exception of a report by Cheng et al. [267] showing transgenerational changes in hormonal control in mice fed a diet supplemented with high levels of Cr(III).

Nickel

Nickel, like the other metals is able to alter the regulation of a variety of genes, including NF-κB [12]. Nickel has also been shown to promote the induction of hypoxia through activation of the transcription factor HIF-1 [268]. Increased levels of HIF-1 correlate with angiogenesis of new vasculature in tumors [269]. Other microarray studies have shown that nickel acetate exposure induces large-scale alterations of gene expression in human lung epithelial cells [270]. Some of the genes most strikingly affected include metallothionein and the heat shock proteins. Similarly, nickel sulfate-induced lung injury in mice showed gene expression patterns representative of both hypoxic and oxidative stress responses [271].

In contrast to the other metals presented in this section, the principle carcinogenic mechanism of nickel appears to be epigenetic in nature (reviewed in [12, 227]). The effects of nickel on DNA methylation were first suggested when it was noted that nickel-immortalized cells could be induced to senesce by demethylation with 5-azacytidine [272]. Since then it has been shown that nickel treatment alters methylation-dependent chromatin condensation [224], causes gene silencing [273], and modifies the activity of DNA methyltransferases [274]. Additionally, when mice are injected with nickel sulfide, the resultant tumors all exhibit hypermethylation of the p16 gene, an important regulator of cell cycle control [275].

More recently, it has been shown that nickel can induce epigenetic changes by both hypoacetylation and localized hypermethylation [276, 277] and that chemical demethylation and deacetylation can reverse gene silencing [278, 279].

Summary

Agents responsible for human carcinogenesis are grossly varied in their properties, and metals are no exception. However, it seems likely that metals share several common means by which to induce cancer. Critically, the most important of these appears to be the generation of oxidative stress and deregulation of key maintenance genes within cells. That said, the nature of the dose of each of these metals, as well as confounding variables required to produce a carcinogenesis, remain at best an unresolved issue. However, with time, and as research progresses, it is likely that a more complete picture will emerge on metal-induced carcinogenesis.

Acknowledgements
This work was supported in part by the Electric Power Research Institute contract no. WOEP-P4898/C2396, the U.S. Environmental Protection Agency's Science to Achieve Results (STAR) program, and the Centre for Cellular and Molecular Biology, School of Biological and Chemical Sciences, Deakin University, Australia. These supporting agencies were not involved in any way in the study design, data collection, or interpretation of the results presented here.

References

1 IARC (1980) *IARC monographs on the evaluation of carcinogenic risks to humans. Arsenic and arsenic compounds. Vol. 23, Lyon, France.* The Agency; Secretariat of the World Health Organization, Geneva
2 IARC (1980) *Some Metals and Metallic Compounds. IARC monographs on the evaluation of carcinogenic risks to humans. Vol. 23.* World Health Organization, Geneva
3 IARC (1990) *Chromium, nickel and welding. IARC monographs on the evaluation of carcinogenic risks to humans, Vol. 49, Lyon, France.* World Health Organization, Geneva
4 IARC (1993) *Beryllium, cadmium, mercury, and exposures in the glass manufacturing industry. IARC monographs on the evaluation of carcinogenic risks to humans, Vol. 58.* World Health Organization, Geneva

5 Snow ET (1992) Metal carcinogenesis: mechanistic implications. *Pharmacol Ther* 53: 31–65
6 Klein CB, Snow ET, Frenkel K (1998) Molecular mechanisms in metal carcinogenesis: role of oxidative stress. In: OI Aruoma, B Halliwell (eds): *Molecular Biology of Free Radicals in Human Diseases.* OICA International, Saint Lucia, London, 80–137
7 Theophanides T, Anastassopoulou J (2002) Copper and carcinogenesis. *Crit Rev Oncol Hematol* 42: 57–64
8 Huang X (2003) Iron overload and its association with cancer risk in humans: evidence for iron as a carcinogenic metal. *Mutat Res* 533: 153–171
9 Cangul H, Broday L, Salnikow K, Sutherland J, Peng W, Zhang Q, Poltaratsky V, Yee H, Zoroddu MA, Costa M (2002) Molecular mechanisms of nickel carcinogenesis. *Toxicol Lett* 127: 69–75
10 Waisberg M, Joseph P, Hale B, Beyersmann D (2003) Molecular and cellular mechanisms of cadmium carcinogenesis. *Toxicology* 192: 95–117
11 Toyokuni S (2002) Iron and carcinogenesis: from Fenton reaction to target genes. *Redox Rep* 7: 189–197
12 Kasprzak KS, Sunderman FW Jr, Salnikow K (2003) Nickel carcinogenesis. *Mutat Res* 533: 67–97
13 O'Brien TJ, Ceryak S, Patierno SR (2003) Complexities of chromium carcinogenesis: role of cellular response, repair and recovery mechanisms. *Mutat Res* 533: 3–36
14 Huang C, Ke Q, Costa M, Shi X (2004) Molecular mechanisms of arsenic carcinogenesis. *Mol Cell Biochem* 255: 57–66
15 Dubins JS, LaVelle JM (1986) Nickel(II) genotoxicity: potentiation of mutagenesis of simple alkylating agents. *Mutat Res* 162: 187–199
16 Ding M, Shi X (2002) Molecular mechanisms of Cr(VI)-induced carcinogenesis. *Mol Cell Biochem* 234–235: 293–300
17 Galaris D, Evangelou A (2002) The role of oxidative stress in mechanisms of metal-induced carcinogenesis. *Crit Rev Oncol Hematol* 42: 93–103
18 Kim JY, Mukherjee S, Ngo LC, Christiani DC (2004) Urinary 8-hydroxy-2'-deoxyguanosine as a biomarker of oxidative DNA damage in workers exposed to fine particulates. *Environ Health Perspect* 112: 666–671
19 Machella N, Regoli F, Cambria A, Santella RM (2004) Application of an immunoperoxidase staining method for detection of 7,8-dihydro-8-oxodeoxyguanosine as a biomarker of chemical-induced oxidative stress in marine organisms. *Aquat Toxicol* 67: 23–32
20 De Flora S, Bagnasco M, Serra D, Zanacchi P (1990) Genotoxicity of chromium compounds. A review. *Mutat Res* 238: 99–172
21 Lee AJ, Hodges NJ, Chipman JK (2004) Modified comet assay as a biomarker of sodium dichromate-induced oxidative DNA damage: optimization and reproducibility. *Biomarkers* 9: 103–115
22 Zhitkovich A (2005) Importance of Chromium-DNA Adducts in Mutagenicity and Toxicity of Chromium(VI). *Chem Res Toxicol* 18: 3–11
23 Klein CB, Costa M (1997) DNA methylation, heterochromatin and epigenetic carcinogens. *Mutat Res* 386: 163–180
24 Costa M (2002) Molecular mechanisms of nickel carcinogenesis. *Biol Chem* 383: 961–967
25 Chen H, Li S, Liu J, Diwan BA, Barrett JC, Waalkes MP (2004) Chronic inorganic arsenic exposure induces hepatic global and individual gene hypomethylation: implications for arsenic hepatocarcinogenesis. *Carcinogenesis* 25: 1779–1786
26 Lee TC, Wei ML, Chang WJ, Ho IC, Lo JF, Jan KY, Huang H (1989) Elevation of glutathione levels and glutathione S-transferase activity in arsenic-resistant Chinese hamster ovary cells. *In Vitro Cell Dev Biol* 25: 442–448
27 Stohs SJ, Bagchi D, Hassoun E, Bagchi M (2001) Oxidative mechanisms in the toxicity of chromium and cadmium ions. *J Environ Pathol Toxicol Oncol* 20: 77–88
28 Ercal N, Gurer-Orhan H, Aykin-Burns N (2001) Toxic metals and oxidative stress part I: mechanisms involved in metal-induced oxidative damage. *Curr Top Med Chem* 1: 529–539
29 Prozialeck WC, Grunwald GB, Dey PM, Reuhl KR, Parrish AR (2002) Cadherins and NCAM as potential targets in metal toxicity. *Toxicol Appl Pharmacol* 182: 255–265
30 Waxman S, Anderson KC (2001) History of the development of arsenic derivatives in cancer therapy. *Oncologist* 6, Suppl 2: 3–10
31 Weider B, Fournier JH (1999) Activation analyses of authenticated hairs of Napoleon Bonaparte confirm arsenic poisoning. *Am J Forensic Med Pathol* 20: 378–382
32 Evens AM, Tallman MS, Gartenhaus RB (2004) The potential of arsenic trioxide in the treatment of malignant disease: past, present, and future. *Leuk Res* 28: 891–900

80 Aust SD, Morehouse LA, Thomas CE (1985) Role of metals in oxygen radical reactions. *J Free Radic Biol Med* 1: 3–25
81 Wardman P, Candeias LP (1996) Fenton chemistry: an introduction. *Radiat Res* 145: 523–531
82 Fridovich I (1978) The biology of oxygen radicals. *Science* 201: 875–880
83 Fantone JC, Ward PA (1982) Role of oxygen-derived free radicals and metabolites in leukocyte-dependent inflammatory reactions. *Am J Pathol* 107: 395–418
84 Perner A, Andresen L, Pedersen G, Rask-Madsen J (2003) Superoxide production and expression of NAD(P)H oxidases by transformed and primary human colonic epithelial cells. *Gut* 52: 231–236
85 Smith KR, Klei LR, Barchowsky A (2001) Arsenite stimulates plasma membrane NADPH oxidase in vascular endothelial cells. *Am J Physiol Lung Cell Mol Physiol* 280: L442–L449
86 McCord JM (2000) The evolution of free radicals and oxidative stress. *Am J Med* 108: 652–659
87 Xiao L, Pimentel DR, Wang J, Singh K, Colucci WS, Sawyer DB (2002) Role of reactive oxygen species and NAD(P)H oxidase in alpha(1)-adrenoceptor signaling in adult rat cardiac myocytes. *Am J Physiol Cell Physiol* 282: C926–C934
88 Cui XL, Douglas JG (1997) Arachidonic acid activates c-jun N-terminal kinase through NADPH oxidase in rabbit proximal tubular epithelial cells. *Proc Natl Acad Sci USA* 94: 3771–3776
89 Lynn S, Gurr JR, Lai HT, Jan KY (2000) NADH oxidase activation is involved in arsenite-induced oxidative DNA damage in human vascular smooth muscle cells. *Circ Res* 86: 514–519
90 Chou WC, Jie C, Kenedy AA, Jones RJ, Trush MA, Dang CV (2004) Role of NADPH oxidase in arsenic-induced reactive oxygen species formation and cytotoxicity in myeloid leukemia cells. *Proc Natl Acad Sci USA* 101: 4578–4583
91 Wang S, Leonard SS, Ye J, Gao N, Wang L, Shi X (2004) Role of reactive oxygen species and Cr(VI) in Ras-mediated signal transduction. *Mol Cell Biochem* 255: 119–127
92 McCord JM, Fridovich I (1969) Superoxide dismutase. An enzymic function for erythrocuprein (hemocuprein). *J Biol Chem* 244: 6049–6055
93 Kitchin KT, Ahmad S (2003) Oxidative stress as a possible mode of action for arsenic carcinogenesis. *Toxicol Lett* 137: 3–13
94 Stohs SJ, Bagchi D (1995) Oxidative mechanisms in the toxicity of metal ions. *Free Radic Biol Med* 18: 321–336
95 Yamanaka K, Okada S (1994) Induction of lung-specific DNA damage by metabolically methylated arsenics via the production of free radicals. *Environ Health Perspect* 102, Suppl 3: 37–40
96 Vahter M (2002) Mechanisms of arsenic biotransformation. *Toxicology* 181–182: 211–217
97 Kitchin KT (2001) Recent advances in arsenic carcinogenesis: modes of action, animal model systems, and methylated arsenic metabolites. *Toxicol Appl Pharmacol* 172: 249–261
98 Keyse SM, Tyrrell RM (1989) Heme oxygenase is the major 32-kDa stress protein induced in human skin fibroblasts by UVA radiation, hydrogen peroxide, and sodium arsenite. *Proc Natl Acad Sci USA* 86: 99–103
99 Zheng XH, Watts GS, Vaught S, Gandolfi AJ (2003) Low-level arsenite induced gene expression in HEK293 cells. *Toxicology* 187: 39–48
100 Ahmad S, Kitchin KT, Cullen WR (2000) Arsenic species that cause release of iron from ferritin and generation of activated oxygen. *Arch Biochem Biophys* 382: 195–202
101 Goering PL, Aposhian HV, Mass MJ, Cebrian M, Beck BD, Waalkes MP (1999) The enigma of arsenic carcinogenesis: role of metabolism. *Toxicol Sci* 49: 5–14
102 Tsai SM, Wang TN, Ko YC (1999) Mortality for certain diseases in areas with high levels of arsenic in drinking water. *Arch Environ Health* 54: 186–193
103 Liu SX, Athar M, Lippai I, Waldren C, Hei TK (2001) Induction of oxyradicals by arsenic: implication for mechanism of genotoxicity. *Proc Natl Acad Sci USA* 98: 1643–1648
104 Lee TC, Ho IC (1995) Modulation of cellular antioxidant defense activities by sodium arsenite in human fibroblasts. *Arch Toxicol* 69: 498–504
105 Hei TK, Liu SX, Waldren C (1998) Mutagenicity of arsenic in mammalian cells: role of reactive oxygen species. *Proc Natl Acad Sci USA* 95: 8103–8107
106 Shi H, Hudson LG, Ding W, Wang S, Cooper KL, Liu S, Chen Y, Shi X, Liu KJ (2004) Arsenite causes DNA damage in keratinocytes via generation of hydroxyl radicals. *Chem Res Toxicol* 17: 871–878
107 Huang H, Huang CF, Wu DR, Jinn CM, Jan KY (1993) Glutathione as a cellular defence against arsenite toxicity in cultured Chinese hamster ovary cells. *Toxicology* 79: 195–204
108 Fan SR, Ho IC, Yeoh FL, Lin CJ, Lee TC (1996) Squalene inhibits sodium arsenite-induced sis-

ter chromatid exchanges and micronuclei in Chinese hamster ovary-K1 cells. *Mutat Res* 368: 165–169
109 Kessel M, Liu SX, Xu A, Santella R, Hei TK (2002) Arsenic induces oxidative DNA damage in mammalian cells. *Mol Cell Biochem* 234–235: 301–308
110 Snow ET, Hu Y, Klein CB, McCluskey KL, Schuliga M, Sykora P (2003) Regulation of redox and DNA repair genes by arsenic: low dose protection against oxidative stress? In: WR Chappell, CO Abernathy, RL Calderon (eds): *Arsenic Exposure and Health Effects*. Elsevier Science Ltd, Amsterdam, 303–317
111 Oya-Ohta Y, Kaise T, Ochi T (1996) Induction of chromosomal aberrations in cultured human fibroblasts by inorganic and organic arsenic compounds and the different roles of glutathione in such induction. *Mutat Res* 357: 123–129
112 Matsui M, Nishigori C, Toyokuni S, Takada J, Akaboshi M, Ishikawa M, Imamura S, Miyachi Y (1999) The role of oxidative DNA damage in human arsenic carcinogenesis: detection of 8-hydroxy-2'-deoxyguanosine in arsenic-related Bowen's disease. *J Invest Dermatol* 113: 26–31
113 An Y, Gao Z, Wang Z, Yang S, Liang J, Feng Y, Kato K, Nakano M, Okada S, Yamanaka K (2004) Immunohistochemical analysis of oxidative DNA damage in arsenic-related human skin samples from arsenic-contaminated area of China. *Cancer Lett* 214: 11–18
114 Lynn S, Shiung JN, Gurr JR, Jan KY (1998) Arsenite stimulates poly(ADP-ribosylation) by generation of nitric oxide. *Free Radic Biol Med* 24: 442–449
115 Liu F, Jan KY (2000) DNA damage in arsenite- and cadmium-treated bovine aortic endothelial cells. *Free Radic Biol Med* 28: 55–63
116 Shi XL, Dalal NS (1990) Evidence for a Fenton-type mechanism for the generation of .OH radicals in the reduction of Cr(VI) in cellular media. *Arch Biochem Biophys* 281: 90–95
117 Luo H, Lu Y, Shi X, Mao Y, Dalal NS (1996) Chromium (IV)-mediated fenton-like reaction causes DNA damage: implication to genotoxicity of chromate. *Ann Clin Lab Sci* 26: 185–191
118 Tsou TC, Yang JL (1996) Formation of reactive oxygen species and DNA strand breakage during interaction of chromium (III) and hydrogen peroxide *in vitro*: evidence for a chromium (III)-mediated Fenton-like reaction. *Chem Biol Interact* 102: 133–153
119 Aiyar J, Berkovits HJ, Floyd RA, Wetterhahn KE (1990) Reaction of chromium (VI) with hydrogen peroxide in the presence of glutathione: reactive intermediates and resulting DNA damage. *Chem Res Toxicol* 3: 595–603
120 Sugden KD, Campo CK, Martin BD (2001) Direct oxidation of guanine and 7,8-dihydro-8-oxoguanine in DNA by a high-valent chromium complex: a possible mechanism for chromate genotoxicity. *Chem Res Toxicol* 14: 1315–1322
121 Martin BD, Schoenhard JA, Sugden KD (1998) Hypervalent chromium mimics reactive oxygen species as measured by the oxidant-sensitive dyes 2',7'-dichlorofluorescin and dihydrorhodamine. *Chem Res Toxicol* 11: 1402–1410
122 Tsou TC, Lai HJ, Yang JL (1999) Effects of mannitol or catalase on the generation of reactive oxygen species leading to DNA damage by Chromium(VI) reduction with ascorbate. *Chem Res Toxicol* 12: 1002–1009
123 Liu S, Dixon K (1996) Induction of mutagenic DNA damage by chromium (VI) and glutathione. *Environ Mol Mutagen* 28: 71–79
124 Ning J, Grant MH (2000) The role of reduced glutathione and glutathione reductase in the cytotoxicity of chromium (VI) in osteoblasts. *Toxicol In Vitro* 14: 329–335
125 Asatiani N, Sapojnikova N, Abuladze M, Kartvelishvili T, Kulikova N, Kiziria E, Namchevadze E, Holman HY (2004) Effects of Cr(VI) long-term and low-dose action on mammalian antioxidant enzymes (an *in vitro* study). *J Inorg Biochem* 98: 490–496
126 Lloyd DR, Phillips DH, Carmichael PL (1997) Generation of putative intrastrand cross-links and strand breaks in DNA by transition metal ion-mediated oxygen radical attack. *Chem Res Toxicol* 10: 393–400
127 Yang CF, Shen HM, Shen Y, Zhuang ZX, Ong CN (1997) Cadmium-induced oxidative cellular damage in human fetal lung fibroblasts (MRC-5 cells). *Environ Health Perspect* 105: 712–716
128 Pourahmad J, O'Brien PJ, Jokar F, Daraei B (2003) Carcinogenic metal induced sites of reactive oxygen species formation in hepatocytes. *Toxicol In Vitro* 17: 803–810
129 Filipic M, Hei TK (2004) Mutagenicity of cadmium in mammalian cells: implication of oxidative DNA damage. *Mutat Res* 546: 81–91
130 Ikediobi CO, Badisa VL, Ayuk-Takem LT, Latinwo LM, West J (2004) Response of antioxidant enzymes and redox metabolites to cadmium-induced oxidative stress in CRL-1439 normal rat

liver cells. *Int J Mol Med* 14: 87–92
131 Casalino E, Calzaretti G, Sblano C, Landriscina C (2002) Molecular inhibitory mechanisms of antioxidant enzymes in rat liver and kidney by cadmium. *Toxicology* 179: 37–50
132 Stohs SJ, Bagchi D, Hassoun E, Bagchi M (2000) Oxidative mechanisms in the toxicity of chromium and cadmium ions. *J Environ Pathol Toxicol Oncol* 19: 201–213
133 Koizumi T, Li ZG (1992) Role of oxidative stress in single-dose, cadmium-induced testicular cancer. *J Toxicol Environ Health* 37: 25–36
134 Casalino E, Sblano C, Landriscina C (1997) Enzyme activity alteration by cadmium administration to rats: the possibility of iron involvement in lipid peroxidation. *Arch Biochem Biophys* 346: 171–179
135 Lin X, Zhuang Z, Costa M (1992) Analysis of residual amino acid—DNA crosslinks induced in intact cells by nickel and chromium compounds. *Carcinogenesis* 13: 1763–1768
136 Shi X, Dalal N, Kasprzak KS (1994) Enhanced generation of hydroxyl radical and sulfur trioxide anion radical from oxidation of sodium sulfite, nickel(II) sulfite, and nickel subsulfide in the presence of nickel(II) complexes. *Environ Health Perspect* 102, Suppl 3: 91–96
137 Chen CY, Wang YF, Huang WR, Huang YT (2003) Nickel induces oxidative stress and genotoxicity in human lymphocytes. *Toxicol Appl Pharmacol* 189: 153–159
138 McNeilly JD, Heal MR, Beverland IJ, Howe A, Gibson MD, Hibbs LR, MacNee W, Donaldson K (2004) Soluble transition metals cause the pro-inflammatory effects of welding fumes *in vitro*. *Toxicol Appl Pharmacol* 196: 95–107
139 Bal W, Liang R, Lukszo J, Lee SH, Dizdaroglu M, Kasprzak KS (2000) Ni(II) specifically cleaves the C-terminal tail of the major variant of histone H2A and forms an oxidative damage-mediating complex with the cleaved-off octapeptide. *Chem Res Toxicol* 13: 616–624
140 Chen CY, Wang YF, Lin YH, Yen SF (2003) Nickel-induced oxidative stress and effect of antioxidants in human lymphocytes. *Arch Toxicol* 77: 123–130
141 Lloyd DR, Phillips DH (1999) Oxidative DNA damage mediated by copper(II), iron(II) and nickel(II) fenton reactions: evidence for site-specific mechanisms in the formation of double-strand breaks, 8-hydroxydeoxyguanosine and putative intrastrand cross-links. *Mutat Res* 424: 23–36
142 Kawanishi S, Inoue S, Oikawa S, Yamashita N, Toyokuni S, Kawanishi M, Nishino K (2001) Oxidative DNA damage in cultured cells and rat lungs by carcinogenic nickel compounds. *Free Radic Biol Med* 31: 108–116
143 Rodriguez RE, Misra M, Diwan BA, Riggs CW, Kasprzak KS (1996) Relative susceptibilities of C57BL/6, (C57BL/6 × C3H/He)F1, and C3H/He mice to acute toxicity and carcinogenicity of nickel subsulfide. *Toxicology* 107: 131–140
144 Rodriguez RE, Misra M, North SL, Kasprzak KS (1991) Nickel-induced lipid peroxidation in the liver of different strains of mice and its relation to nickel effects on antioxidant systems. *Toxicol Lett* 57: 269–281
145 Herrero MC, Alvarez C, Cartana J, Blade C, Arola L (1993) Nickel effects on hepatic amino acids. *Res Commun Chem Pathol Pharmacol* 79: 243–248
146 Salnikow K, Gao M, Voitkun V, Huang X, Costa M (1994) Altered oxidative stress responses in nickel-resistant mammalian cells. *Cancer Res* 54: 6407–6412
147 Kunkel TA (2004) DNA replication fidelity. *J Biol Chem* 279: 16895–16898
148 Harfe BD, Jinks-Robertson S (2000) DNA mismatch repair and genetic instability. *Annu Rev Genet* 34: 359–399
149 Wood RD, Shivji MK (1997) Which DNA polymerases are used for DNA-repair in eukaryotes? *Carcinogenesis* 18: 605–610
150 Friedberg EC, Walker GC, Siede W (1995) *DNA repair and mutagenesis.* ASM Press, Washington, D.C, xvii, 698
151 Mitchell RJ, Farrington SM, Dunlop MG, Campbell H (2002) Mismatch repair genes hMLH1 and hMSH2 and colorectal cancer: A HuGE review. *Am J Epidemiol* 156: 885–902
152 Evans MD, Cooke MS (2004) Factors contributing to the outcome of oxidative damage to nucleic acids. *Bioessays* 26: 533–542
153 Jin YH, Clark AB, Slebos RJ, Al-Refai H, Taylor JA, Kunkel TA, Resnick MA, Gordenin DA (2003) Cadmium is a mutagen that acts by inhibiting mismatch repair. *Nat Genet* 34: 326–329
154 Thilly WG (2003) Have environmental mutagens caused oncomutations in people? *Nat Genet* 34: 255–259
155 Wood RD (1996) DNA repair in eukaryotes. *Annu Rev Biochem* 65: 135–167
156 Christmann M, Tomicic MT, Roos WP, Kaina B (2003) Mechanisms of human DNA repair: an

update. *Toxicology* 193: 3–34
157 Reynolds M, Peterson E, Quievryn G, Zhitkovich A (2004) Human nucleotide excision repair efficiently removes chromium-DNA phosphate adducts and protects cells against chromate toxicity. *J Biol Chem* 279: 30419–30424
158 Costa RMA, Chigancas V, Galhardo RD, Carvalho H, Menck CFM (2003) The eukaryotic nucleotide excision repair pathway. *Biochimie* 85: 1083–1099
159 Okui T, Fujiwara Y (1986) Inhibition of human excision DNA repair by inorganic arsenic and the co-mutagenic effect in V79 Chinese hamster cells. *Mutat Res* 172: 69–76
160 Hartwig A, Groblinghoff UD, Beyersmann D, Natarajan AT, Filon R, Mullenders LH (1997) Interaction of arsenic(III) with nucleotide excision repair in UV-irradiated human fibroblasts. *Carcinogenesis* 18: 399–405
161 Li JH, Rossman TG (1989) Inhibition of DNA ligase activity by arsenite: a possible mechanism of its comutagenesis. *Mol Toxicol* 2: 1–9
162 Lee-Chen SF, Yu CT, Jan KY (1992) Effect of arsenite on the DNA repair of UV-irradiated Chinese hamster ovary cells. *Mutagenesis* 7: 51–55
163 Lynn S, Lai HT, Gurr JR, Jan KY (1997) Arsenite retards DNA break rejoining by inhibiting DNA ligation. *Mutagenesis* 12: 353–358
164 Hu Y, Su L, Snow ET (1998) Arsenic toxicity is enzyme specific and its affects on ligation are not caused by the direct inhibition of DNA repair enzymes. *Mutat Res* 408: 203–218
165 Lee-Chen SF, Wang MC, Yu CT, Wu DR, Jan KY (1993) Nickel chloride inhibits the DNA repair of UV-treated but not methyl methanesulfonate-treated Chinese hamster ovary cells. *Biol Trace Elem Res* 37: 39–50
166 Hartwig A, Mullenders LH, Schlepegrell R, Kasten U, Beyersmann D (1994) Nickel(II) interferes with the incision step in nucleotide excision repair in mammalian cells. *Cancer Res* 54: 4045–4051
167 Hartwig A, Schlepegrell R, Dally H, Hartmann M (1996) Interaction of carcinogenic metal compounds with deoxyribonucleic acid repair processes. *Ann Clin Lab Sci* 26: 31–38
168 Bal, W, Schwerdtle T, Hartwig A (2003) Mechanism of nickel assault on the zinc finger of DNA repair protein XPA. *Chem Res Toxicol* 16: 242–248
169 Asmuss M, Mullenders LH, Eker A, Hartwig A (2000) Differential effects of toxic metal compounds on the activities of Fpg and XPA, two zinc finger proteins involved in DNA repair. *Carcinogenesis* 21: 2097–2104
170 Schwerdtle T, Seidel A, Hartwig A (2002) Effect of soluble and particulate nickel compounds on the formation and repair of stable benzo[a]pyrene DNA adducts in human lung cells. *Carcinogenesis* 23: 47–53
171 Hu W, Feng Z, Tang MS (2004) Nickel (II) enhances benzo[a]pyrene diol epoxide-induced mutagenesis through inhibition of nucleotide excision repair in human cells: a possible mechanism for nickel (II)-induced carcinogenesis. *Carcinogenesis* 25: 455–462
172 Wiederhold L, Leppard JB, Kedar P, Karimi-Busheri F, Rasouli-Nia A, Weinfeld M, Tomkinson AE, Izumi T, Prasad R, Wilson SH et al. (2004) AP endonuclease-independent DNA base excision repair in human cells. *Mol Cell* 15: 209–220
173 Izumi T, Wiederhold LR, Roy G, Roy R, Jaiswal A, Bhakat KK, Mitra S, Hazra TK (2003) Mammalian DNA base excision repair proteins: their interactions and role in repair of oxidative DNA damage. *Toxicology* 193: 43–65
174 Slupphaug G, Kavli B, Krokan HE (2003) The interacting pathways for prevention and repair of oxidative DNA damage. *Mutat Res* 531: 231–251
175 Krokan HE, Nilsen H, Skorpen F, Otterlei M, Slupphaug G (2000) Base excision repair of DNA in mammalian cells. *FEBS Lett* 476: 73–77
176 Nealon K, Nicholl ID, Kenny MK (1996) Characterization of the DNA polymerase requirement of human base excision repair. *Nucleic Acids Res* 24: 3763–3770
177 Fortini P, Pascucci B, Parlanti E, Sobol RW, Wilson SH, Dogliotti E (1998) Different DNA polymerases are involved in the short- and long-patch base excision repair in mammalian cells. *Biochemistry* 37: 3575–3580
178 Caldecott KW, McKeown CK, Tucker JD, Ljungquist S, Thompson LH (1994) An interaction between the mammalian DNA repair protein XRCC1 and DNA ligase III. *Mol Cell Biol* 14: 68–76
179 Tomkinson AE, Levin DS (1997) Mammalian DNA ligases. *Bioessays* 19: 893–901
180 Schwerdtle T, Walter I, Hartwig A (2003) Arsenite and its biomethylated metabolites interfere

with the formation and repair of stable BPDE-induced DNA adducts in human cells and impair XPAzf and Fpg. *DNA Repair (Amst)* 2: 1449–1463
181 Snow ET, Schuliga M, Chouchane S, Hu Y (2001) Sub-toxic arsenite induces a multi-component protective response against oxidative stress in human cells. In: WR Chappel, CO Abernathy, R Calderon R (eds): *Arsenic Exposure and Health Effects, IV*. Elsevier Science, Ltd, San Diego, 265–275
182 Andrew AS, Karagas MR, Hamilton JW (2003) Decreased DNA repair gene expression among individuals exposed to arsenic in United States drinking water. *Int J Cancer* 104: 263–268
183 Chou WC, Hawkins AL, Barrett JF, Griffin CA, Dang CV (2001) Arsenic inhibition of telomerase transcription leads to genetic instability. *J Clin Invest* 108: 1541–1547
184 Zhang TC, Schmitt MT, Mumford JL (2003) Effects of arsenic on telomerase and telomeres in relation to cell proliferation and apoptosis in human keratinocytes and leukemia cells *in vitro*. *Carcinogenesis* 24: 1811–1817
185 Fatur T, Lah TT, Filipic M (2003) Cadmium inhibits repair of UV-, methyl methanesulfonate- and N-methyl-N-nitrosourea-induced DNA damage in Chinese hamster ovary cells. *Mutat Res* 529: 109–116
186 Hartwig A, Asmuss M, Ehleben I, Herzer U, Kostelac D, Pelzer A, Schwerdtle T, Burkle A (2002) Interference by toxic metal ions with DNA repair processes and cell cycle control: molecular mechanisms. *Environ Health Perspect* 110, Suppl 5: 797–799
187 Zharkov DO, Rosenquist TA (2002) Inactivation of mammalian 8-oxoguanine-DNA glycosylase by cadmium(II): implications for cadmium genotoxicity. *DNA Repair (Amst)* 1: 661–670
188 Potts RJ, Watkin RD, Hart BA (2003) Cadmium exposure down-regulates 8-oxoguanine DNA glycosylase expression in rat lung and alveolar epithelial cells. *Toxicology* 184: 189–202
189 McNeill DR, Narayana A, Wong, HK, Wilson, DM, 3rd (2004) Inhibition of Ape1 nuclease activity by lead, iron, and cadmium. *Environ Health Perspect* 112: 799–804
190 Hodges NJ, Chipman JK (2002) Down-regulation of the DNA-repair endonuclease 8-oxo-guanine DNA glycosylase 1 (hOGG1) by sodium dichromate in cultured human A549 lung carcinoma cells. *Carcinogenesis* 23: 55–60
191 Schroeder M, Mass MJ (1997) CpG methylation inactivates the transcriptional activity of the promoter of the human p53 tumor suppressor gene. *Biochem Biophys Res Commun* 235: 403–406
192 Scicchitano DA, Pegg AE (1987) Inhibition of O6-alkylguanine-DNA-alkyltransferase by metals. *Mutat Res* 192: 207–210
193 Takahashi K, Imaeda T, Kawazoe Y (1988) Effect of metal ions on the adaptive response induced by N-methyl-N-nitrosourea in Escherichia coli. *Biochem Biophys Res Commun* 157: 1124–1130
194 Hartmann M, Hartwig A (1998) Disturbance of DNA damage recognition after UV-irradiation by nickel(II) and cadmium(II) in mammalian cells. *Carcinogenesis* 19: 617–621
195 Menzel DB, Hamadeh HK, Lee E, Meacher DM, Said V, Rasmussen RE, Greene H, Roth RN (1999) Arsenic binding proteins from human lymphoblastoid cells. *Toxicol Lett* 105: 89–101
196 Simons SS Jr, Chakraborti PK, Cavanaugh AH (1990) Arsenite and cadmium(II) as probes of glucocorticoid receptor structure and function. *J Biol Chem* 265: 1938–1945
197 Bodwell JE, Kingsley LA, Hamilton JW (2004) Arsenic at very low concentrations alters glucocorticoid receptor (GR)-mediated gene activation but not GR-mediated gene repression: complex dose-response effects are closely correlated with levels of activated GR and require a functional GR DNA binding domain. *Chem Res Toxicol* 17: 1064–1076
198 Walton FS, Harmon AW, Paul DS, Drobna Z, Patel YM, Styblo M (2004) Inhibition of insulin-dependent glucose uptake by trivalent arsenicals: possible mechanism of arsenic-induced diabetes. *Toxicol Appl Pharmacol* 198: 424–433
199 Kapahi P, Takahashi T, Natoli G, Adams SR, Chen Y, Tsien RY, Karin M (2000) Inhibition of NF-kappa B activation by arsenite through reaction with a critical cysteine in the activation loop of Ikappa B kinase. *J Biol Chem* 275: 36062–36066
200 Doussiere J, Bouzidi F, Poinas A, Gaillard J, Vignais PV (1999) Kinetic study of the activation of the neutrophil NADPH oxidase by arachidonic acid. Antagonistic effects of arachidonic acid and phenylarsine oxide. *Biochemistry* 38(49): 16394–16406
201 Bremnes RM, Veve R, Hirsch FR, Franklin WA (2002) The E-cadherin cell-cell adhesion complex and lung cancer invasion, metastasis, and prognosis. *Lung Cancer* 36: 115–124
202 Takeichi M (1990) Cadherins: a molecular family important in selective cell-cell adhesion. *Annu Rev Biochem* 59: 237–252
203 Wheelock MJ, Johnson KR (2003) Cadherin-mediated cellular signaling. *Curr Opin Cell Biol* 15:

509–514
204 Pearson CA, Prozialeck WC (2001) E-Cadherin, beta -Catenin and cadmium carcinogenesis. *Med Hypotheses* 56: 573–581
205 Prozialeck WC (2000) Evidence that E-cadherin may be a target for cadmium toxicity in epithelial cells. *Toxicol Appl Pharmacol* 164: 231–249
206 Hirohashi S, Kanai Y (2003) Cell adhesion system and human cancer morphogenesis. *Cancer Sci* 94: 575–581
207 Gouveia MA (1988) The testes in cadmium intoxication: morphological and vascular aspects. *Andrologia* 20: 225–231
208 Prozialeck WC, Niewenhuis RJ (1991) Cadmium (Cd^{2+}) disrupts intercellular junctions and actin filaments in LLC-PK1 cells. *Toxicol Appl Pharmacol* 107: 81–97
209 Boveri M, Pazos P, Gennari A, Casado J, Hartung T, Prieto P (2004) Comparison of the sensitivity of different toxicological endpoints in Caco-2 cells after cadmium chloride treatment. *Arch Toxicol* 78: 201–206
210 Prozialeck WC, Lamar PC, Lynch SM (2003) Cadmium alters the localization of N-cadherin, E-cadherin, and beta-catenin in the proximal tubule epithelium. *Toxicol Appl Pharmacol* 189: 180–195
211 Prozialeck WC, Lamar PC (1999) Interaction of cadmium (Cd(2+)) with a 13-residue polypeptide analog of a putative calcium-binding motif of E-cadherin. *Biochim Biophys Acta* 1451: 93–100
212 Niewenhuis RJ, Dimitriu C, Prozialeck WC (1997) Ultrastructural characterization of the early changes in intercellular junctions in response to cadmium (Cd^{2+}) exposure in LLC-PK1 cells. *Toxicol Appl Pharmacol* 142: 1–12
213 Stearns DM, Kennedy LJ, Courtney KD, Giangrande PH, Phieffer LS, Wetterhahn KE (1995) Reduction of chromium(VI) by ascorbate leads to chromium-DNA binding and DNA strand breaks *in vitro*. *Biochemistry* 34: 910–919
214 Zhitkovich A, Song Y, Quievryn G, Voitkun V (2001) Non-oxidative mechanisms are responsible for the induction of mutagenesis by reduction of Cr(VI) with cysteine: role of ternary DNA adducts in Cr(III)-dependent mutagenesis. *Biochemistry* 40: 549–560
215 Arakawa H, Ahmad R, Naoui M, Tajmir-Riahi HA (2000) A comparative study of calf thymus DNA binding to Cr(III) and Cr(VI) ions. Evidence for the guanine N-7-chromium-phosphate chelate formation. *J Biol Chem* 275: 10150–10153
216 Kowara R, Salnikow K, Diwan BA, Bare RM, Waalkes MP, Kasprzak KS (2004) Reduced Fhit protein expression in nickel-transformed mouse cells and in nickel-induced murine sarcomas. *Mol Cell Biochem* 255: 195–202
217 Predki PF, Harford C, Brar P, Sarkar B (1992) Further characterization of the N-terminal copper(II)- and nickel(II)-binding motif of proteins. Studies of metal binding to chicken serum albumin and the native sequence peptide. *Biochem J* 287 (Pt 1): 211–215
218 Bush GL, Tassin AM, Friden H, Meyer DI (1991) Secretion in yeast. Purification and *in vitro* translocation of chemical amounts of prepro-alpha-factor. *J Biol Chem* 266: 13811–13814
219 Vosters AF, Evans DB, Tarpley WG, Sharma SK (1992) On the engineering of rDNA proteins for purification by immobilized metal affinity chromatography: applications to alternating histidine-containing chimeric proteins from recombinant Escherichia coli. *Protein Expr Purif* 3: 18–26
220 Kondo K, Ozaki T, Nakamura Y, Sakiyama S (1995) DAN gene product has an affinity for Ni^{2+}. *Biochem Biophys Res Commun* 216: 209–215
221 Zoroddu MA, Schinocca L, Kowalik-Jankowska T, Kozlowski H, Salnikow K, Costa M (2002) Molecular mechanisms in nickel carcinogenesis: modeling Ni(II) binding site in histone H4. *Environ Health Perspect* 110, Suppl 5: 719–723
222 Bal W, Lukszo J, Kasprzak KS (1997) Mediation of oxidative DNA damage by nickel(II) and copper(II) complexes with the N-terminal sequence of human protamine HP2. *Chem Res Toxicol* 10: 915–921
223 Misra M, Olinski R, Dizdaroglu M, Kasprzak KS (1993) Enhancement by L-histidine of nickel(II)-induced DNA-protein cross-linking and oxidative DNA base damage in the rat kidney. *Chem Res Toxicol* 6: 33–37
224 Costa M (1995) Model for the epigenetic mechanism of action of nongenotoxic carcinogens. *Am J Clin Nutr* 61 (3 Suppl): 666S–669S
225 Abernathy CO, Liu YP, Longfellow D, Aposhian HV, Beck B, Fowler B, Goyer R, Menzer R, Rossman T, Thompson C et al. (1999) Arsenic: health effects, mechanisms of actions, and

research issues. *Environ Health Perspect* 107: 593–597
226 Hughes MF (2002) Arsenic toxicity and potential mechanisms of action. *Toxicol Lett* 133: 1–16
227 Sutherland JE, Costa M (2003) Epigenetics and the environment. *Ann N Y Acad Sci* 983: 151–160
228 Takiguchi M, Achanzar WE, Qu W, Li G, Waalkes MP (2003) Effects of cadmium on DNA-(Cytosine-5) methyltransferase activity and DNA methylation status during cadmium-induced cellular transformation. *Exp Cell Res* 286: 355–365
229 Chen H, Liu J, Merrick BA, Waalkes MP (2001) Genetic events associated with arsenic-induced malignant transformation: applications of cDNA microarray technology. *Mol Carcinog* 30: 79–87
230 Rea MA, Gregg JP, Qin Q, Phillips MA, Rice RH (2003) Global alteration of gene expression in human keratinocytes by inorganic arsenic. *Carcinogenesis* 24: 747–756
231 Kuo TT, Hu S, Lo SK, Chan HL (1997) p53 expression and proliferative activity in Bowen's disease with or without chronic arsenic exposure. *Hum Pathol* 28: 786–790
232 Vogt BL, Rossman TG (2001) Effects of arsenite on p53, p21 and cyclin D expression in normal human fibroblasts – a possible mechanism for arsenite's comutagenicity. *Mutat Res* 478: 159–168
233 Yih LH, Lee TC (2000) Arsenite induces p53 accumulation through an ATM-dependent pathway in human fibroblasts. *Cancer Res* 60: 6346–6352
234 Hamadeh HK, Trouba KJ, Amin RP, Afshari CA, Germolec D (2002) Coordination of altered DNA repair and damage pathways in arsenite-exposed keratinocytes. *Toxicol Sci* 69: 306–316
235 Cavigelli M, Li WW, Lin A, Su B, Yoshioka K, Karin M (1996) The tumor promoter arsenite stimulates AP-1 activity by inhibiting a JNK phosphatase. *Embo J* 15: 6269–6279
236 Simeonova PP, Wang S, Toriuma W, Kommineni V, Matheson J, Unimye N, Kayama F, Harki D, Ding M, Vallyathan V et al. (2000) Arsenic mediates cell proliferation and gene expression in the bladder epithelium: association with activating protein-1 transactivation. *Cancer Res* 60: 3445–3453
237 Hu Y, Jin X, Snow ET (2002) Effect of arsenic on transcription factor AP-1 and NF-kappaB DNA binding activity and related gene expression. *Toxicol Lett* 133: 33–45
238 Shaulian E, Karin M (2002) AP-1 as a regulator of cell life and death. *Nat Cell Biol* 4: E131–E136
239 Eferl R, Wagner EF (2003) AP-1: a double-edged sword in tumorigenesis. *Nat Rev Cancer* 3: 859–868
240 Karin M (1998) The NF-kappa B activation pathway: its regulation and role in inflammation and cell survival. *Cancer J Sci Am* 4, Suppl 1: S92–S99
241 Karin M, Ben-Neriah Y (2000) Phosphorylation meets ubiquitination: the control of NF-kappaB activity. *Annu Rev Immunol* 18: 621–663
242 Barchowsky A, Dudek EJ, Treadwell MD, Wetterhahn KE (1996) Arsenic induces oxidant stress and NF-kappa B activation in cultured aortic endothelial cells. *Free Radic Biol Med* 21: 783–990
243 Wijeweera JB, Gandolfi AJ, Parrish A, Lantz RC (2001) Sodium arsenite enhances AP-1 and NFkappaB DNA binding and induces stress protein expression in precision-cut rat lung slices. *Toxicol Sci* 61: 283–294
244 Zhao CQ, Young MR, Diwan BA, Coogan TP, Waalkes MP (1997) Association of arsenic-induced malignant transformation with DNA hypomethylation and aberrant gene expression. *Proc Natl Acad Sci USA* 94: 10907–10912
245 Mass MJ, Wang L (1997) Arsenic alters cytosine methylation patterns of the promoter of the tumor suppressor gene p53 in human lung cells: a model for a mechanism of carcinogenesis. *Mutat Res* 386: 263–277
246 Boonchai W, Walsh M, Cummings M, Chenevix-Trench G (2000) Expression of p53 in arsenic-related and sporadic basal cell carcinoma. *Arch Dermatol* 136: 195–198
247 Sciandrello G, Caradonna F, Mauro M, Barbata G (2004) Arsenic-induced DNA hypomethylation affects chromosomal instability in mammalian cells. *Carcinogenesis* 25: 413–417
248 Templeton DM, Wang Z, Miralem T (1998) Cadmium and calcium-dependent c-fos expression in mesangial cells. *Toxicol Lett* 95: 1–8
249 Wang Z, Templeton DM (1998) Induction of c-fos proto-oncogene in mesangial cells by cadmium. *J Biol Chem* 273: 73–79
250 Lee MJ, Nishio H, Ayaki H, Yamamoto M, Sumino K (2002) Upregulation of stress response mRNAs in COS-7 cells exposed to cadmium. *Toxicology* 174: 109–117
251 Misra UK, Gawdi G, Akabani G, Pizzo SV (2002) Cadmium-induced DNA synthesis and cell

proliferation in macrophages: the role of intracellular calcium and signal transduction mechanisms. *Cell Signal* 14: 327–340
252 Zheng H, Liu J, Choo KH, Michalska AE, Klaassen CD (1996) Metallothionein-I and -II knockout mice are sensitive to cadmium-induced liver mRNA expression of c-jun and p53. *Toxicol Appl Pharmacol* 136: 229–235
253 Cherian MG, Jayasurya A, Bay BH (2003) Metallothioneins in human tumors and potential roles in carcinogenesis. *Mutat Res* 533: 201–209
254 Bhave MR, Wilson MJ, Waalkes MP (1988) Methylation status and organization of the metallothionein-I gene in livers and testes of strains of mice resistant and susceptible to cadmium. *Toxicology* 50: 231–245
255 McKenna IM, Bare RM, Waalkes MP (1996) Metallothionein gene expression in testicular interstitial cells and liver of rats treated with cadmium. *Toxicology* 107: 121–130
256 Ren XY, Zhou Y, Zhang JP, Feng WH, Jiao BH (2003) Expression of metallothionein gene at different time in testicular interstitial cells and liver of rats treated with cadmium. *World J Gastroenterol* 9: 1554–1558
257 Liu J, Kadiiska MB, Corton JC, Qu W, Waalkes MP, Mason RP, Liu Y, Klaassen CD (2002) Acute cadmium exposure induces stress-related gene expression in wild-type and metallothionein-I/II-null mice. *Free Radic Biol Med* 32: 525–535
258 Uchida M, Teranishi H, Aoshima K, Katoh T, Kasuya M, Inadera H (2004) Reduction of erythrocyte catalase and superoxide dismutase activities in male inhabitants of a cadmium-polluted area in Jinzu river basin, Japan. *Toxicol Lett* 151: 451–457
259 Compere SJ, Palmiter RD (1981) DNA methylation controls the inducibility of the mouse metallothionein-I gene lymphoid cells. *Cell* 25: 233–240
260 Stallings RL, Crawford BD, Tobey RA, Tesmer J, Hildebrand CE (1986) 5-Azacytidine-induced conversion to cadmium resistance correlates with early S phase replication of inactive metallothionein genes in synchronized CHO cells. *Somat Cell Mol Genet* 12: 423–432
261 Poirier LA, Vlasova TI (2002) The prospective role of abnormal methyl metabolism in cadmium toxicity. *Environ Health Perspect* 110, Suppl 5: 793–795
262 Ye J, Shi X (2001) Gene expression profile in response to chromium-induced cell stress in A549 cells. *Mol Cell Biochem* 222: 189–197
263 Wang S, Chen F, Zhang Z, Jiang BH, Jia L, Shi X (2004) NF-kappaB prevents cells from undergoing Cr(VI)-induced apoptosis. *Mol Cell Biochem* 255: 129–137
264 Cheng RY, Alvord WG, Powell D, Kasprzak KS, Anderson LM (2002) Microarray analysis of altered gene expression in the TM4 Sertoli-like cell line exposed to chromium(III) chloride. *Reprod Toxicol* 16: 223–236
265 Izzotti A, Cartiglia C, Balansky R, D'Agostini F, Longobardi M, De Flora S (2002) Selective induction of gene expression in rat lung by hexavalent chromium. *Mol Carcinog* 35: 75–84
266 Gao N, Jiang BH, Leonard SS, Corum L, Zhang Z, Roberts JR, Antonini J, Zheng JZ, Flynn DC, Castranova V et al. (2002) p38 Signaling-mediated hypoxia-inducible factor 1alpha and vascular endothelial growth factor induction by Cr(VI) in DU145 human prostate carcinoma cells. *J Biol Chem* 277: 45041–45048
267 Cheng RY, Alvord WG, Powell D, Kasprzak KS, Anderson LM (2002) Increased serum corticosterone and glucose in offspring of chromium(III)-treated male mice. *Environ Health Perspect* 110: 801–804
268 Salnikow K, Davidson T, Kluz T, Chen H, Zhou D, Costa M (2003) GeneChip analysis of signaling pathways effected by nickel. *J Environ Monit* 5: 206–209
269 Choi KS, Bae MK, Jeong JW, Moon HE, Kim KW (2003) Hypoxia-induced angiogenesis during carcinogenesis. *J Biochem Mol Biol* 36: 120–127
270 Cheng RY, Zhao A, Alvord WG, Powell DA, Bare RM, Masuda A, Takahashi T, Anderson LM, Kasprzak KS (2003) Gene expression dose-response changes in microarrays after exposure of human peripheral lung epithelial cells to nickel(II). *Toxicol Appl Pharmacol* 191: 22–39
271 McDowell SA, Gammon K, Bachurski CJ, Wiest JS, Leikauf JE, Prows DR, Leikauf GD (2000) Differential gene expression in the initiation and progression of nickel-induced acute lung injury. *Am J Respir Cell Mol Biol* 23: 466–474
272 Klein CB, Conway K, Wang XW, Bhamra RK, Lin XH, Cohen MD, Annab L, Barrett JC, Costa M (1991) Senescence of nickel-transformed cells by an X chromosome: possible epigenetic control. *Science* 251: 796–799
273 Lee YW, Klein CB, Kargacin B, Salnikow K, Kitahara J, Dowjat K, Zhitkovich A, Christie NT,

Costa M (1995) Carcinogenic nickel silences gene expression by chromatin condensation and DNA methylation: a new model for epigenetic carcinogens. *Mol Cell Biol* 15: 2547–2557
274 Lee YW, Broday L, Costa M (1998) Effects of nickel on DNA methyltransferase activity and genomic DNA methylation levels. *Mutat Res* 415: 213–218
275 Govindarajan B, Klafter R, Miller MS, Mansur C, Mizesko M, Bai X, LaMontagne K Jr, Arbiser JL (2002) Reactive oxygen-induced carcinogenesis causes hypermethylation of p16(Ink4a) and activation of MAP kinase. *Mol Med* 8: 1–8
276 Forzani C, Loulergue C, Lobreaux S, Briat JF, Lebrun M (2001) Nickel resistance and chromatin condensation in Saccharomyces cerevisiae expressing a maize high mobility group I/Y protein. *J Biol Chem* 276: 16731–16738
277 Broday L, Cai J, Costa M (1999) Nickel enhances telomeric silencing in Saccharomyces cerevisiae. *Mutat Res* 440: 121–130
278 Yan Y, Kluz T, Zhang P, Chen HB, Costa M (2003) Analysis of specific lysine histone H3 and H4 acetylation and methylation status in clones of cells with a gene silenced by nickel exposure. *Toxicol Appl Pharmacol* 190: 272–277
279 Zhang Q, Salnikow K, Kluz T, Chen LC, Su WC, Costa M (2003) Inhibition and reversal of nickel-induced transformation by the histone deacetylase inhibitor trichostatin A. *Toxicol Appl Pharmacol* 192: 201–211

Cancer: Cell Structures, Carcinogens and Genomic Instability
Edited by Leon P. Bignold

Actions of ultraviolet light on cellular structures

David I. Pattison and Michael J. Davies

The Heart Research Institute, 145 Missenden Rd, Camperdown, NSW 2050, Australia

Summary. Solar radiation is the primary source of human exposure to ultraviolet (UV) radiation. Overexposure without suitable protection (i.e., sunscreen and clothing) has been implicated in mutagenesis and the onset of skin cancer. These effects are believed to be initiated by UV-mediated cellular damage, with proteins and DNA as primary targets due to a combination of their UV absorption characteristics and their abundance in cells. UV radiation can mediate damage *via* two different mechanisms: (a) direct absorption of the incident light by the cellular components, resulting in excited state formation and subsequent chemical reaction, and (b) photosensitization mechanisms, where the light is absorbed by endogenous (or exogenous) sensitizers that are excited to their triplet states. The excited photosensitizers can induce cellular damage by two mechanisms: (a) electron transfer and hydrogen abstraction processes to yield free radicals (Type I); or (b) energy transfer with O_2 to yield the reactive excited state, singlet oxygen (Type II). Direct UV absorption by DNA leads to dimers of nucleic acid bases including cyclobutane pyrimidine species and pyrimidine (6-4) pyrimidone compounds, together with their Dewar isomers. These three classes of dimers are implicated in the mutagenicity of UV radiation, which is typified by a high level of CC→TT and C→T transversions. Single base modifications can also occur *via* sensitized reactions including Type 1 and Type II processes. The main DNA product generated by $^1O_2$ is 8-oxo-Gua; this is a common lesion in DNA and is formed by a range of other oxidants in addition to UV. The majority of UV-induced protein damage appears to be mediated by $^1O_2$, which reacts preferentially with Trp, His, Tyr, Met, Cys and cystine side chains. Direct photo-oxidation reactions (particularly with short-wavelength UV) and radicals can also be formed *via* triplet excited states of some of these side chains. The initial products of $^1O_2$-mediated reactions are endoperoxides with the aromatic residues, and zwitterions with the sulfur-containing residues. These intermediates undergo a variety of further reactions, which can result in radical formation and ring-opening reactions; these result in significant yields of protein cross-links and aggregates, but little protein fragmentation. This review discusses the formation of these UV-induced modifications and their downstream consequences with particular reference to mutagenesis and alterations in protein structure and function.

Key words: DNA, free radicals, photoproducts, protein, singlet oxygen, ultraviolet.

Introduction

Nature of UV and solar radiation

UV light is defined as the region of the electromagnetic spectrum with wavelengths from 200 to 400 nm. This light is broken down into three distinct wavelength bands, known as UVC (*ca.* 200–280 nm), UVB (*ca.* 280–320 nm) and UVA (*ca.* 320–400 nm). As with all electromagnetic radiation, the shortest wavelength radiation (UVC) is the most energetic, and has the greatest potential for biological damage.

The major source of human exposure to UV light is *via* the sun. Solar radiation contains all three forms of UV radiation, but UV radiation with wavelengths below 295 nm (i.e., the entire UVC region) is absorbed by the Earth's upper atmosphere, and does not reach the Earth's surface. The UV light that does reach the Earth's surface comprises primarily (*ca.* 95%) UVA wavelengths, with the remainder (*ca.* 5%) comprising the shorter wavelength (295–320 nm) UVB radiation. This chapter will focus on the biological effects of UVB and UVA radiation, as these are the most biologically relevant [1, 2].

Mechanisms of UV-induced biological damage

UV radiation is capable of inducing biological damage *via* two discreet mechanisms [1–4]. Firstly, there is direct absorption of UV photons by the cellular material (particularly DNA or proteins) that can lead to photo-induced reactions. Secondly, there is the possibility of photosensitized processes, where UV light is absorbed by an endogenous or exogenous (in the case of therapeutic methods such as photodynamic therapy) sensitizer. The electronically excited sensitizer (typically in the triplet state) can harmlessly revert back to its ground state *via* intramolecular decay processes, or can damage other cellular material. Cellular damage can occur by two major pathways often called Type I and Type II mechanisms. Type I damage involves one electron oxidation or hydrogen atom abstraction from cellular targets, resulting in free radical formation. The Type II mechanism involves energy transfer from the molecule that originally absorbed the UV light (the sensitizer) to molecular oxygen, with the consequent formation of an excited state of oxygen – singlet oxygen. The latter species is a powerful oxidant that can undergo further reactions with cellular material.

Singlet oxygen is the first excited singlet state ($^1\Delta_g$ $^1O_2$) of molecular oxygen. This state, which has both electrons in the same molecular orbital with paired spins, is formed readily, being only *ca.* 94 kJ mol^{-1} above the ground triplet state ($^3\Sigma$), and has a relatively long lifetime of a few microseconds (reviewed in [2, 5, 6]). Other excited states can also be formed (e.g., the $^1\Sigma_g$ state, which has the two highest energy electrons in different orbitals with paired spins), but these are of higher energy and much shorter lived, and usually undergo rapid decay rather than chemical reaction [5]. The $^1\Delta_g$ state (henceforth denoted simply as $^1O_2$) is therefore the most important excited state of oxygen in biological systems. The transfer of energy from the original chromophore to molecular oxygen to give $^1O_2$ is rarely 100% efficient, and hence energy transfer to oxygen (Type II) and electron/hydrogen transfer (Type I) reactions usually occur simultaneously, and competitively. Thus, many photosensitizers give both $^1O_2$ and radicals such as $O_2^{-}\cdot$ [5]; the yields of species generated by these two processes are known to be dependent on the sensitizer, the excitation wavelength and the reaction conditions [5]. A number of reviews of the yields of $^1O_2$ and the rate constants for the reactions of these species are available [7, 8].

UVB radiation is responsible for the majority of damage resulting from direct absorption of UV light by cellular structures. The primary macromolecular structures that absorb UVB light are DNA and proteins, but a number of other low-molecular-weight materials (e.g. heme groups, carotenoids and vitamin A, eumelanin and phaeomelanin, pyridoxamine, urocanic acid; reviewed in [2]) can also be major chromophores for UVB radiation. The significance, and extent, of absorption by these other compounds is obviously tissue and situation dependent, with some cells and organs containing higher levels of these chromophores (either endogenously or as a result of exogenous addition). Thus, melanocytes in the skin contain particularly high levels of melanins and the retina particularly high levels of carotenoids and related species. The levels of particular chromophores within tissues can also be artificially raised by, for example, exposure to heme compounds or their precursors (e.g. 5-aminolaevulinic acid) in photodynamic therapy [2]. Other macromolecular structures such as lipids and polysaccharides do not have any major absorption bands in the UVB region, and thus do not undergo direct damage by UVB radiation.

It is well established (reviewed in [9–11]) that UVB radiation is both genotoxic and mutagenic, and plays a key role in DNA damage and skin cancer. Thus, there have been considerable efforts to develop sunscreens that protect the skin from UVB radiation. However, in the past many sunscreens have not offered protection against UVA radiation, and this has led to increased exposure of the population to UVA radiation, as the dose-limiting side effects of UVB exposure (e.g. sunburn) are attenuated [10]. The extent of human exposure to UVA radiation has also increased due to the use of sun beds and tanning salons as an alternative to sun exposure. These artificial tanning methods use light that is almost exclusively in the UVA region to induce skin tanning.

UVA radiation is typically less damaging than UVB light, but is directly absorbed to some extent by DNA and proteins. However, the majority of damage induced by UVA radiation is *via* photosensitization and Type I- and II-mediated mechanisms (reviewed in [4, 10–12]), resulting in free radical and $^1O_2$ formation, and ultimately oxidative damage to DNA, proteins and lipids. The processes that give rise to such damage, and its consequences, are reviewed below.

Consequences of UV irradiation on DNA

The UV absorption spectrum for DNA comprises a rising band in the far UV (<220 nm) with a further peak at 260 nm, which tails into the near-UVB wavelengths at around 300 nm [13]. The purine and pyrimidine bases are responsible for the absorption maximum at 260 nm [2], and it is these bases that dominate the UV photochemistry of DNA.

DNA photoproducts induced by direct UV absorption

The majority of products detected in DNA following UVB irradiation are pyrimidine products, as these bases absorb further into the near-UV region than the purine bases. Most of these products are photoadducts between adjacent bases to yield dimers (reviewed in [9, 12, 14, 15]). Typically these occur on the same DNA strand, but there is also evidence for inter-strand dimer formation in double-stranded DNA under certain conditions [16]. These lesions are described in detail below.

Cyclobutane pyrimidine dimers

Cyclobutane pyrimidine dimers (P<>P; Scheme 1a) are formed when a pyrimidine base in the triplet excited state undergoes a [2 + 2] addition to the C5–C6 double bond of a second pyrimidine base (reviewed in [12]). In free solution a

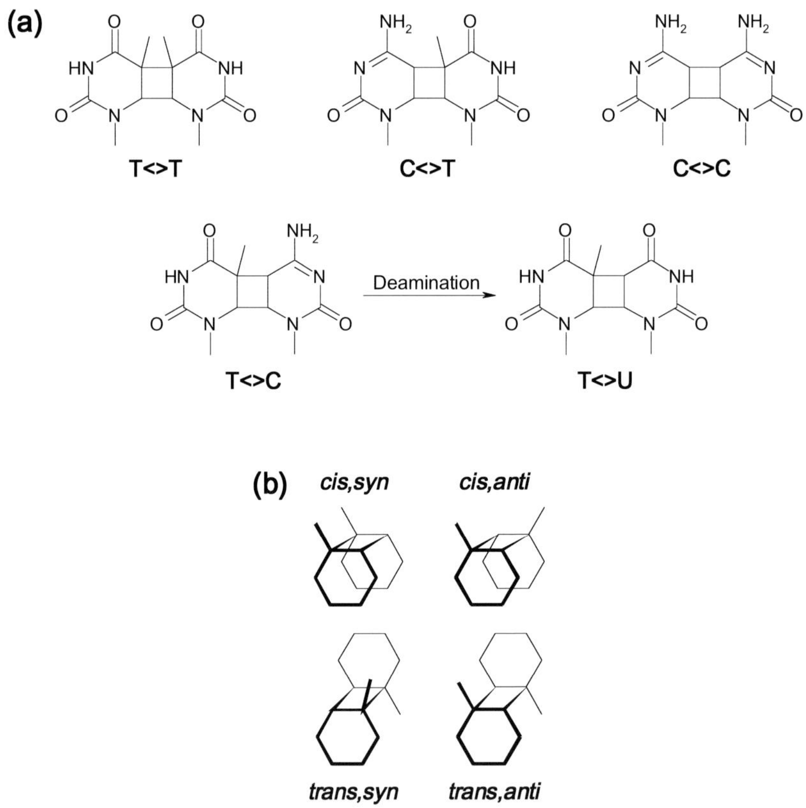

Scheme 1. Structures (a) of the cyclobutane pyrimidine dimers (P<>P) and the possible diastereoisomers (b) of T<>T.

mixture of diastereoisomers (Scheme 1b) is generated that differ in the orientation of the two pyrimidine rings relative to the cyclobutane ring, and on the relative orientations of the C5–C6 bonds in each pyrimidine base. In double-stranded DNA in its natural configuration (i.e. in the B form), where the dimer involves two adjacent pyrimidine bases on the same strand, only the *syn* isomers can be generated, and the *cis* isomer is greatly preferred over the *trans* isomer [12]. In single-stranded or denatured DNA, the *trans,syn* isomer becomes more prevalent due to the increased flexibility of the DNA backbone. Dimer formation (typically the *cis,syn* or *trans,anti* isomers) between the two strands of double-stranded DNA can also be detected in trace amounts in aqueous solutions with UVC irradiation [16]. However, in situations where a different DNA conformation is adopted (e.g., in 80% ethanol or in the dry state) the incidence of inter-strand dimers dramatically increases, as evidenced by the increased proportion of *anti* isomers [16].

The P<>P dimers themselves are only mildly photoactive but under irradiation with UVC light the dimerization can be reversed by photo-induced splitting of the cyclobutane ring, to yield the original monomer bases [17, 18]. There are also a number of DNA repair processes invoked *in vivo*; in many organisms these include photolyase enzymes that are activated upon exposure to UV light (reviewed in [9, 13]).

For P<>P dimers that contain cytosine a further reaction can occur; due to the saturation of the C5–C6 bond in these products, they undergo deamination *via* hydrolysis of the C4 amino group to yield a carbonyl function (Scheme 1) [19, 20]. This results in the formation of uracil-containing products, for example, T<>C becomes a T<>U dimer. This has implications in their mutagenic properties, as discussed in the section ‘Mutagenicity of the DNA lesions’ [21].

Pyrimidine (6-4) pyrimidone dimers and their Dewar valence isomers

The formation of pyrimidine (6-4) pyrimidone dimers [(6-4)PP; Scheme 2] occurs when a pyrimidine base in its singlet excited state reacts *via* a [2 + 2] cycloaddition with a second pyrimidine base (reviewed in [12]). This process is similar to the formation of P<>P dimers, but the (6-4)PP adducts are generated by cyclization between the C5–C6 bond of a pyrimidine and the C4 carbonyl or imino groups (for thymine or cytosine, respectively) of its 3' neighbor. The resulting oxetane and azetidine products are unstable and rapidly rearrange to yield the (6-4)PP adducts, in which the carbonyl or imino group of the 3' base is transferred to the C5 position of the 5' base.

The UV spectra of the (6-4)PP dimers exhibit an absorption band that is shifted by *ca.* 50 nm from the native bases into the near UV region of the spectrum. Thus, upon exposure to UVB or UVA light, the (6-4)PP adducts are readily converted into their Dewar valence isomers (Scheme 2) [22, 23]. These Dewar isomers are only moderately photoactive, but can undergo reversion to the (6-4)PP adducts upon exposure to short-wavelength UV radiation [13].

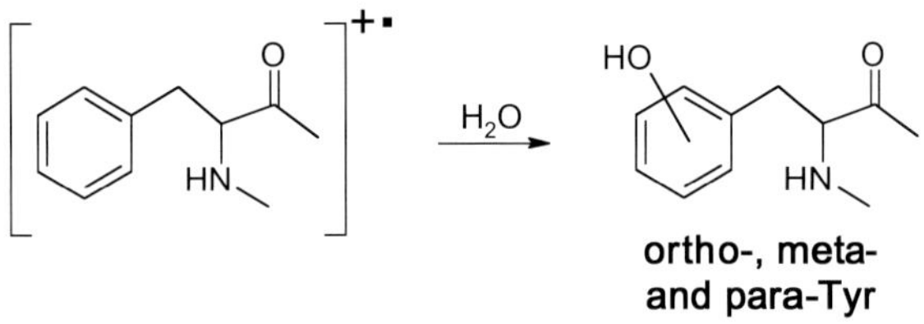

Scheme 8. Hydration products of the Phe radical cation,

zyl radicals [53]. With Tyr, direct cleavage of the phenolic -O-H bond can occur, yielding the phenoxyl radical, a proton and a hydrated electron [2].

The hydrated electron (e^-_{aq}) produced by direct photo-ionization can add rapidly to O_2, to give $O_2^{-\cdot}$, which can, in turn, induce further protein damage. e^-_{aq}-mediated addition to free carboxyl groups (e.g., the C terminus, or Asp/Glu side chains) and amine groups (e.g., the N terminus, or Lys side chains) results in deamination and $H^\cdot$ elimination. Hydrated electrons also react with cystine to give the disulfide radical anion ($RSSR^{-\cdot}$), and with peptide backbone carbonyl groups yielding a radical anion that can subsequently give rise to backbone cleavage [4, 55, 61, 62].

Protein photoproducts induced by $^1O_2$ reactions

As the rate constants for reaction of $^1O_2$ with protein side chains are higher than those with most other cellular targets [7], and proteins are present in most biological systems at particularly high concentrations, these are major targets for $^1O_2$ (reviewed in [3, 4, 6]). The majority of reactions of $^1O_2$ with proteins occur *via* reactions that result in chemical change, rather than quenching pathways that result in relaxation to ground state O_2 without inducing protein damage. Of the common amino acids present in proteins, Trp, His, Tyr, Met and Cys, react with $^1O_2$ at significant rates at physiological pH values [3–7]. At high pH, where Arg and Lys are in their neutral (unprotonated) forms, photo-oxidation also occurs at these residues [63]. Other amino acids can also be consumed as a result of indirect photo-oxidation processes, due to further reactions of $^1O_2$-induced intermediates at the above residues. These reactions have been implicated in cross-linking/aggregation of proteins [5]. The mechanisms and products that arise from $^1O_2$ reaction with these reactive side chains are reviewed below.

Reaction of $^1O_2$ with tryptophan residues

Reaction of $^1O_2$ with Trp gives both *N*-formylkynurenine (and hence kynurenine *via* hydrolysis) and 3α-hydroxypyrroloindoles, *via* the initial formation of either a dioxetane across the C2–C3 double bond, or a hydroperoxide at C3 (Scheme 9) [64, 65]. Subsequent decomposition of these intermediates gives *N*-formylkynurenine, whereas ring closure yields 3α-hydroxypyrroloindole. Decomposition of these peroxides may involve non-radical reactions, or thermal homolysis of the -O–O- bond to give radicals. Metal ions, heat and UV light have been shown to catalyze peroxide decomposition to radicals; the mechanism and products of such reactions has been reviewed [65]. The overall process appears to be common to both the free amino acid and *N*-blocked Trp, and hence analogous materials are formed on oxidation of Trp oxidation in proteins. Interestingly, *N*-formylkynurenine and kynurenine are more effective photo-sensitizing agents than the parent amino acid [8], thus the formation of these materials on UV-exposed proteins may lead to enhanced photo-oxidation due to further generation of reactive species. This may be of particular significance in the human lens where it is known that free Trp oxidation products (e.g., kynurenine, 3-hydroxykynurenine, 3-hydroxykynurenine *O*-β-D-glucoside and 4-(2-amino-3-hydroxyphenyl)-4-oxobutanoic acid *O*-β-D-glucoside) can become bound to lens proteins (e.g., [66–68]) and subsequently act as photosensitizers of peroxide formation and further damage [69].

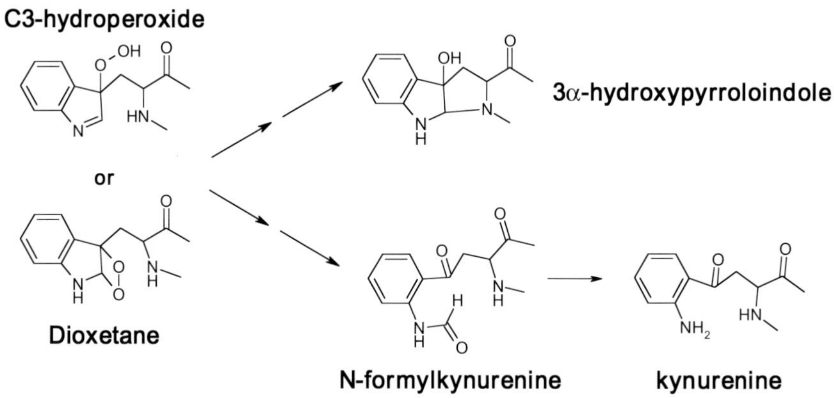

Scheme 9. Products of $^1O_2$-mediated oxidation of Trp.

Reaction of $^1O_2$ with tyrosine residues

With free Tyr, the primary products formed are unstable endoperoxides (Scheme 10), which react *via* ring-opening mechanisms to give a C1 hydroperoxide, and cyclized products involving nucleophilic addition of the α-amino group [70–73]. The endoperoxide species are unstable and rapidly decompose. This process can be catalyzed by metal ions and UV light to give radicals; the ultimate products of these reactions are unclear [72, 74]. Thermal decay gives rise to a cyclized indolic product, 3a-hydroxy-6-oxo-2,3,3a,6,7,7a-hexahydro-

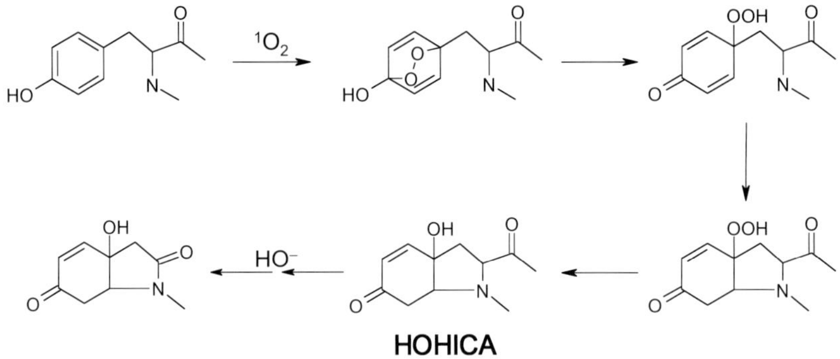

Scheme 10. Products of $^1O_2$-mediated oxidation of Tyr.

1H-indol-2-carboxylic acid (HOHICA; Scheme 10) [70–73]. This product can be oxidized further in basic conditions, giving rise to a decarboxylated keto compound [70, 71].

In peptides and proteins, the nucleophilic ring closure reactions of the α-amino group are less favorable (due to its incorporation in the peptide bond) allowing other nucleophilic additions to occur (A. Wright, C.L. Hawkins and M.J. Davies, unpublished data); these processes could play an important role in protein cross-linking. Previous studies have reported that 3,4-dihydroxyphenylalanine and di-tyrosine are not major products of $^1O_2$-mediated oxidation of Tyr on proteins [75, 76]; more recent data have suggested the opposite [69].

Studies with Tyr-containing peptides have shown that there is an increase in the yield of free amino groups on reaction with the tri-peptide Tyr-Tyr-Tyr, little change for Tyr-Tyr, and a loss of amine groups for Tyr itself [70]. These observations allude to the occurrence of peptide bond cleavage [70]. This may arise as a result of radical formation on peroxide decomposition, with these species subsequently abstracting a hydrogen atom from the backbone α-carbon site; α-carbon radicals are known intermediates in peptide bond cleavage [55]. Direct evidence for intermolecular hydrogen atom abstraction by similar peroxide-derived radicals has been obtained by EPR spectroscopy (A. Wright, C.L. Hawkins and M.J. Davies, unpublished data).

Reaction of $^1O_2$ with histidine residues

Oxidation of His by $^1O_2$ occurs *via* the initial formation of one or more endoperoxides (Scheme 11) and the consumption of a single molecule of O_2 per mole His [77]. The structure of these endoperoxides has been determined by low temperature NMR in organic solvents [78, 79]. At higher temperatures they undergo a series of ill-defined reactions to give a complex mixture of products which include aspartic acid, asparagines and urea (Scheme 11) [77,

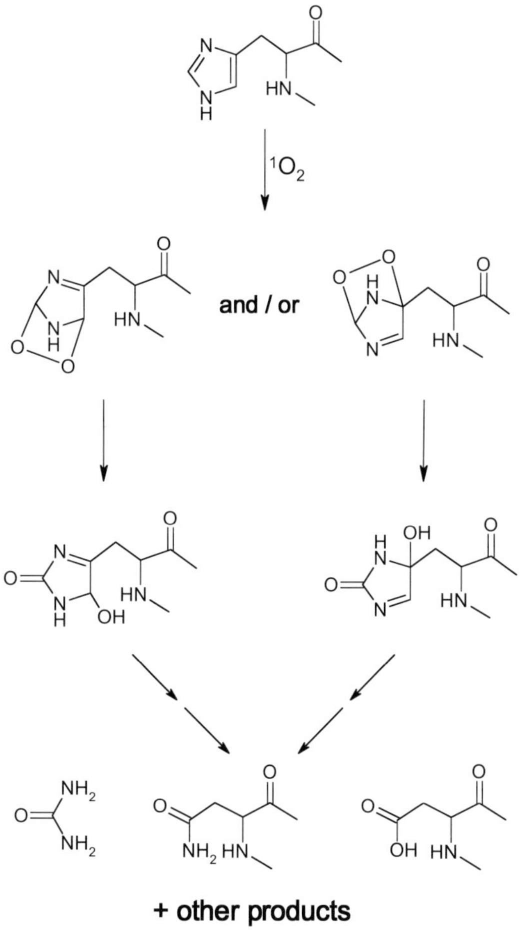

Scheme 11. Products of $^1O_2$-mediated oxidation of His.

80]. The mechanisms probably involve radical processes and ring-opening reactions [74, 81]. The final products of His oxidation can undergo further reactions [82, 83], including the formation of His-His and His-Lys cross-links *via* the reaction of a nucleophilic nitrogen on one His ring, or a Lys side chain, with a keto group on a second oxidized His. Endoperoxide formation appears to occur regardless of whether the α-amino group is free or blocked, so it is likely that similar materials are generated on proteins.

Reaction of $^1O_2$ with methionine residues

Reaction of free Met with $^1O_2$ occurs *via* the formation of a zwitterionic species ($R_2S^+{-}OO^-$), which undergoes subsequent reaction with a second molecule of the parent to give two moles of the sulfoxide ($R_2S{=}O$) [84] (reviewed in [5]). With some sensitizers, other intermediates including a stable nitrogen-sulfur cyclic intermediate have been reported (reviewed in [5]). Subsequent hydrolysis of this species yields 1 mol of sulfoxide and 1 mol of H_2O_2 [84]. This type of reaction may only be of significance with free Met, due to the involvement of the free amino group.

With free Met, the stoichiometry of molecular O_2:Met consumption is pH dependent (reviewed in [6]). The stoichiometry and intermediates involved in the corresponding reactions on proteins remain to be fully elucidated, although it is clear that methionine sulfoxide can be a major product.

Reaction of $^1O_2$ with cysteine and cystine residues

Rapid, but non-quantitative generation of the disulfide (RSSR) occurs when free Cys reacts with $^1O_2$ [5, 85]. Other products are formed, probably including cysteic acid (RSO_3H), but these have not been fully elucidated [5].

It has been suggested that reaction of $^1O_2$ with free cystine occurs *via* a zwitterion ($RS^+{-}OO^-$), similar to that with Met [86]. This species probably reacts with a further molecule of cystine yielding two molecules of the mono-sulfoxide RSS(=O)R. The occurrence of these reactions in proteins, where the 3-D structure is likely to constrain the cystine molecules, remains to be determined.

Physical and chemical consequences of photo-oxidation of proteins

A range of enzymes have long been known to be inactivated on exposure to UV light [87], and a considerable number of studies have reported similar results (e.g., [2, 5, 6]). These studies have shown that much of the photochemistry described above for free amino acids and peptides, also occurs with proteins. Thus, the major amino acids consumed on photo-oxidation are His, Trp, Met, Tyr and Cys, although a few studies have also reported the loss of Phe, Arg and Lys residues. With proteins, which contain all of these residues, His and Trp are usually the most susceptible (e.g., [5]), although there are some exceptions, as would be expected on the basis of the rate constants. It has been shown that photo-oxidation of proteins can give rise to a number of reactive species (reviewed in [3]) including protein peroxides, with these probably localized on His, Trp and Tyr residues. Similar protein-bound peroxides have been detected in a number of cell types exposed to visible light and sensitizers (e.g., Rose Bengal) that generate $^1O_2$ ([88], Policarpio, V. and Davies, M.J., unpublished data). Thus it is likely that exposure of cells to UVA will also generate intracellular protein (and other) peroxides.

The formation of high-molecular-weight aggregates (dimers and higher species) is a common consequence of photo-oxidation of proteins [89]. Some of these aggregates may arise from radical-radical termination reactions of two Tyr-derived phenoxyl radicals to give di- (or bi-) tyrosine [69, 90], although other reports have suggested that di-tyrosine is not formed [75, 76]. Other reactions probably play a key role in the formation of aggregates, and it has been proposed that many cross-links arise as a result of "dark" reactions that occur after the cessation of light exposure [91]. It has been suggested that cross-links are formed as a result of the oxidation of His residues to products, which then react with Lys, Cys or other His residues [75, 82, 83, 92, 93]. Studies on His derivatives have shown that carbonyl-containing materials are formed [77], and an increase in the yield of (unspecified) protein carbonyls has been reported on photo-oxidation of albumin by porphyrin sensitizers [94, 95]. It is possible that these cross-links arise from reaction of carbonyl compounds, formed by photo-oxidation, with Lys, Arg and Cys side chains. The formation of such cross-links may explain the observed loss of Lys and Arg residues in some photo-oxidized proteins (reviewed in [5]). His residues may be of particular importance in cross-link generation, as such links have been reported to be absent from photo-oxidized proteins that lack this amino acid [75, 76].

In contrast to the prevalence of reports on protein aggregation, there are relatively few reports of backbone cleavage (i.e., fragmentation), at least with short UV exposure times [5, 96]. Photo-oxidation of lysozyme in the presence of a sensitizer has been reported to give peptide fragments as a result of backbone rupture; this has been proposed to occur either *via* oxidation of Trp residues and/or the formation of radicals [97]. Backbone α-carbon radicals are known to be key intermediates in backbone cleavage in the presence of O_2 (reviewed in [4, 59, 62]), and peroxides formed on His and Tyr residues have been shown to yield such radicals in small peptides (Wright, A., Hawkins, C.L. and Davies, M.J., unpublished data) [72, 74].

Photo-oxidized proteins can behave in a markedly different manner to their non-exposed counterparts. Changes include: an increase in susceptibility of the oxidized protein to proteolytic enzymes; alterations in mechanical properties (e.g., of silk and collagen); an increased extent, or susceptibility to, unfolding; changes in conformation; an increase in hydrophobicity; altered light scattering properties and optical rotation; and changes in binding of co-factors and metal ions (e.g., [5, 75, 93, 98, 99]). The physico-chemical bases for these changes are not fully understood, although studies on apo-horseradish peroxidase, which is highly resistant to photo-oxidation, have given useful information. The 3-D protein structure appears to shield most of the susceptible residues from damage [98, 100]; in the holo enzyme the heme iron also appears to play a protective role.

It has been demonstrated that the photo-oxidation of DNA and proteins are not ends in themselves, and that damage to these macromolecules can initiate or exacerbate damage to other molecules. Thus, although DNA and proteins may be the major initial sites of UV-induced damage, the formation of oxidation products on these targets can give rise to secondary damage. It has, for example, been shown that peroxides formed on peptides and proteins can subsequently oxidize susceptible residues on other proteins (and thereby give rise to enzyme inactivation [101, 102]), can deplete low-molecular-mass antioxidants [103], and can also give rise to the formation of oxidized DNA bases, strand breaks and DNA-protein adducts [104–106]. Decomposition of these peroxides to radicals may also initiate lipid oxidation chain reactions and thereby result in significant membrane damage.

Summary and conclusions

The interaction of UV radiation with cells leads primarily to the modification of DNA and proteins, due to a combination of their UV absorption characteristics and abundance in cells. In DNA, several dimeric products between nucleic acid bases are generated by direct UV absorption, together with single base modifications resulting from photo-ionization and radical reactions. A series of products is also generated *via* sensitized reactions, resulting in radical formation or $^1O_2$ generation. The primary product from reaction of $^1O_2$ with DNA is 8-oxo-Gua; this is a common product following DNA oxidation by a variety of oxidants. The dimeric products are strongly implicated in the mutagenicity observed in UV-irradiated DNA, which is typified by a high level of CC→TT and C→T transversions.

In proteins, the majority of UV-induced damage appears to be mediated by $^1O_2$, which reacts preferentially with the side chains of Trp, His, Tyr, Met, Cys and cystine residues, although direct photo-oxidation reactions and radical species may also play a role, particularly with short wavelength UV. The initial products of $^1O_2$-mediated reactions are endoperoxides with the aromatic residues, and zwitterions with the sulfur-containing residues. These species undergo a variety of further reactions, including radical processes and ring-opening reactions that ultimately lead to disruption of protein structure. Cross-linking and aggregation of proteins predominates over fragmentation processes for $^1O_2$-mediated oxidation.

The chemical manifestations of UV radiation on proteins and DNA are now reasonably well characterized, although there is still considerable work to be done in elucidating the mechanisms of these reactions and their relative importance in different scenarios (e.g., different UV wavelengths, cellular *versus* isolated environments, pH). It is also clear that initial damage to one molecule can subsequently give rise to significant secondary damage and have major biological ramifications (e.g., DNA mutations in the early stages of carcinogenesis, disruption of cellular and tissue function by, for example, protein cross-linking and enzyme inactivation); these aspects have yet to be fully elucidated.

Acknowledgements
Work in the authors' laboratories is supported by the Australian Research Council and the National Health and Medical Research Council.

References

1 Bensasson RV, Land EJ, Truscott TG (1983) *Pulse radiolysis and flash photolysis: Contributions to the chemistry of biology and medicine*. Pergamon Press, Oxford

2 Bensasson RV, Land EJ, Truscott TG (1993) *Excited states and free radicals in biology and medicine: Contributions from flash photolysis and pulse radiolysis*. Oxford University Press, Oxford

3 Davies MJ (2004) Reactive species formed on proteins exposed to singlet oxygen. *Photochem Photobiol Sci* 3: 17–25

4 Davies MJ, Truscott RJW (2001) Photo-oxidation of protein and its role in cataractogenesis. *J Photochem Photobiol B Biol* 63: 114–125

5 Straight RC, Spikes JD (1985) Photosensitized oxidation of biomolecules. In: AA Frimer (ed.): *Singlet O_2, Vol. 4*. CRC Press, Boca Raton, 91–143

6 Davies MJ (2003) Singlet oxygen-mediated damage to proteins and its consequences, *Biochem Biophys Res Commun 305*: 761–770

7 Wilkinson F, Helman WP, Ross AB (1995) Rate constants for the decay and reactions of the lowest electronically excited state of molecular oxygen in solution. An expanded and revised compilation. *J Phys Chem Ref Data 24*: 663–1021

8 Redmond RW, Gamlin JN (1999) A compilation of singlet oxygen yields from biologically relevant molecules, *Photochem Photobiol* 70: 391–475

9 Sinha RP, Hader DP (2002) UV-induced DNA damage and repair: A review. *Photochem Photobiol Sci* 1: 225–236

10 Matsumura Y, Ananthaswamy HN (2004) Toxic effects of ultraviolet radiation on the skin. *Toxicol Appl Pharmacol* 195: 298–308

11 Ichihashi M, Ueda M, Budiyanto A, Bito T, Oka M, Fukunaga M, Tsuru K, Horikawa T (2003) UV-induced skin damage. *Toxicology* 189: 21–39

12 Ravanat JL, Douki T, Cadet J (2001) Direct and indirect effects of UV radiation on DNA and its components. *J Photochem Photobiol B* 63: 88–102

13 Taylor JS (1994) Unraveling the molecular pathway from sunlight to skin-cancer. *Acc Chem Res* 27: 76–82

14 Cadet J, Anselmino C, Douki T, Voituriez L (1992) Photochemistry of nucleic-acids in cells, *J Photochem Photobiol B* 15: 277–298

15 Black HS, deGruijl FR, Forbes PD, Cleaver JE, Ananthaswamy HN, deFabo EC, Ullrich SE, Tyrrell RM (1997) Photocarcinogenesis: An overview. *J Photochem Photobiol B* 40: 29–47

16 Douki T, Laporte G, Cadet J (2003) Inter-strand photoproducts are produced in high yield within A-DNA exposed to UVC radiation, *Nucleic Acids Res* 31: 3134–3142

17 Cadet J, Vigny P (1990) The photochemistry of nucleic acids. In: H Morrison (ed.): *Bioorganic photochemistry, Vol. 1*. Wiley and Sons, New York, 1–272

18 Cadet J, Voituriez L, Hruska FE, Kan L-S, de Leeuw FAA, Altona C (1985) Characterization of thymidine ultraviolet photoproducts. Cyclobutane dimers and 5,6-dihydrothymines. *Can J Chem* 63: 2861–2868

19 Lemaire DGE, Ruzsicska BP (1993) Kinetic analysis of the deamination reactions of cyclobutane dimers of thymidyl-3',5'-2'-deoxycytidine and 2'-deoxycytidyl-3',5'-thymidine, *Biochemistry* 32: 2525–2533

20 Setlow RB, Carrier WL, Bollum FJ (1965) Pyrimidine dimers in UV-irradiated poly dI:dC. *Proc Natl Acad Sci USA* 53: 1111–1118

21 Peng W, Shaw BR (1996) Accelerated deamination of cytosine residues in UV-induced cyclobutane pyrimidine dimers leads to CC-TT transitions. *Biochemistry* 35: 10172–10181

22 Taylor JS, Cohrs MP (1987) DNA, light, and Dewar pyrimidones – the structure and biological significance of TpT3. *J Am Chem Soc* 109: 2834–2835

23 Taylor JS, Lu HF, Kotyk JJ (1990) Quantitative conversion of the (6-4) photoproduct of TpdC to its Dewar valence isomer upon exposure to simulated sunlight. *Photochem Photobiol* 51: 161–167

24 Douki T, Cadet J (1992) Far-UV photochemistry and photosensitization of 2'-deoxycytidylyl-(3'-5')-thymidine - isolation and characterization of the main photoproducts. *J Photochem Photobiol B Biol* 15: 199–213

25 Wierzchowski KL, Shugar D (1961) Photochemistry of cytosine nucleosides and nucleotides II. *Acta Biochim Pol* 8: 219–234

26 Boorstein RJ, Hilbert TP, Cunningham RP, Teebor GW (1990) Formation and stability of repairable pyrimidine photohydrates in DNA. *Biochemistry* 29: 10455–10460

27 Douki T, Vadesne-Bauer G, Cadet J (2002) Formation of 2 '-deoxyuridine hydrates upon exposure of nucleosides to gamma radiation and UVC-irradiation of isolated and cellular DNA. *Photochem Photobiol Sci* 1: 565–569
28 Kumar S, Joshi PC, Sharma ND, Bose SN, Davies RJH, Takeda N, McCloskey J A (1991) Adenine photodimerization in deoxyadenylate sequences - elucidation of the mechanism through structural studies of a major D(ApA) photoproduct. *Nucleic Acids Res* 19: 2841–2847
29 Kumar S, Sharma ND, Jeremy R, Davies H, Phillipson DW, McCloskey JA (1987) The isolation and characterization of a new type of dimeric adenine photoproduct in UV-irradiated deoxyadenylates. *Nucleic Acids Res* 15: 1199–1216
30 Zhao XD, Nadji S, Kao JLF, Taylor JS (1996) The structure of D(TpA)*, the major photoproduct of thymidylyl-(3'-5')-deoxyadenosine. *Nucleic Acids Res* 24: 1554–1560
31 Wang YS, Taylor JS, Gross ML (2001) Isolation and mass spectrometric characterization of dimeric adenine photoproducts in oligodeoxynucleotides. *Chem Res Toxicol* 14: 738–745
32 Sharma ND, Davies RJH (1989) Extent of formation of a dimeric adenine photoproduct in polynucleotides and DNA. *J Photochem Photobiol B* 3: 247–258
33 Cabrera-Juarez E, Setlow JK (1977) Formation of a thymine photoproduct in transforming DNA by near ultraviolet irradiation. *Arch Biochem Biophys* 475: 315–322
34 Donnellan JE, Setlow RB (1965) Thymine photoproducts but not thymine dimers are found in ultraviolet irradiated bacterial spores. *Science* 149: 308–310
35 Douki T, Cadet J (2003) Formation of the spore photoproduct and other dimeric lesions between adjacent pyrimidines in UVC-irradiated dry DNA. *Photochem Photobiol Sci* 2: 433–436
36 Steenken S (1989) Purine-bases, nucleosides, and nucleotides – aqueous-solution redox chemistry and transformation reactions of their radical cations and e^- and OH adducts. *Chem Rev* 89: 503–520
37 Cadet J, Teoule R (1978) Comparative study of oxidation of nucleic acid components by hydroxyl radicals, singlet oxygen and superoxide anion radical. *Photochem Photobiol* 28: 661–667
38 Sheu C, Foote CS (1993) Endoperoxide formation in a guanosine derivative. *J Am Chem Soc* 115: 10446–10447
39 Douki T, Court M, Sauvaigo S, Odin F, Cadet J (2000) Formation of the main UV-induced thymine dimeric lesions within isolated and cellular DNA as measured by high performance liquid chromatography-tandem mass spectrometry, *J Biol Chem* 275: 11678–11685
40 Douki T, Cadet J (2001) Individual determination of the yield of the main UV-induced dimeric pyrimidine photoproducts in DNA suggests a high mutagenicity of CC photolesions. *Biochemistry* 40: 2495–2501
41 Yoon JH, Lee CS, O'Connor TR, Yasui A, Pfeifer GP (2000) The DNA damage spectrum produced by simulated sunlight. *J Mol Biol* 299: 681–693
42 Kielbassa C, Roza L, Epe B (1997) Wavelength dependence of oxidative DNA damage induced by UV and visible light. *Carcinogenesis* 18: 811–816
43 Douki T, Reynaud-Angelin A, Cadet J, Sage E (2003) Bipyrimidine photoproducts rather than oxidative lesions are the main type of DNA damage involved in the genotoxic effect of solar UVA radiation, *Biochemistry* 42: 9221–9226
44 Rochette PJ, Therrien JP, Drouin R, Perdiz D, Bastien N, Drobetsky EA, Sage E (2003) UVA-induced cyclobutane pyrimidine dimers form predominantly at thymine-thymine dipyrimidines and correlate with the mutation spectrum in rodent cells. *Nucleic Acids Res* 31: 2786–2794
45 Brash DE (1997) Sunlight and the onset of skin cancer. *Trends Genet* 13: 410–414
46 Sage E, Lamolet B, Brulay E, Moustacchi E, Chateauneuf A, Drobetsky EA (1996) Mutagenic specificity of solar UV light in nucleotide excision repair-deficient rodent cells. *Proc Natl Acad Sci USA* 93: 176–180
47 Nairn RS, Mitchell DL, Adair GM, Thompson LH, Siciliano MJ, Humphrey RM (1989) UV mutagenesis, cyto-toxicity and split-dose recovery in a human CHO cell hybrid having intermediate (6-4) photoproduct repair. *Mutat Res* 217: 193–201
48 You YH, Lee DH, Yoon JH, Nakajima S, Yasui A, Pfeifer GP (2001) Cyclobutane pyrimidine dimers are responsible for the vast majority of mutations induced by UVB irradiation in mammalian cells. *J Biol Chem* 276: 44688–44694
49 Shibutani S, Takeshita M, Grollman AP (1991) Insertion of specific bases during DNA-synthesis past the oxidation-damaged base 8-oxodG. *Nature* 349: 431–434
50 Meloun B, Moravek L, Kostka V (1975) Complete amino acid sequence of human serum albumin. *FEBS Lett* 58: 134–137

51 Creed D (1984) The photophysics and photochemistry of the near-UV absorbing amino acids. II. Tyrosine and its simple derivatives. *Photochem Photobiol* 39: 363–375
52 Creed D (1984) The photophysics and photochemistry of the near-UV absorbing amino acids. I. Tryptophan and its simple derivatives. *Photochem Photobiol* 39: 537–562
53 Davies MJ, Gilbert BC (1991) Free radical reactions. Fragmentation and rearrangements in aqueous solution. *Adv Detailed React Mech* 1: 35–81
54 Bent DV, Hayon E (1975) Excited state chemistry of aromatic amino acids and related peptides. II. Phenylalanine. *J Phys Chem* 97: 2606–2612
55 Davies MJ, Dean RT (1997) *Radical-mediated protein oxidation: From chemistry to medicine*. Oxford University Press, Oxford
56 Candeias LP, Wardman P, Mason RP (1997) The reaction of oxygen with radicals from oxidation of tryptophan and indole-3-acetic acid. *Biophys J* 67: 229–237
57 Wardman P, von Sonntag C (1995) Kinetic factors that control the fate of thiyl radicals in cells. *Meth Enzymol* 251: 31–45
58 Schoneich C (1995) Kinetics of thiol reactions. *Meth Enzymol* 251: 45–55
59 Hawkins CL, Davies MJ (2001) Generation and propagation of radical reactions on proteins. *Biochim Biophys Acta* 1504: 196–219
60 Nikogosyan DN, Gorner H (1992) Photolysis of aromatic amino acids in aqueous solution by nanosecond 248 and 193 nm laser light. *J Photochem Photobiol B Biol* 13: 219–234
61 von Sonntag C (1987) *The chemical basis of radiation biology*. Taylor and Francis, London
62 Garrison WM (1987) Reaction mechanisms in the radiolysis of peptides, polypeptides, and proteins. *Chem Rev* 87: 381–398
63 Papeschi G, Monici M, Pinzauti S (1982) pH effect on dye sensitized photooxidation of amino acids and albumins. *Med Biol Environ* 10: 245–250
64 Nakagawa M, Watanabe H, Kodato S, Okajima H, Hino T, Flippen JL, Witkop B (1977) A valid model for the mechanism of oxidation of tryptophan to formylkynurenine – 25 years later. *Proc Natl Acad Sci USA* 74: 4730–4733
65 Saito I, Matsuura T, Nakagawa M, Hino T (1977) Peroxidic intermediates in photosensitized oxygenation of tryptophan derivatives. *Acc Chem Res* 10: 346–352
66 Wood AM (1993) *Tryptophan metabolism in the human lens*. PhD Thesis, University of Wollongong, Australia, 43–44
67 Hood BD, Garner B, Truscott RJ (1999) Human lens coloration and aging. Evidence for crystallin modification by the major ultraviolet filter, 3-hydroxy-kynurenine O-beta-D-glucoside. *J Biol Chem* 274: 32547–32550
68 Garner B, Shaw DC, Lindner RA, Carver JA, Truscott RJ (2000) Non-oxidative modification of lens crystallins by kynurenine: A novel post-translational protein modification with possible relevance to ageing and cataract. *Biochim Biophys Acta* 1476: 265–278
69 Parker NR, Jamie JF, Davies MJ, Truscott RJW (2004) Protein-bound kynurenine is a photosensitizer of oxidative damage. *Free Radic Biol Med*: 37: 1479–1489
70 Criado S, Soltermann AT, Marioli JM, Garcia NA (1998) Sensitized photooxidation of di- and tripeptides of tyrosine. *Photochem Photobiol* 68: 453–458
71 Katsuya E, Seya K, Hikino H (1988) Photo-oxidation of L-tyrosine, an efficient, 1,4-chirality transfer reaction. *J Chem Soc Chem Commun* 934–935
72 Wright A, Bubb WA, Hawkins CL, Davies MJ (2002) Singlet oxygen-mediated protein oxidation: Evidence for the formation of reactive side-chain peroxides on tyrosine residues, *Photochem Photobiol* 76: 35–46
73 Jin FM, Leitich J, von Sonntag C (1995) The photolysis ($\lambda = 254$ nm) of tyrosine in aqueous solutions in the absence and presence of oxygen - the reaction of tyrosine with singlet oxygen. *J Photochem Photobiol A Chem* 92: 147–153
74 Wright A, Hawkins CL, Davies MJ (2000) Singlet oxygen-mediated protein oxidation: Evidence for the formation of reactive peroxides. *Redox Rep* 5: 159–161
75 Balasubramanian D, Du X, Zigler JSJ (1990) The reaction of singlet oxygen with proteins, with special reference to crystallins. *Photochem Photobiol* 52: 761–768
76 Balasubramanian D, Kanwar R (2002) Molecular pathology of dityrosine cross-links in proteins: Structural and functional analysis of four proteins. *Mol Cell Biochem* 234: 27–38
77 Tomita M, Irie M, Ukita T (1969) Sensitized photooxidation of histidine and its derivatives. Products and mechanism of the reaction. *Biochemistry* 8: 5149–5160
78 Kang P, Foote CS (2000) Synthesis of a C-13, N-15 labeled imidazole and characterization of the

2,5-endoperoxide and its decomposition. *Tet Lett* 41: 9623–9626
79 Kang P, Foote CS (2002) Photosensitized oxidation of C-13,N-15-labeled imidazole derivatives. *J Am Chem Soc* 124: 9629–9638
80 Kai S, Suzuki M (1996) Dye-sensitized photooxidation of 2,4-disubstituted imidazoles - the formation of isomeric imidazolinones. *Heterocycles* 43: 1185–1188
81 Chang SH, Teshima GM, Milby T, Gillece-Castro B, Canova-Davis E (1997) Metal-catalyzed photooxidation of histidine in human growth hormone. *Anal Biochem* 244: 221–227
82 Shen HR, Spikes JD, Kopecekova P, Kopecek J (1996) Photodynamic crosslinking of proteins. I. Model studies using histidine- and lysine-containing N-(2-hydroxypropyl)methacrylamide copolymers. *J Photochem Photobiol B Biol* 34: 203–210
83 Shen H-R, Spikes JD, Smith CJ, Kopecek J (2000) Photodynamic cross-linking of proteins IV. Nature of the His-His bond(s) formed in the Rose Bengal-photosensitized cross-linking of *N*-benzoyl-L-histidine. *J Photochem Photobiol A Chem* 130: 1–6
84 Sysak PK, Foote CS, Ching T-Y (1977) Chemistry of singlet oxygen - XXV. Photooxygenation of methionine. *Photochem Photobiol* 26: 19–27
85 Rougee M, Bensasson RV, Land EJ, Pariente R (1988) Deactivation of singlet molecular oxygen by thiols and related compounds, possible protectors against skin photosensitivity. *Photochem Photobiol* 47: 485–489
86 Foote CS, Peters JW (1971) Chemistry of singlet oxygen. XIV. A reactive intermediate in sulfide photooxidation. *J Am Chem Soc* 93: 3795–3796
87 von Tappeiner H (1903) Über die Wirkung fluoreszierender Substanzen auf Fermente und Toxine. *Ber Deut Chem Ges* 36: 3035–3038
88 Wright A, Hawkins CL, Davies MJ (2003) Photo-oxidation of cells generates long-lived intracellular protein peroxides. *Free Radic Biol Med* 34: 637–647
89 Goosey JD, Zigler JS Jr, Kinoshita JH (1980) Cross-linking of lens crystallins in a photodynamic system: A process mediated by singlet oxygen. *Science* 208: 1278–1280
90 Shen H, Spikes JD, Smith CJ, Kopecek J (2000) Photodynamic cross-linking of proteins V. Nature of the tyrosine-tyrosine bonds formed in the FMN-sensitized intermolecular cross-linking of *N*-acetyl-L-tyrosine. *J Photochem Photobiol A Chem* 133: 115–122
91 Dubbelman TM, Haasnoot C, van Steveninck J (1980) Temperature dependence of photodynamic red cell membrane damage., *Biochim Biophys Acta* 601: 220–227
92 Verweij H, van Steveninck J (1982) Model studies on photodynamic crosslinking, *Photochem Photobiol* 35: 265–267
93 Dillon J, Chiesa R, Wang RH, McDermott M (1993) Molecular changes during the photooxidation of alpha-crystallin in the presence of uroporphyrin. *Photochem Photobiol* 57: 526–530
94 Silvester JA, Timmins GS, Davies MJ (1998) Protein hydroperoxides and carbonyl groups generated by porphyrin-induced photo-oxidation of bovine serum albumin. *Arch Biochem Biophys* 350: 249–258
95 Silvester JA, Timmins GS, Davies MJ (1998) Photodynamically-generated bovine serum albumin radicals: Evidence for damage transfer and oxidation at cysteine and tryptophan residues. *Free Radic Biol Med* 24: 754–766
96 Michaeli A, Feitelson J (1994) Reactivity of singlet oxygen toward amino acids and peptides. *Photochem Photobiol* 59: 284–289
97 Gomyo T, Fujimaki M (1970) Studies on changes in protein by dye sensitized photooxidation. Part 3. On the photodecomposition products of lysozyme. *Agric Biol Chem* 34: 302–309
98 Silva E, De Landea C, Edwards AM, Lissi E (2000) Lysozyme photo-oxidation by singlet oxygen: Properties of the partially inactivated enzyme. *J Photochem Photobiol B* 55: 196–200
99 Prinsze C, Dubbelman TM, Van Steveninck J (1990) Protein damage, induced by small amounts of photodynamically generated singlet oxygen or hydroxyl radicals. *Biochim Biophys Acta* 1038: 152–157
100 Kang YJ, Spikes JD (1976) The porphyrin-sensitized photooxidation of horseradish apoperoxidase. *Arch Biochem Biophy.* 172: 565–573
101 Morgan PE, Dean RT, Davies MJ (2002) Inhibition of glyceraldehyde-3-phosphate dehydrogenase by peptide and protein peroxides generated by singlet oxygen attack. *Eur J Biochem* 269: 1916–1925
102 Hampton MB, Morgan PE, Davies MJ (2002) Inactivation of cellular caspases by peptide-derived tryptophan and tyrosine peroxides. *FEBS Lett* 527: 289–292
103 Morgan PE, Dean RT, Davies MJ (2004) Protective mechanisms against peptide and protein per-

oxides generated by singlet oxygen. *Free Radic Biol Med* 36: 484–496
104 Luxford C, Dean RT, Davies MJ (2000) Radicals derived from histone hydroperoxides damage nucleobases in RNA and DNA. *Chem Res Toxicol* 13: 665–672
105 Luxford C, Dean RT, Davies MJ (2002) Induction of DNA damage by oxidised amino acids and proteins. *Biogerentology* 3: 95–102
106 Luxford C, Morin B, Dean RT, Davies MJ (1999) Histone H1- and other protein- and amino acid-hydroperoxides can give rise to free radicals which oxidize DNA. *Biochem J* 344: 125–134

Cancer: Cell Structures, Carcinogens and Genomic Instability
Edited by Leon P. Bignold

Actions of radiation on living cells in the "post-bystander" era

Carmel Mothersill and C. B. Seymour

Medical Physics and Applied Radiation Sciences Unit, McMaster University, Hamilton, Ontario L8S 4K1, Canada

Abstract. Over the past 20 years there has been increasing evidence that cells and the progeny of cells surviving a dose of ionizing radiation can exhibit a wide range of effects inconsistent with the level of dose received. Recently, the cause of these delayed effects has been ascribed to so-called bystander effects, occurring in cells not directly hit by an ionizing track, but which are influenced by signals from irradiated cells. These effects are not necessarily deleterious, although most of the literature deals with adverse delayed effects. What is important to consider is what, if anything, these effects mean for what is still the central dogma of radiobiology and radiation protection, i.e., that DNA double-strand breaks are the primary radiation-induced lesion that can be quantifiably related to received dose, and which determine the probability that a cancer will result from a radiation exposure. In this chapter we review the history of radiation biology which led to the DNA paradigm. We explore the issues and the evidence which are now challenging the view that dose deposition in DNA is all important. We conclude that in the low-dose region, the primary determinant of radiation exposure outcome is the genetic and epigenetic background of the individual and not the dose. This effectively dissociates dose from effect as a quantitative relationship, but it does not necessarily mean that the effect is unrelated to DNA damage somewhere in the system.

Key words: Bystander effects, genomic instability, radiation, radiation carcinogenesis, radiobiology.

Historical evidence for a central role of DNA and other candidate "targets" in the production and expression of radiation damage

This section reviews the older literature concerning action of radiation on living cells, principally mammalian ones. It must be remembered that much of the early radiation biology literature was concerned with yeast, bacteria or fruit flies and many early theories were formalized using these model systems, and then applied to mammalian systems. It should also be noted as an introductory remark that radiation may be either particulate or non-particulate. X and gamma rays are non-particulate, whereas neutrons, electrons and protons are particulate. Particulate radiations tend to be more densely ionizing, producing a Bragg-Peak ionization deposition. The dose deposition in tissues and cell culture systems varies considerably depending not only on the type of radiation but also on the energy of the radiation.

There is a famous quote that those not aware of history are condemned to repeat it. In this context it is important to be aware of how this field developed, in order to understand how the prevailing paradigms that exist today originated. Unconditional acceptance of paradigms can lead to an uncritical acceptance of dogma. Similarly, unless we regard the scientists responsible for developing the theories as charlatans, we must acknowledge they formed their conclusions based on solid, painstaking research. Therefore, we shall attempt to review the field critically after giving a brief historical perspective.

Classical radiobiology really started in the 1940s with the almost simultaneous publication of two books on mechanisms of action of ionizing radiation on living cells. Lea published "Actions of radiations on living cells" in 1946 [1] and Timofeeff-Ressovsky and Zimmer published "Das Trefferprinzip in der Biologie" in 1947 [2]. Both authors noted that there had been much radiation research on the chemical effects of ionizing radiation on pure substances in the gaseous, liquid and solid states, and on aqueous solutions. These experiments of necessity involved what would be considered now as extremely high doses. In the absence of our modern sophisticated knowledge of molecular and cellular biology, books on radiobiology wrote about the actions of radiation on the chemical constituents of living systems. Lea decided that the actions of radiation on dilute aqueous solutions could be described as indirect actions since most of the molecules of solute that reacted had not been excited or ionized directly by the radiation, but their reaction followed excitation or ionization of the solvent molecules. Lea noted that, although the total energy dissipated by the radiation per gram of solution did not vary with the concentration, the energy dissipated in the solute per gram of solution was proportional to the concentration, and with dilute solutions was only a minute fraction of the total energy dissipated per gram of solution. Lea concluded "we see that the weight of solute reacting is proportional not to the energy dissipated directly in the solute alone, but to the energy dissipated in the solution altogether". However, Lea also noted that, at sufficiently low concentrations, the ionic yield did not remain constant but diminished with diminution of the dose.

At this time there was a hypothesis of activated water following irradiation. This was an intermediary body of finite life, which caused reactions in many solutes. With regard to the total reaction, Lea was able to state that one of the characteristics of indirect action was that it was always proportional to dose.

Early radiobiologists took it for granted that the biological effects of ionizing radiation were due to the chemical changes induced by the radiation, but noted there was a problem in explaining why marked biological changes were produced by doses of radiation that produced only small degrees of chemical change.

There were various hypotheses to explain this, including cell poisons (products of cellular decomposition) and activated water mechanisms (depending on the sensitivity of enzymes in a cell being greater than enzymes in a concentrated solution). One hypothesis was that the direct action of radiation was inversely proportional to the molecular weight, and that this could be related to the localization of the radiation damage (e.g., in chromosomes).

This led into the concept of target theory, also initially known as "Treffertheorie" and discussed in the book by Timofeeff-Ressovsky and

Zimmer. It was a mechanistic theory to explain the different ionic efficiencies of different radiations, and involved a calculation on the size of the target molecule or structure from the proportion of effected organisms after a given dose of radiation.

However, even using target theory, it was necessary to postulate the spread of the ionization. To quote Lea, "transference of energy from one part of a molecule to another is a process known to occur, and capable of interpretation on current quantum-mechanical theory". This is the part of target theory that tends to be forgotten in modern texts.

One attempt to explain this was to regard radiation as a point source of heat, although Gray [3] favored the production of active radicals (OH and H).

At this stage target theory was still a theory, and only applicable to certain aspects of radiation damage, such as mutation or chromosome breakage. As Lea stated, "There are many actions of radiation on living organisms which are not to be interpreted on the target theory". Lea considered it more of a working hypothesis, and accepted there were many confounding factors, such as surrounding tissue effects. Target theory would apply where the biological effect was due to a single inactivation (viruses); chromosome aberrations (if the particle was densely ionizing); and if a large number of ionizations must be produced within the target (bacteria, bean seeds, yeast). Many scientists were not enamored of this "multi-hit" target theory, mainly because target theory was based on the exponential cell killing of radiation. However, a multi-hit model would explain a shouldered survival curve. Equally, a shouldered survival curve could be explained as a technical artifact due, for example, to clumping of cells, different intrinsic radiosensitivities, or the activation of different methods of cell kill at higher doses. The one-ionization event that Lea favored meant that time was irrelevant, in that further radiation events were immaterial.

Having a target theory hypothesis meant that differing energies and types of radiation could be used to identify the target size and hence the target identity. Much work was done on direct and indirect actions of radiation causing the inactivation of viruses. For example, Lea noted that when the mosaic tobacco virus was irradiated in several different concentrations and also dry, the inactivation dose was the same in dry and in concentrated solution, but was less in dilute solution. The addition of gelatin to the dilute solution resulted in the activation dose being raised almost to the value for the dry preparation. Lea concluded that in dilute solutions the indirect action of radiation predominated, but that it could be inhibited by a sufficient concentration of protein (virus or gelatin).

At that time the structure of DNA was not known, but the stage was set for its adoption as the critical "target" for radiation induced biological effects by the formalization of target theory, which basically said that energy deposition had to occur in a target in the cell, and that if the energy was deposited in a critical target it caused a biological effect. Chromosomes were known to be damaged by radiation and it was logical therefore to look at DNA as a candidate target. Equally logical was the idea that reproductive failure had to mean that the genome was the target. Thus, the nucleus was assumed to contain or be the target and many experiments supported this view (reviewed in [4]). However, as Tikvah Alper pointed out [5], it is more correct to consider the entire replicative machinery in the cell as a target, in that an energy deposition in any sensitive organelle essential for cell replication will cause reproductive death. The focus on DNA and later on DNA double-strand breaks (DSBs) can be traced to mathematical interpretation and formalism of target theory by Kellerer and Rossi [6] and by Chadwick and Leenhouts [7] among others, and also to classic experimental evidence correlating DSBs in a quantitative fashion with dose [8]. This correlation still forms the mechanistic justification for the use of a linear, no-threshold model in radiation protection. Later investigations of repair of DSBs served to consolidate the evidence that these were the critical lesion caused by radiation, which, if left unrepaired or if mis-repaired, led to the observable and dose-dependent biological consequences [9]. The central importance of DSBs dominated radiobiology during the 1970s and 1980s, although other "targets" were suggested such as membranes [10, 11]. These are reviewed in the next section. Critically, for the development of the field in general, unrepaired DNA DSBs were considered lethal and cells containing them did not reproduce. Quantitative assays of cell survival deemed that cells that reproduced five to six times were "survivors", in that they did not contain a lethal lesion [12]. How the conceptual leap from "survivor" to "undamaged, perfectly normal cell" occurred is not clear, but the assumption is made in Elkind and Whitmore's book [12] that cells that have survived to form a colony are normal, and will not show any effect of the progenitor irradiation. The evidence was there that this was not as simple as it seemed, but it appears to have been largely disregarded [13–18]. The only serious challenge to the DNA DSB paradigm came from proponents of various types of repair models or pool models [19–21]. These suggested that the final expression of the dose-response relationship was determined more by how the cells coped with the damage than with the quantitative amount of energy deposition or consequent DSBs. Cellular "fitness" was an issue and proponents of these models placed great emphasis on the low-dose effects, which predominated in the shoulder region of the semi-log plot of the traditional dose-response curve [22]. These models still regarded the cell survival curve as an "inactivation curve", in that increasing radiation doses inactivated increasing numbers of cells. The theory was that inactivation could be reduced by various intervention strategies.

Review of the evidence for radiobiological effects due to cellular damage or responses outside the nucleus

By far the most radiobiological work in this area deals with radiation effects in the cell nucleus. Thousands of papers have been published compared with only a few hundred dealing with membrane or mitochondrial damage, and less than 20 dealing with effects involving other organelles such as lysosomes, Golgi apparatus, ribosomes and endoplasmic reticulum. When reviewing radiation effects in cellular organelles, it is important to stress that the cell is an integrated functional unit, operating in multicellular organisms, within a tissue structure. Experiments looking at specific effects in specific organelles must be analyzed in context. What happens in one situation cannot be taken as necessarily being indicative of a global mechanism. This is particularly true at low doses of exposure, where hierarchical controls of survival and response at the tissue level, dictate much of what happens at the level of the individual cell. This is discussed later in the chapter.

Most work on cellular organelles other than the nucleus relates to radiation-induced membrane damage or compromised membrane function (reviewed in [22–25]). Often, however, membrane damage is seen as an indirect effect. Membranes are considered to be an important site of radical formation due to the lipids, which enable peroxy-radical formation [11]. These radicals are extremely toxic, causing DNA breakage. More recently, there has been consideration of the role of membrane channels and membrane-bound proteins in the modulation of radiation damage [25]. Again however, DNA damage is what is being modulated. Some possible indicators that membrane damage *per se* is determining cellular outcome following exposure are the reports by Gulbins and Kolesnick [26] and Lucero et al. [27] that ionizing radiation among other stimuli can affect raft formation in cell membranes, leading to transmembrane signaling. Benderitter et al. [28] show multiple changes in the phospholipids content of cell membranes after irradiation and consider that it must be considered as a critical target. Differences in radiation response between cells that can communicate through gap junctions and those that cannot [29] also support a direct role for cell membrane damage in determining response. "Response" here is tissue response, not individual cell response, thus the argument is semantic since it pivots on how radiation damage is defined–as a targeted dose deposition or as a final cellular outcome.

The literature concerning mitochondria is interesting. Mitochondria are crucial for oxidative metabolism, and thus have roles in energy production for repair, but are also the major site for the generation of oxidative stress. There are thousands of mitochondria in cells, which are derived almost exclusively from the maternal contribution to the zygote. Mitochondria contain their own DNA and have their own genes controlling key cellular functions but do not repair DNA damage very well [30, 31]. They have been well studied as candidates for expression of ionizing radiation damage and apoptosis. Most of the studies are concerned with the oxidative stress aspects of mitochondrial radiation damage, for reviews see [31–33] but some studies have actually looked at mitochondrial DNA damage [34]. The problem is that with so many mitochondria in a cell, propagation of mutations in mitochondrial DNA is considered unlikely as a mechanism of fixation of DNA damage at the cellular level. It is very likely that mitochondrial responses, including initiation of apoptosis and generation of reactive oxygen species (ROS), are key determinants of radiation outcome, but whether mitochondria are "targets" in the classical sense, in that direct energy deposition within their structure is required, is not clear.

Lysosomes are important because they release degradative enzymes following cellular injury. Only six published studies of lysosomal activity following irradiation could be found and these dealt mainly with increased numbers of lysosomes occurring after high-dose exposure [10, 35–37]. Direct damage to lysosomes has not been reported but the generation of lipid peroxy radicals or inflammatory responses involving lysosomes has been seen [38, 39].

In the very few studies of radiation effects involving the Golgi apparatus, the emphasis is on the response of Golgi-related enzymes to radiation [40–45]. The Golgi apparatus is critical for intracellular trafficking of repair proteins and the trans-Golgi-network (TGN) has important functions in controlling repair, cell cycle progression, cytokinesis and genome stability, and it is likely that studies of the role of the Golgi apparatus in these processes will increase. The fundamental role of the TGN in the cytoskeleton, particularly the assembly and disassembly of the microtubules, means that compromised Golgi function will impact on outcome following radiation exposure [43]. This is likely to be a secondary effect of oxidative stress, changed energy budgeting or induction of apoptosis pathways, and not due to primary energy deposition in Golgi structures [44].

Another cellular organelle that shows changed activity following irradiation is the endoplasmic reticulum (ER) [46–52]. This organelle is probably continuous with the nuclear membrane, and is a site of protein synthesis, assembly and degradation [51]. The ER is the organelle in which newly synthesized secretory and transmembrane proteins form their proper tertiary structure by post-translational modification, folding, and oligomerization. However, many of these proteins are unfolded or misfolded by extracellular or intracellular stimuli. The accumulation of misfolded proteins constitutes a risk for living cells. Eukaryotic cells possess at least three different mechanisms to adapt to ER stress and thereby survive: (1) translational attenuation to limit further accumulation of misfolded proteins; (2) transcriptional activation of genes encoding ER-resident chaperones; and (3) the ER-associated degradation (ERAD) pathway to restore the folding capacity. If the cells are exposed to prolonged or strong ER stress, the cells are destroyed by apoptosis. Recent evidence indicates that ER stress signaling pathways are mediated in part by several protein kinases and play an important role in the pathogenesis of neurodegenerative disorders. There has been considerable interest recently in ER stress, thought to arise from the physical overload of the ER with misfolded proteins after exposure to cellular insults requiring repair or activation of defense response pathways [52]. Papers considering ionizing radiation as the cellular insult are few [49–52], but clearly this site is important because mutations, particularly point mutations induced by ionizing radiation damage in DNA, are likely to lead to protein misfolding and consequent ER stress and apoptosis–a type of "cellular constipation". This again raises the issue of the nature of primary and secondary types of radiation damage. It could be argued that virtually all "damage" is secondary, and that effects such as DNA strand breaks that we ascribe to energy deposition in a target are actually resulting from apoptosis, triggered by cellular stress responses. An interesting point is that ER stress resulting from protein misfolding, while admittedly due to DNA mutations, is independent of the locus of the mutation. This again dissociates the specific site of energy deposition from the consequences.

New evidence for radiation effects in the absence of DNA strand break induction in the cell showing the effect

The fundamental paradigm shift from DNA-centered to response-driven radiobiology really started with the growing realization that all survivors are not equally normal or healthy, and that progeny of irradiated survivors exhibit many differences when compared to their unirradiated counterparts. The literature has references to this as early as 1964 [13–18], but the major lines of evidence came from the late 1980s and early 1990s when demonstrations by Seymour et al., Gorgogo and Little, Born and Trott, Mendonca et al., Streffer et al., Kadhim et al. and Marder and Morgan [53–59] using entirely different systems and endpoints, all showed a high frequency of non-clonal effects in the distant progeny of irradiated cells. These were variously called lethal mutations, genomic instability, delayed reproductive death or chromosomal instability, but basically all pointed to a persistence of expression of radiation damage in distant progeny of irradiated cells, deemed by conventional dogma of the time to have survived the dose. A further challenge to the neat association of DNA DSBs with outcome occurred in the 1990s with several papers showing that unirradiated cells that received signals from, or were in proximity to, irradiated cells, but in which no energy deposition had occurred, could demonstrate effects similar to those seen in irradiated cells [60–66]. The effects were also transmissible to progeny [67–69]. These changes in unirradiated cells known (perhaps incorrectly) as "bystander effects", make it difficult to talk of targets for radiation interaction unless the response to the energy deposition can be assessed at the level of the population of cells rather than the individual cell. This whole field has been extensively reviewed [70–74], but it is important to point out that, again, the historical evidence for what were then known as abscopal and clastogenic effects was in the literature since 1954 [75–89]. The current popularity of all these "non-targeted" or "bystander" effects is attributable to the fact that they predominate in the low-dose region of the survival curve [90], and thus may have major implications for the understanding of mechanisms of radiation action following low-dose exposures. It is also because modern molecular biological techniques and sophisticated cell culture methods permit the detection of effects at these low doses, which could not be seen with less sensitive assays or the clonogenic cell lines available. That being said, it is clear that in the "pre genomic instability/bystander" era very good and careful research was done both using cell cultures and animal models (reviewed in textbooks such as Lea, Timofeeff-Ressovsky and Zimmer, Elkind and Whitmore, Bacq and Alexander, Alper, and Hall [1, 2, 5, 12, 91, 92]. Critical analysis of the data is required to try to integrate the "old" with the "new". In particular, primary sources of data need to be reexamined rather that reliance on reviews, because often there is a tendency to go from the particular to the general if it fits current theories. A good example of this is the dogma that there is an equal effect per fraction, i.e., that between doses there is full recovery of cells so that they respond to the next dose as if never irradiated. Very rigorous work by Elkind and Sutton [12, 93] defined this split-dose effect, which became known as "Elkind Recovery". The theory fitted many of the mathematical predictions and confirmed prevailing ideas that cells accumulated sub-lethal (single-strand breaks) DNA damage that, if time was allowed, could be completely repaired, restoring cells to their pre-irradiation state. Higher doses led to increased frequencies of lethal DSBs, and thus the recovery effect was confined to the shoulder region of the curve and did not alter the terminal slope of the curve [94]. Logically and experimentally, this was a generalization containing many assumptions from particular data sets generated from particular cell lines. The initial experiments were done with high doses and with CHO cells, which have a high plating efficiency. Many data did not fit, e.g., McNally et al. [95] showed that the iso-effect per fraction broke down after more than five fractions, Bryant [96] showed decreased sensitivity after split-dose irradiation and Alper suggested that *in vivo* the survival of clonogenic stem cells in tissues cannot be measured at all [5]. An alternative interpretation of the iso-effect per fraction data could be that a residual damage effect is cancelled out by an adaptive response involving availability of induced repair enzymes. This would obviously reach a limit, whereas the residual damage would continue to accumulate, leading to apparent initial adherence to an iso-effect law followed by increasing deviation as fraction number increased. This example is presented merely to suggest that the data are the data, it is the interpretation of the data that needs critical review.

Points of commonality: DNA damage induced by radicals, bystander effects in repair-deficient cell lines, induction of instability by DNA break-inducing chemicals, and prevention by radical scavengers

Following on from the previous paragraph, it is sometimes thought that the "new" radiobiology replaces the "old", but actually one of the most keenly argued points is whether DNA damage (some would say a DSB) is necessary somewhere in the system to trigger non-targeted effects. This has been addressed in several ways using cytoplasmic targeting with microbeams, medium transfer protocols, and endonucleases instead of radiation to induce breaks [62, 97, 98]. The data are controversial because even with cytoplasmic targeting, the possibility of some nucleic acid being irradiated cannot be excluded as mitochondria have DNA [99]. In the medium transfer experiments, broad field irradiation of cells is used to produce the signal, so DNA damage cannot be excluded as a cause of signal production here. Here clearly DNA damage is not necessary to make cells respond to the signal by showing increased mutations and induced genomic instability since signal recipients were not in the radiation field at all. The use of endonucleases produced contradictory data, Chang and Little [100] concluded that endonucleases could result in delayed effects, while Limoli et al. [101] cut DNA at several different sites and got no induction of chromosomal instability. They concluded that cutting DNA *per se* was not sufficient to produce delayed effects in their system. Work with Auger electrons [102] suggested that these strand-breaking electrons were operating via indirect mechanisms. Radical scavengers that prevent radical damage to DNA also reduce delayed effects [103, 104].

Some of the most convincing evidence that DNA damage is necessary, however remotely, comes from experiments by Nagazawa and Little and by Mothersill et al. [105, 106] showing that bystander signals from DNA repair-deficient cells are more toxic, in the sense that they produces greater effects than their wild-type line. Morgan's group also have evidence that chromosomally unstable colonies derived from cells with rearrangements produce a more toxic DIE factor than their wild-type relatives [107]. Other evidence comes from the ability of DNA strand-breaking chemicals such as bleomycin, but not DNA intercalating agents such as cis platinum, to produce delayed effects [108]. Radical scavengers also reduce delayed effects [109–111]. Work reported very recently by Suzuki et al. [112] using the γH_2AX assay shows that signals form irradiated cells can induce γH_2AX activity in unirradiated bystander cells. Thus, it would appear that DNA damage is associated with the production of delayed effects. The question is whether this damage is just a manifestation of free radical damage [112–116] or whether among the DNA damage specific lesions such as DSBs or clustered damage is required [117, 118]. This question has not been satisfactorily resolved.

Types of delayed and bystander-induced effects and their time of occurrence in the context of damage processing by cells

One of the most often reported delayed or bystander signal-associated consequence of low-dose radiation exposure of cells is apoptosis or other forms of programmed cell death, such as terminal differentiation [119]. Apoptosis is often referred to as a protective mechanism, in that it removes dead or damaged cells from the population [120]. This shifts the emphasis in the delayed effect field, and requires consideration of the possibility that damage management at the cell population level is actually accounting for many of the low-dose effects being observed as discussed in the next section. Such management would imply that cell communication via signaling pathways is coordinating the radiation response at low doses. This type of response is well accepted in many areas of biology and involves, for example, cytokines, ceramides and other molecules associated with intercellular communication in tissues. This is outside the scope of the present review but is discussed extensively in [121] and is considered in [122] in relation to radiation response. The idea has not gained much favor in radiobiology, probably because of the association of dose (energy deposition) with damage in a one-to-one relationship, but it makes sense that at low doses where cellular energy supplies are adequate to cope with the damage caused by energy deposition, there would be coordination among the cells in a population or tissue to prevent energy wastage on non-important or unrepairable lesions, and a shuttling of cellular energy into the most productive areas. Again this type of preferential distribution of available resources is not novel and is well established in other areas of biology [123]. Obviously, once the damage burden exceeds a certain level in a tissue or cell population, other mechanisms are needed to salvage tissue or population functionality. Figures 1 and 2 represent an attempt to conceptualize these ideas. This analysis would predict late effects as a consequence of population tolerance of a certain amount of damage, which could be dealt with later at a time when population reserves of energy and substrates needed for apoptosis or repair were replenished. Thus, the time component of response becomes established in the context of energy supply and the kinetics of the "coping mecha-

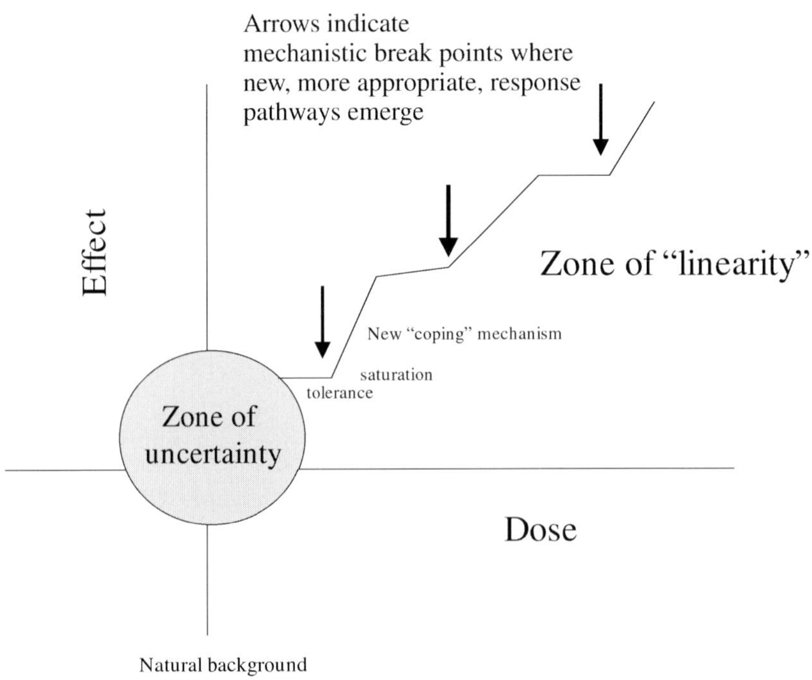

Figure 1. Proposed dose-response relationship for radiation-induced effects.

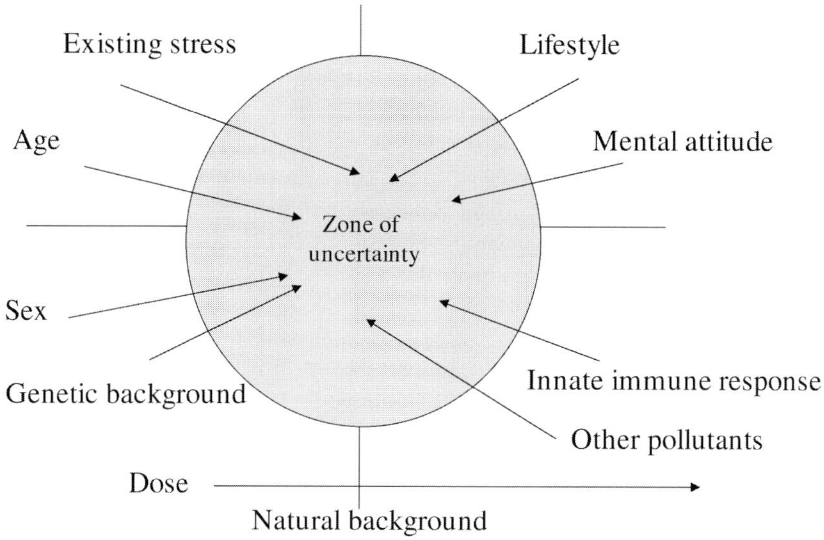

Figure 2. Factors influencing outcome in the zone of uncertainty.

nisms". It is often forgotten that repair processes take time. Protein synthesis is complex and substrate mobilization is not instantaneous [124]. Cells have a limited capacity at the purely physical level of space in cellular organelles, such as the ER as discussed previously and in reviews [125, 126]. At the population level, disposal of dead cells and general "clean-up" is an issue. Effects in progeny or transgenerational effects would occur when cell division occurred before repair was complete.

Hormetic effects of radiation: adaptive responses and positive effects induced by DNA damage

Many theories have been advanced to explain how genomic instability effects, such as delayed death, chromosomal instability or mutation, occur *de novo*, several generations after exposure. These include induction of mutations in mutator phenotype genes or repair genes [127], but mechanisms relying on gene mutations induced at the time of exposure, cannot explain the persistent, high frequency of the event or its species wide occurrence. They also cannot explain transgenerational instability occurring in the offspring of irradiated male mice [128, 129]. The most convincing mechanism suggested initially by Wright's group [113] and later confirmed by many other laboratories is that elevation of oxidative stress occurs after exposure to low levels of toxins and that this persists in progeny for several generations. The ROS damage DNA causing breaks, which can in certain circumstances become fixed and lead to deleterious mutations. In any case the repair of these breaks places a burden on cellular energy and protein production systems. The alteration in energy utilization patterns and the associated enhanced demand for protein synthesis carries with it further generation of ROS due to the higher metabolic rate. Questions remain concerning the driver of the oxidative stress, how it persists in distant progeny and most importantly, why it is not selected against in systems where enhanced cell death, deleterious mutations and reduced growth occur. Clearly, the ROS driver must be continuously present and must be acting at the population level not at the cell level. A candidate driver is the bystander signal pathway. Most of the evidence comes from radiobiology because only in this system, can cells be uniquely targeted. The important point is that bystander signals induce genomic instability-like effects, and appear to do this by elevating ROS in the recipient cells [130, 131]. The recipient cells in turn produce bystander signals, which could both temporarily and spatially distribute and maintain the instability phenotype in a population. The key question of why this mechanism exists at all and why genomic instability persists at roughly the same level in distant progeny is without an answer at present. The phenomenon occurs at doses of radiation, which do not normally kill cells. The bystander effect means that perfectly healthy, non-exposed cells are showing effects which should only occur if they received a dose of radiation. This has prompted us to consider the possibility that the process is beneficial at some level of organization that supercedes the individual cell. Many of the features of genomic instability suggest a permanently enhanced tolerance for mutations or plasticity in the progeny leading to enhanced frequencies of chromosome damage or death in the population [132]. This hypothesis was and still is very popular with Russian radiobiologists, and unfortunately the discussion has not been very accessible to scientists in the West as much of it was secret and, if published, was in Russian. Timofeeff-Ressovsky and colleagues [133] showed for the first time that the frequency of surviving cells carrying chromosome abnormalities did not increase, while the mitotic index did increase, as a result of exposing resting cells to low-dose irradiation, suggesting perhaps that an enhanced division rate was compensating for death (by apoptosis?) of cells carrying mutations. They also defined "cascade mutagenesis" in yeast, which is very similar to what we call genomic instability and is again evidence of plasticity of the genome occurring after exposure to low doses [134–139]. We suggest that genomic instability may represent part of the mechanism by which adaptation to altered environmental conditions is achieved at the population level. This "natural selection" is assumed to result from selection among the biodiverse population for those which happen to be best suited to the new challenge. We extend this hypothesis to suggest that the change in environmental conditions actively liberates the exposed population from the tight controls needed to maintain the stability of the DNA and consequent fitness of the population in its previous environment. We suggest a two-stage mechanism; part 1 involves the release of cells from the tight control which previously terminated any showing abnormalities in the genome, thus allowing a higher tolerance of mutation in the population (i.e., genomic instability). This method of inducing variety is well known in plant breeding where radiation is used to generate novel varieties of plants [140]. Stage two then involves selection from among the wider range of available phenotypes for those best suited to the new environmental conditions. While this discussion and hypothesis has been generated based on information mainly produced from research on cells, it can obviously hold for organisms as well. Genomic instability and bystander signals are known to be produced *in vivo* and have been described in mammals, teleost fish and crustaceans, suggesting a very widespread distribution in the animal kingdom [141–144].

Is the central dogma challenged or is the system just more complex than previously recognized?

To summarize, this paper reviews the historical data leading to the conclusion that DNA damage, and, more particularly, DSBs are the critical site of energy deposition that causes death, mutation and chromosome aberration. It then considers whether is conclusion, upon which our radiation protection legislation is based, is challenged by the growing realization that at low doses much of the damage ascribed to radiation does not require deposition of ionizing radiation in the cell showing the effect. We conclude that at low doses, it is likely, but by no means certain, that direct DNA damage is necessary somewhere in the system, to trigger the non-targeted effects measured in unexposed cells, but the evidence that a DBS is required is weak. We stress the importance of radiation response and of the emergent properties of tissues and cell populations in determining the final outcome. The consequences of this more complex situation are that the uncertainty at low doses is probably not resolvable, and that, as shown in Figures 1 and 2, any outcome has a probability of occurring, which cannot be defined given our present understanding of these mechanisms.

Acknowledgements
We wish to acknowledge our funding agencies in Ireland (Science Foundation Ireland and Irish Cancer Research), and in Canada, (the Canada Chair program and the National Science and Engineering Research Council). We also wish to acknowledge the late Tikvah Alper, whose inspiration and willingness to look at and analyze data which did not fit laid the way for much of this field of research.

References

1 Lea DE (1962) *Actions of Radiations on Living Cells. 2nd edn.* Cambridge University Press, Cambridge, UK
2 Timofeeff-Ressovsky NW, Zimmer KG (1947) *Das Trefferprinzip in der Biologie.* S. Herzel Verlag, Leipzig
3 Gray LH (1954) Conditions which affect the biologic damage resulting from exposure to radiation. *Acta Radiol* 41: 63–83

4 Savage JRK (2002) Reflections and meditations upon complex chromosomal changes. *Mutation Res* 512: 93–109
5 Alper T (1979) *Cellular Radiobiology*. Cambridge University Press, Cambridge, UK
6 Kellerer AM, Rossi HH (1972) The Theory of Dual Radiation Action. *Curr Topics Radiat Res* 8: 85–158
7 Chadwick KH, Leenhouts HP (1973) A molecular theory of cell survival. *Phys Med Biol* 18: 78–87
8 Barendsen GW (1994) RBE-LET relationships for different types of lethal radiation damage in mammalian cells: comparison with DNA dsb and an interpretation of differences in radiosensitivity. *Int J Radiat Biol* 66: 433–436
9 Pieffer P (1988) The mutagenic potential of DNA double-strand break repair. *Toxicol Lett* 96–97: 119–129
10 AlperT (1977) The role of membrane damage in radiation-induced cell death. *Adv Exp Med Biol* 84: 139–165
11 Freeman BA, Crapo JD (1982) Biology of disease: free radicals and tissue injury. *Lab Invest* 47: 412–426
12 Elkind MM, Whitmore GF (1967) *The Radiobiology of Cultured Mammalian Cells*. Gordon and Breach, New York
13 Sinclair WK (1964) X-ray-induced heritable damage (small-colony formation) in cultured mammalian cells. *Radiat Res* 21: 584–611
14 Beer JZ, Szumiel I (1975) Heritable cell cycle disturbances and late recovery in x-irradiated murine lymphoma L5178Y-S cell populations *in vitro*. *Adv Exp Med Biol* 53: 497–509
15 Beer JZ (1968) Hereditary radiation-induced lesions in somatic mammalian cells. *Postepy Biochem* 14: 463–472
16 Joshi GP, Nelson WJ, Revell SH, Shaw CA (1982) Discrimination of slow growth from non-survival among small colonies of diploid Syrian hamster cells after chromosome damage induced by a range of x-ray doses. *Int J Radiat Biol Relat Stud Phys Chem Med* 42: 283–296
17 Todd P (1968) Defective mammalian cells isolated from x-irradiated cultures, *Mutat Res* 5: 173–183
18 Todd PW (1975) Heavy-ion irradiation of human and Chinese hamster cells *in vitro*. *Radiat Res* 61: 288–297
19 Powers EL (1962) Considerations of Survival Curves and Target Theory. *Physics Med Biol* 7: 3–28
20 Orr JS, Hope CS, Wakerley SE, Stark JM (1966) A metabolic theory of cell survival curves. *Phys Med Biol* 11: 103–108
21 Laurie J, Orr JS, Foster CJ (1972) Repair processes and cell survival. *Br J Radiol* 45: 362–368
22 Alper T, Gilles NE, Elkind MM (1960) The sigmoid survival curve in radiobiology. *Nature (Lond)* 186: 1062–1063
23 Leyko W, Bartosz G (1986) Membrane effects of ionizing radiation and hyperthermia. *Int J Radiat Biol Relat Stud Phys Chem Med* 49: 743–770
24 Radford IR (1999) Initiation of ionizing radiation-induced apoptosis: DNA damage-mediated or does ceramide have a role? *Int J Radiat Biol* 75: 521–528
25 Dainiak N, Tan BJ (1995) Utility of biological membranes as indicators for radiation exposure: alterations in membrane structure and function over time. *Stem Cells* 13 (Suppl 1): 142–152
26 Gulbins E, Kolesnick R (2003) Raft ceramide in molecular medicine. *Oncogene* 22: 7070–7077
27 Lucero H, Gae D, Tacioli GE (2003) Novel localization of the DNA-PK complex in lipid rafts: a putative role in the signal transduction pathway of the ionizing radiation response. *J Biol Chem* 278: 22136–22243
28 Bendritter M, Vincent-Genod L, Pouget JP, Voisin P (2003) The cell membrane as a biosensor of oxidative stress induced by radiation exposure: a multiparameter investigation. *Radiat Res* 159: 471–483
29 Azzam EI, de Toledo SM, Little JB (2003) Oxidative metabolism, gap junctions and the ionizing radiation-induced bystander effect. *Oncogene* 22: 7050–7057
30 Koehler CM, Bauer, Matthias F (eds): (2004) *Mitochondrial Function and Biogenesis Series*. Topics in Current Genetics, Vol. 8
31 Gaziev AI, Podlutskii AI (2003) Low efficiency of DNA repair system in mitochondria. *Tsitologiia* 45: 403–417
32 Ferlini C, D'Amelio R, Scambia G (2002) Apoptosis induced by ionizing radiation. The biological basis of radiosensitivity. *Subcell Biochem* 36: 171–186

33 Janssen YM, Van Houten B, Borm PJ, Mossman BT (1993) Cell and tissue responses to oxidative damage. *Lab Invest* 69: 261–274
34 Kang D, Hamasaki N (2003) Mitochondrial oxidative stress and mitochondrial DNA. *Clin Chem Lab Med* 41: 1281–1288
35 Hume SP, Field SB (1977) Acid phosphatase activity following hyperthermia of mouse spleen and its implication in heat potentiation of X-ray damage. *Radiat Res* 72: 145–153
36 Abok K, Rundquist I, Forsberg B, Brunk U (1984) Dimethylsulfoxide increases the survival and lysosomal stability of mouse peritoneal macrophages exposed to low-LET ionizing radiation and/or ionic iron in culture. *Virchows Arch B Cell Pathol Incl Mol Pathol* 46: 307–320
37 Fatemi SH, Antosh M, Cullan GM, Sharp JG (1985) Late ultrastructural effects of heavy ions and gamma irradiation in the gastrointestinal tract of the mouse. *Virchows Arch B Cell Pathol Incl Mol Pathol* 48: 325–340
38 Lorimore SA, Coates PJ, Scobie GE, Milne G, Wright EG (2001) Inflammatory-type responses after exposure to ionizing radiation *in vivo*: a mechanism for radiation-induced bystander effects? *Oncogene* 20: 7085–7095
39 Tappel A (2005) Lysosomal enzymes and initiation of breast cancer. *Med Hypotheses* 64: 288–289
40 Yang VV, Stearner SP, Dimitrievich GS, Griem ML (1977) Radiation damage to the microvasculature in the rabbit ear chamber. An electron microscope study. *Radiat Res* 70: 107–117
41 Rich KA, Kerr JB, de Kretser DM (1979) Evidence for Leydig cell dysfunction in rats with seminiferous tubule damage. *Mol Cell Endocrinol* 13: 123–135
42 Zhang LL, Collier PA, Ashwell KW (1995) Mechanisms in the induction of neuronal heterotopiae following prenatal cytotoxic brain damage. *Neurotoxicol Teratol* 17: 297–311
43 Behrend L, Stoter M, Kurth M, Rutter G, Heukeshoven J, Deppert W, Knippschild U (2000) Interaction of casein kinase 1 delta (CK1delta) with post-Golgi structures, microtubules and the spindle apparatus. *Eur J Cell Biol* 79: 240–251
44 Bennett CB, Lewis LK, Karthikeyan G, Lobachev KS, Jin YH, Sterling JF (2001) Snipe JR, Resnick MA. Genes required for ionizing radiation resistance in yeast. *Nat Genet* 29: 426–434
45 Aonuma M, Nasu M, Iwata H, Yosue T (2004) Radioprotection of the murine submandibular gland by isoproterenol: autoradiography study with ^{3}H-leucine. *Odontology* 92: 14–21
46 Arendarcik J, Stanikova A, Rajtova V, Molnarova M (1983) Changes in the hypothalamus in continuously irradiated sheep. *Vet Med (Praha)* 28: 519–527
47 Khizhniak SV, Voitsitskii VM, Ostapchenko SG, Kucherenko NE (1990) The effect of ionizing radiation on Ca^{2+}-ATPase activity from the sarcoplasmic reticulum of rabbit skeletal muscles. *Ukr Biokhim Zh* 62: 58–63
48 Ueda T, Toyoshima Y, Moritani T, Ri K, Otsuki N, Kushihashi T, Yasuhara H, Hishida T (1996) Protective effect of dipyridamole against lethality and lipid peroxidation in liver and spleen of the ddY mouse after whole-body irradiation. *Int J Radiat Biol* 69: 199–204
49 Koufen P, Brdiczka D, Stark G (2000) Inverse dose-rate effects at the level of proteins observed in the presence of lipids. *Int J Radiat Biol* 76: 625–631
50 Schroder M, Kaufman RJ (2005) ER stress and the unfolded protein response. *Mutat Res* 569: 29–63
51 Rudner J, Lepple-Wienhues A, Budach W, Berschauer J, Friedrich B, Wesselborg S, Schulze-Osthoff K, Belka C (2001) Wild-type, mitochondrial and ER-restricted Bcl-2 inhibit DNA damage-induced apoptosis but do not affect death receptor-induced apoptosis. *J Cell Sci* 114: 4161–4172
52 Oloumi A, Lam W, Banath JP, Olive PL (2002) Identification of genes differentially expressed in V79 cells grown as multicell spheroids. *Int J Radiat Biol* 78: 483–492
53 Seymour CB, Mothersill C, Alper T (1986) High yields of lethal mutations in somatic mammalian cells that survive ionizing radiation. *Int J Radiat Biol* 50: 167–179
54 Gorgojo L, Little JB (1989) Expression of lethal mutations in progeny of irradiated mammalian cells. *Int J Radiat Biol* 55: 619–630
55 Born R, Trott KR (1988) Clonogenicity of the progeny of surviving cells after irradiation. *Int J Radiat Biol Relat Stud Phys Chem Med* 53: 319–330
56 Mendonca MS, Kurohara W, Antoniono R, Redpath JL (1989) Plating efficiency as a function of time postirradiation: evidence for the delayed expression of lethal mutations. *Radiat Res* 119: 387–393
57 Pampfer S, Streffer C (1989) Increased chromosome aberration levels in cells from mouse fetuses after zygote X-irradiation. *Int J Radiat Biol* 55: 85–92

58 Kadhim MA, MacDonald DA, Goodhead DT, Lorimore SA, Marsden SJ, Wright EG (1992) Transmission of chromosomal instability after plutonium -particle irradiation. *Nature* 355: 738–740
59 Marder BA, Morgan WF (1993) Delayed chromosomal instability induced by DNA damage. *Mol Cell Biol* 13: 6667–6677
60 Prise KM, Belyakov OV, Folkard M, Michael BD (1998) Studies of bystander effects in human fibroblasts using a charged particle microbeam. *Int J Radiat Biol* 74: 793–798
61 Nagasawa H, Little JB (1992) Induction of sister-chromatid exchanges by extremely low doses of alpha particles: *Cancer Res* 52: 6394–6396
62 Deshpande A, Goodwin EH, Bailey SM, Marrone BL, Lehnert BE (1996) Alpha-particle induced sister chromatid exchange in normal human lung fibroblasts: Evidence for an extranuclear target. *Radiat Res* 145: 260–267
63 Mothersill C, Seymour C (1997) Medium from irradiated human epithelial cells but not human fibroblasts reduces the clonogenic survival of unirradiated cells. *Int J Radiat Biol* 71: 421–427
64 Mothersill C, Seymour CB (1998) Cell-cell contact during gamma irradiation is not required to induce a bystander effect in normal human keratinocytes: evidence for release during irradiation of a signal controlling survival into the medium. *Radiat Res* 149: 256–262
65 Lorimore SA, Kadhim MA, Pocock DA, Papworth D, Stevens DL, Goodhead DT, Wright EG (1998) Chromosomal instability in the descendants of unirradiated surviving cells after alpha-particle irradiation: *Proc Natl Acad Sci USA* 95: 5730–5733
66 Zhou H, Randers-Pehrson G, Waldren CA, Vannais D, Hall EJ, Hei TK (2000) Induction of a bystander mutagenic effect of alpha-particles in mammalian cells: *Proc Natl Acad Sci USA* 97: 2099–2104
67 Seymour CB, Mothersill C (1997) Delayed expression of lethal mutations and genomic instability in the progeny of human epithelial cells which survived in a bystander killing environment: *Radiat Oncol Invest* 5: 106–110
68 Lewis DA, Mayhugh BM, Qin Y, Trott K, Mendonca MS (2001) Production of delayed death and neoplastic transformation in CGL1 cells by radiation-induced bystander effects. *Radiat Res* 156: 251–258
69 Watson GE, Lorimore SA, Macdonald DA, Wright EG (2000) Chromosomal instability in unirradiated cells induced *in vivo* by a bystander effect of ionizing radiation. *Cancer Res* 60: 5608–5611
70 Mothersill C, Seymour CB (2001) Radiation-induced bystander effects: past history and future directions. *Radiat Res* 155: 759–767
71 Morgan WF (2003) Non-targeted and delayed effects of exposure to ionizing radiation: I. Radiation-induced genomic instability and bystander effects *in vitro*. *Radiat Res* 159: 567–580
72 Mothersill C, Seymour C (2003) Low-dose radiation effects: experimental hematology and the changing paradigm. *J Exper Haematol* 31: 437–445
73 Lorimore SA, Wright EG. (2003) Radiation-induced genomic instability and bystander effects: related inflammatory-type responses to radiation-induced stress and injury? A review. *Internat J Radiat Biol* 79: 15–25
74 Little JB, Morgan WF (2003) Guest editors of special issue. Oncogene 13: 6977
75 Parsons WB, Watkins CH, Pease GL, Childs DS (1954) Changes in sternal bone marrow following roentgen-ray therapy to the spleen in chronic granulocytic leukemia. *Cancer* 7: 179–189
76 Souto J (1962) Tumour development in the rat induced by the blood of irradiated animals. *Nature* 195: 1317–1318
77 Hollowell JG, Littlefield LG (1967) Chromosome aberrations induced by plasma from irradiated patients. A brief report. *JSC Med Assoc* 63: 437–442
78 Goh K, Sumner H (1968) Breaks in normal human chromosomes: Are they induced by a transferable substance in the plasma of persons exposed to total-body irradiation? *Radiat Res* 35: 171–181
79 Hollowell JG, Littlefield LG (1968) Littlefield. Chromosome damage induced by plasma of X-rayed patient: An indirect effect of radiation. *Proc Soc Exp Biol Med* 129: 240–244
80 Littlefield LG, Hollowell JG, Pool WH (1969) Chromosomal aberrations induced by plasma from irradiated patients: An indirect effect of X-radiation. Radiology 93: 879–886
81 Scott D (1969) The effect of irradiated plasma on normal human chromosomes and its relevance to the long-lived lymphocyte hypothesis. *Cell Tissue Kinet* 2: 295–305
82 Goyanes-Villaescusa V (1971) Chromosomal abnormalities in lymphocytes of children and baby rabbits born from mothers treated by X-irradiation before pregnancy. A transplacentary plasmatic chromosome damage factor? *Blut* 22: 93–96

83 Demoise CF, Conard RA (1972) Effects of age and radiation exposure on chromosomes in a Marshall island population. *J Gerontol* 27: 197–201
84 Faguet GB, Reichard SM, Welter DA (1984) Radiation-induced clastogenic factors. *Cancer Genet Cytogenet* 12: 73–83
85 Pant GS, Kamada N (1977) Chromosome aberrations in normal leukocytes induced by the plasma of exposed individuals. *Hiroshima J Med Sci* 26: 149–154
86 Emerit I, Cerutti P (1981) Clastogenic activity from Bloom syndrome fibroblast cultures. *Proc Natl Acad Sci USA* 78: 1868–1872
87 Kahn SH, Emerit I (1985) Lipid peroxidation products and clastogenic in culture media of human leukocytes exposed to the tumour promoter phorbol-myristate-acetate. *Free Radic Biol Med* 1: 443–449
88 Emerit I (1994) Reactive oxygen species, chromosome mutation and cancer: A possible role of clastogenic factors in carcinogenesis. *Free Radic Biol Med* 16: 985–991
89 Emerit I, Arutyunyan R, Oganesian N, Levy A, Cerniavsky L, Sarkisian T, Pogossian A, Asrian K (1995) Radiation-induced clastogenic factors; Anticlastogenic effect of Ginkgo biloba extract. *Free Radic Biol Med* 18: 985–991
90 Seymour CB, Mothersill C (2000) Relative contribution of bystander and targeted cell killing to the low-dose region of the radiation dose-response curve. *Radiat Res* 153: 508–511
91 Bacq ZM, Alexander P (1955) *Fundamentals in Radiobiology.* Butterworth, London
92 Hall EJ (2000) *Radiobiology for the Radiologist.* 5th edn, Lippincott Williams and Wilkins, Philadelphia
93 Elkind MM, Sutton H (1959) X-ray damage and recovery in mammalian cells in culture. *Nature* 184: 1293–1295
94 Bender MA, Gooch PC (1962) The kinetics of x-ray survival of mammalian cells *in vitro*. *Int J Radiat Biol* 5: 133–145
95 McNally MJ, de Ronde J (1976) The effect of repeated small doses of radiation on recovery from sub-lethal damage by Chinese hamster cells irradiated in the plateau phase of growth. *Int J Radiat Biol* 29: 221–234
96 Bryant PE (1975) Decrease in sensitivity after split-dose recovery: evidence for the involvement of protein synthesis (letter). *Int J Radiat Biol Relat Stud Phys Chem Med* 27: 95–102
97 Wu LJ, Randers-Pehrson G, Xu A, Waldren CA, Geard CR, Yu Z, Hei TK (1999) Targeted cytoplasmic irradiation with alpha particles induces mutations in mammalian cells. *Proc Natl Acad Sci USA* 96: 4959–4964
98 Shao C, Folkard M, Michael BD, Prise KM (2004) Targeted cytoplasmic irradiation induces bystander responses. *Proc Natl Acad Sci USA* 101: 13495–13500
99 Singh KK (2004) Mitochondria damage checkpoint in apoptosis and genome stability. *FEMS Yeast Res* 5: 127–132
100 Chang WP, Little JB (1992) Persistently elevated frequency of spontaneous mutations in progeny of CHO clones surviving X-irradiation: association with delayed reproductive death phenotype. *Mutat Res* 270:191–199
101 Limoli CL, Kaplan MI, Phillips JW, Adair GM, Morgan WF (1997) Differential induction of chromosomal instability by DNA strand-breaking agents. *Cancer Res* 57: 4048–4056
102 Walicka MA, Adelstein SJ, Kassis AI (1998) Indirect mechanisms contribute to biological effects produced by decay of DNA-incorporated iodine-125 in mammalian cells *in vitro*: clonogenic survival. *Radiat Res* 149: 142–146
103 Suzuki K, Ojima M, Kodama S, Watanabe M (2003) Radiation-induced DNA damage and delayed induced genomic instability. *Oncogene* 22: 6988–6993
104 Kashino G, Prise KM, Schettino G, Folkard M, Vojnovic B, Michael BD, Suzuki K, Kodama S, Watanabe M (2004) Evidence for induction of DNA double strand breaks in the bystander response to targeted soft X-rays in CHO cells. *Mutat Res* 556: 209–215
105 Nagasawa H, Huo L, Little JB (2003), Increased bystander mutagenic effect in DNA double-strand break repair-deficient mammalian cells. *Int J Radiat Biol* 79: 35–41
106 Mothersill C, Seymour RJ, Seymour CB (2004) Bystander effects in repair deficient cell lines. *Radiat Res* 161:256–263
107 Nagar S, Smith LE, Morgan WF (2003) Characterization of a novel epigenetic effect of ionizing radiation: the death-inducing effect. *Cancer Res* 63: 324–328
108 Seymour CB, Mothersill C (1991) Chemotherapy agents and the induction of late lethal defects. *Anticancer Res* 11: 1605–1608

109 Bishayee A, Hill HZ, Stein D, Rao DV, Howell RW (2001) Free radical-initiated and gap junction-mediated bystander effect due to nonuniform distribution of incorporated radioactivity in a three-dimensional tissue culture model. *Radiat Res* 155: 335–344
110 Zhou H, Randers-Pehrson G, Waldren CA, Vannais D, Hall EJ, Hei TK (2000) Induction of a bystander mutagenic effect of alpha particles in mammalian cells. *Proc Natl Acad Sci USA* 97: 2099–2104
111 Przybyszewski WM, Widel M, Szurko A, Lubecka B, Matulewicz L, Maniakowski Z, Polaniak R, Birkner E, Rzeszowska-Wolny J (2004) Multiple bystander effect of irradiated megacolonies of melanoma cells on non-irradiated neighbours. *Cancer Lett* 214: 91–102
112 Suzuki K, Ojima M, Kodama S, Watanabe M (2005) Delayed activation of a DNA damage checkpoint and radiation-induced genomic instability. *Mut Res*; *in press*
113 Clutton SM, Townsend KM, Walker C, Ansell JD, Wright EG (1996) Radiation-induced genomic instability and persisting oxidative stress in primary bone marrow cultures. *Carcinogenesis* 17: 1633–1639
114 Albanese J, Dainiak N (2000) Ionizing radiation alters Fas antigen ligand at the cell surface and on exfoliated plasma membrane-derived vesicles: Implications for apoptosis and intercellular signaling. *Radiat Res* 153: 49–610
115 Limoli CL, Hartmann A, Shephard L, Yang CR, Boothman DA, Bartholomew J, Morgan WF (1998) Apoptosis, reproductive failure, and oxidative stress in Chinese hamster ovar cells with compromised genomic integrity. *Cancer Res* 58: 3712–3718
116 Rugo RE, Schiestl RH (2004) Increases in oxidative stress in the progeny of X-irradiated cells. *Radiat Res* 162: 416–425
117 Singleton BK, Griffin CS, Thacker J (1999) Clustered DNA damage leads to complex changes in irradiated human cells. *Cancer Res* 62: 6263–6269
118 Sutherland BM, Bennett PV, Sutherland JC, Laval J (2002) Clustered DNA damages induced by x-rays in human cells. *Radiat Res* 157: 611–616
119 Belyakov OV, Folkard M, Mothersill C, Prise KM, Michael BD (2003) A proliferation-dependent bystander effect in primary porcine and human urothelial explants in response to targeted irradiation. *Br J Cancer* 88: 767–774
120 Pollycove M, Feinendegen LE (2003) Radiation-induced *versus* endogenous DNA damage: possible effect of inducible protective responses in mitigating endogenous damage. *Hum Exp Toxicol* 22: 290–306
121 Atta-ur-Rahman, Harvey K, Siddiqui RA (1999) Interleukin-8: An autocrine inflammatory mediator. *Curr Pharm Des* 5: 241–253
122 Lorimore SA, Coates PJ, Scobie GE, Milne G, Wright EG (2001) Inflammatory-type responses after exposure to ionizing radiation *in vivo*: a mechanism for radiation-induced bystander effects? *Oncogene* 20: 7085–7095
123 Hanawalt PC (1990) Selective DNA repair in active genes. *Acta Biol Hung* 41: 77–91
124 Guillet C, Boirie Y, Walrand S (2004) An integrative approach to *in vivo* protein synthesis measurement: from whole tissue to specific proteins. *Curr Opin Clin Nutr Metab Care* 7: 531–538
125 Johnson ES (2004) Protein modification by SUMO. *Annu Rev Biochem* 73: 355–382
126 van Laar T, van der Eb AJ, Terleth C (2002) A role for Rad23 proteins in 26S proteasome-dependent protein degradation? *Mutat Res* 499: 53–61
127 Bignold LP (2004) Chaotic genomes make chaotic cells: the mutator phenotype theory of carcinogenesis applied to clinicopathological relationships of solid tumors. *Cancer Invest* 22: 338–343
128 Lord BI (1999) Transgenerational susceptibility to leukaemia induction resulting from preconception, paternal irradiation. *Int J Radiat Biol* 75: 801–810
129 Dubrova YE (2003) Radiation-induced transgenerational instability. *Oncogene* 22: 7087–7093
130 Lyng FM, Seymour CB, Mothersill C (2002) Early events in the apoptotic cascade initiated in cells treated with medium from the progeny of irradiated cells. *Radiat Prot Dosimetry* 99: 169–172
131 Lyng FM, Seymour CB, Mothersill C (2002) Initiation of apoptosis in cells exposed to medium from the progeny of irradiated cells: a possible mechanism for bystander-induced genomic instability? *Radiat Res* 157: 365–370
132 Bennett PM (2004) Genome plasticity: insertion sequence elements, transposons and integrons, and DNA rearrangement. *Methods Mol Biol* 266: 71–113
133 Timofeeff-Ressovsky NW, Poryadkova NA, Preobrazhenskaya EI (1950-4) *Influence of low-dose*

irradiation on development of plants, Reports, 1950–1954. Funds of Ural Department of USSR Acad Scie
134 Timofeeff-Ressovsky NV, Ivanov II, Korogodin VI (1972) *Die Anwendung des Trefferprinzips in der Strahlengbiologie*. VEB Gustav Fischer Verlag, Jena
135 Korogodin VI, KM Blisnik. (1972) Regularities of radiorace formation in yeasts. Comm.1. Radioraces of diploid yeasts Saccharomyces ellipsoideus vini. *Radiobiologiya* 12: 267–271
136 Korogodin VI., Blisnik KM, Kapultzevich YG (1977) Regularities of radiorace formation in yeasts. Comm.11. Facts and hypotheses, *Radiobiologiya* 17: 492–499
137 Korogodin VI, Korogodina VL, Fajszi CS (1990) Mutability of genes depends on their functional state - a hypothesis. *Biol Zentbl* 109: 447–451
138 Korogodin VI, Korogodina VL, Fajsz CS, Chepurnoy AI, Mikhova-Tzenova N, Simonyan NV (1991) On the dependence of Spontaneous Mutation Rates on the Functional State of Genes. *Yeast* 7: 105–118
139 Korogodin VI (1993) The study of post-irradiation recovery of yeast: the 'premolecular period'. *Mutat Res* 289: 17–26
140 Sandhu SS, Bastos CR, Azini LE, Tulmann Neto A, Colombo C (2002) RAPD analysis of herbicide-resistant Brasilian rice lines produced via mutagenesis. *Genet Mol Res* 1: 359–370
141 Lyng FM, Lyons-Alcantara M, Olwell P, Shuilleabhain SN, Seymour C, Cottell DC, Mothersill C (2004) Ionizing radiation induces a stress response in primary cultures of rainbow trout skin. *Radiat Res* 162: 226–232
142 Dowling K, Seymour CB, Mothersill C (2005) Delayed cell death and bystander effects in the progeny of Chinook Salmon Embryo cells exposed to radiation and a range of aquatic pollutants. *Int J Radiat Biol; in press*
143 O'Reilly JP, Mothersill C (1997) Comparative effects of UV A and UV B on clonogenic survival and delayed cell death in skin cell lines from humans and fish. *Int J Radiat Biol* 72: 111–119
144 Mothersill C, Lyng F, Mulford A, Seymour C, Cottell D, Lyons M, Austin B (2001) Effect of low doses of ionizing radiation on cells cultured from the hematopoietic tissue of the Dublin Bay prawn, Nephrops norvegicus. *Radiat Res* 156: 241–250

Edited by Leon P. Bignold

Viral carcinogenesis and genomic instability

Karl Münger[1] Hiroyuki Hayakawa[1], Christine L. Nguyen[1], Nadja V. Melquiot[1], Anette Duensing[2] and Stefan Duensing[2]

[1] *The Channing Laboratory, Brigham and Women's Hospital, 181 Longwood Avenue, Boston, MA 02115, USA*
[2] *Molecular Virology Program, University of Pittsburgh Cancer Institute, Pittsburgh, PA 15213, USA*

Abstract. Oncogenes encoded by human tumor viruses play integral roles in the viral conquest of the host cell by subverting crucial and relatively non-redundant regulatory circuits that regulate cellular proliferation, differentiation, apoptosis and life span. Human tumor virus oncoproteins can also disrupt pathways that are necessary for the maintenance of the integrity of host cellular genome. Some viral oncoproteins act as powerful mutator genes and their expression dramatically increases the incidence of host cell mutations with every round of cell division. Others subvert cellular safeguard mechanisms intended to eliminate cells that have acquired abnormalities that interfere with normal cell division. Viruses that encode such activities can contribute to initiation as well as progression of human cancers.

Key words: Aneuploidy, centrosomes, cervical cancer, human papillomavirus, tumor suppressor, viral oncogene.

Viruses and cancer

Viruses are obligatory intracellular parasites and hence their life cycles are irrevocably coupled to that of their host cells. Due to the limited coding capacity of viral genomes that is imposed by packaging limits, viruses have developed strategies to target host cellular regulatory structures and reprogram them for their own purposes. The interplay between a virus and its host cell is a fascinating area of study; during co-evolution host cells have developed intricate defense strategies to restrict viral replication, whereas the intruding viruses have evolved to thwart host cellular antiviral defense mechanisms (reviewed in [1]). Non-productive viral infections can arise if a host cell is intrinsically incapable of supporting the viral life cycle or if the viral genome is mutated and rendered replication defective. Under such conditions viral functions are aberrantly expressed but no infectious progeny is produced, which can have perilous consequences for the host cell (reviewed in [2]).

Approximately 20% of all human cancers may have a viral etiology (reviewed in [3]). The concept of viral carcinogenesis was originally derived from studies with animals where infectious entities, many of which were later identified as retroviruses, were shown to cause formation of malignant tumors. In some rare cases, retroviruses can contribute to carcinogenesis by insertional mutagenesis, where integration of the provirus causes high-level dysregulated expression of a cellular proto-oncogene or disruption of a tumor suppressor. More frequently, however, retroviruses "pick up" cellular proto-oncogene sequences during their replication cycles. Since this process is generally associated with concomitant deletion of viral coding sequences, many oncogenic retroviruses are intrinsically defective for completing the infectious life cycle, and require normal "helper" viruses for replication. Replication of retroviral genomes involves the viral reverse transcriptase enzyme that lacks proofreading mechanisms, and thus is considerably more error-prone than host chromosome replication. Moreover, the acquired host cell-derived sequences do not contribute to the viral life cycle and are not subject to the same degree of mutational restriction as the viral genome. Hence they will accumulate mutations at a significantly higher rate than the remainder of the retroviral genome. In rare cases, the resulting expression of specific mutated versions of such retrovirally transduced cellular genes can endow the infected host cell with a growth advantage relative to the surrounding uninfected cells, and a tumor may form (reviewed in [4]).

Even though the concept of oncogene activation and transmission by retroviruses has not been clearly documented in human cancers, the recognition that retrovirally transmitted oncogenes represent specifically altered versions of cellular proto-oncogenes had a major impact on our understanding of carcinogenic mechanisms. In human cancers, proto-oncogenes are frequently mutated and activated through cell intrinsic mechanisms, including point mutations, gene amplification, gene fusion, or alterations that lead to increased mRNA or protein stability. Moreover, activating mutations of oncogenes isolated from human cancers are often identical to those originally discovered with retrovirally activated oncogenes [5].

Oncogenes of human tumor viruses are virally encoded genes that play integral roles for the viral life cycle. To fulfill their roles in the viral life cycle, human tumor virus oncogenes target critical cellular regulatory circuits, including cellular proto-oncogenes and tumor suppressor pathways, and cause their activation or inactivation, respectively.

Some viral oncogenes also subvert cellular processes that are necessary for maintaining genomic integrity of the host cell. Hence, some human tumor viruses also contribute to human carcinogenesis by creating a cellular milieu that is conducive for the generation and accumulation of activating mutations of cellular oncogenes and/or inactivating mutations of tumor suppressors in the host genome. Such viruses contribute not only to initiation but also to progression of human cancer.

Human papillomaviruses and carcinogenesis

Papillomaviruses are ubiquitous non-enveloped viruses with small 8-kb double-stranded DNA genomes. Only one of the DNA strands is actively transcribed, and the complementary strand does not contain any coding information. The papillomavirus genome can be divided into three parts. The early coding region encodes approximately seven open reading frames (ORFs). Individual early ORFs are denoted by the letter "E" and a number according to their relative molecular size. The lower the number, the longer the corresponding ORF. The late coding region consists of two "L" ORFs, which encode the viral capsid proteins. The non-coding region contains multiple *cis* regulatory elements that modulate viral transcription and genome replication (Fig. 1A) (reviewed in [6]). Approximately 200 different human papillo-

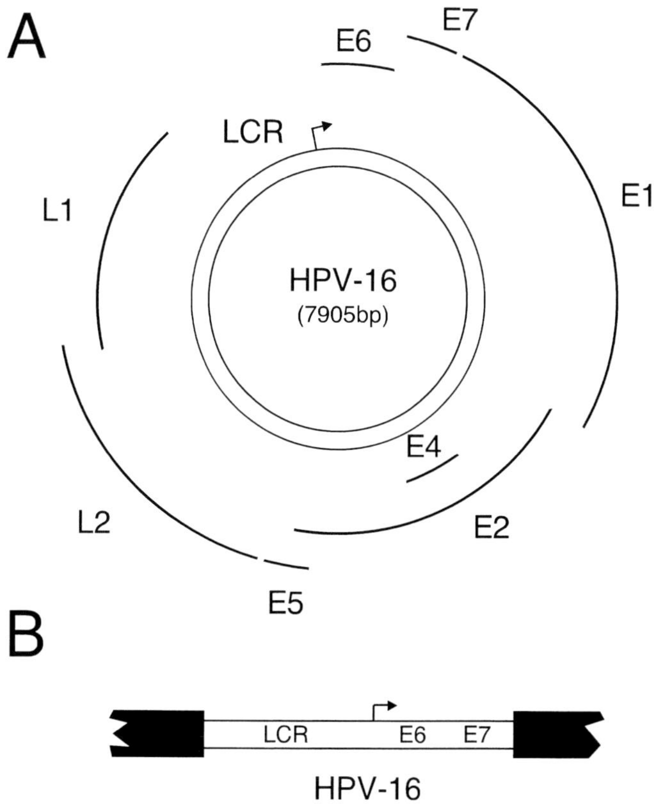

Figure 1. (A) Schematic depiction of the double-stranded circular genome of high-risk HPV-16. Only one of the two DNA strands is actively transcribed and contains all the coding information. The different early (E) and late (L) open reading frames are encoded using each of the different possible phases of translation as indicated by the concentric arcs. The long control region (LCR) does not contain extensive coding potential but contains various *cis* elements that are necessary for the regulation of viral transcription and replication. The position of the major early promoter within the LCR is indicated by an arrow. See text for details. (B) Representation of integrated HPV LCR/E6/E7 sequences in cervical cancer lines. The HPV genes are expressed from the viral promoter within the LCR (indicated by an arrow) expression is dysregulated due to transcriptional and non-transcriptional mechanisms. See text for details.

maviruses (HPVs) have been identified, and additional types likely exist (reviewed in [7]). HPVs display a pronounced tropism for squamous epithelial cells, and approximately 30 HPVs specifically infect mucosal epithelia. The mucosal associated HPVs are classified as "high-risk" and "low-risk" according to the propensity for malignant progression of the lesions that they cause. Low-risk HPVs cause benign warts, which have an extremely low risk for malignant progression. In contrast, infections with high-risk HPVs account for more than 99% of all cervical carcinoma. Worldwide, in excess of 470,000 cervical cancer cases are newly diagnosed each year, and cervical cancer remains a leading cause of cancer death in young women. Since no HPV-specific therapies exist, there are very limited regimens for treatment of late stage invasive cervical cancer, and the death rate has remained unacceptably high at approximately 30% (reviewed in [8]). Cervical cancer incidence is much lower in countries where there is broad access to preventive cytology-based screening programs that allow for detection of potentially pre-cancerous high-risk HPV-associated squamous intraepithelial lesions (SILs). In the US cervical carcinoma accounts for approximately 6% of all cancer cases (13500 per year), and remains frequent in medically underserved segments of the populations. Approximately 20% of human oral cancers, particularly oropharyngeal carcinomas, are also high-risk HPV positive [9]. A fraction of other anogenital tract malignancies such as penile cancer in males and vulvovaginal cancers in females (reviewed in [10]), as well as anal carcinomas that frequently occur in AIDS patients (reviewed in [11]), are also associated with high-risk HPV infections. Even though preventive vaccination strategies using recombinant empty capsid particles yielded promising results [12], it will be decades before they might have a major impact on the incidence of HPV-associated disease (reviewed in [13]).

Due to the small size of their genomes, HPVs do not encode key, rate-limiting replication enzymes, and thus these viruses have adopted a parasitic replication strategy to exploit the cellular DNA replication machinery. Establishing and maintaining a cellular environment conducive for viral genome synthesis is paramount since the HPV life cycle is tightly linked to the differentiation status of the infected keratinocyte. The squamous epithelium is a multilayered organ and only the basal layer contains undifferentiated, actively dividing cells. These cells are the initial targets for infection, and HPVs gain access to these cells either through an injury or at the squamo-columnar transformation zone where basal-like epithelial cells are more readily accessible. Expression of late genes and production of viral capsids, however, only occurs in differentiated epithelial cells. In a normal squamous epithelium, cellular differentiation and proliferation are tightly coupled processes, and cells terminally withdraw from the cell division cycle when they undergo differentiation. To allow for viral genome replication in these growth-arrested cells, HPVs encode regulatory proteins that can uncouple these processes (reviewed in [6]). Consistent with this notion, high-risk HPV E6 and E7 proteins functionally compromise the p53 and retinoblastoma (pRB) tumor suppressors, respectively [14, 15]. In addition, high-risk HPV E6 can activate transcription of hTERT, the catalytic protein subunit of human telomerase [16].

Infection of the anogenital tract with high-risk HPVs is through sexual contact and is quite frequent in the sexually active population [17]. Most infections with high-risk HPVs are transient and do not cause any clinical symptoms. Persistent high-risk HPV infections, however, can cause potentially pre-malignant SILs. Malignant progression of such lesions is an overall rare event that can take decades to occur (reviewed in [18]). An important hallmark of malignant progression of lesions caused by high-risk HPV infection is the frequent integration of the viral sequences into the host cellular genome. HPV genome integration does not cause insertional mutagenesis, and even though HPVs frequently integrate near common chromosomal fragile sites [19], there are no specific integration HPV sites in the human genome [20]. Integration disrupts the integrity of the viral genome and causes a cellular growth advantage due to increased mRNA stability and expression levels of the remaining viral transcripts [21]. Only the HPV E6 and E7 genes remain consistently expressed in HPV-positive cervical cancers (Fig. 1B). HPV E6 and E7 encode small proteins of approximately 100 and 160 amino acid residues, respectively. They lack enzymatic or specific DNA binding activities and appear to function by associating with cellular protein complexes, thereby subverting their biological functions (reviewed in [22]). Ectopic expression of high-risk HPV E6 and E7 in primary human epithelial cells causes life span extension and permits immortalization [23, 24]. When grown under organotypic conditions, HPV E6/E7-expressing keratinocytes form structures that are reminiscent of high-grade pre-cancerous lesions [25]. Moreover, when transgenic mice with expression of HPV-16 E6 and E7 targeted to basal epithelial cells are exposed to continuous low doses of estrogen, they develop cervical cancers that mirror the human disease [26]. Sustained expression of E6 and E7 is necessary for maintenance of the transformed phenotype of human cervical cancer-derived cell lines even after they have been in culture for decades and have accumulated a plethora of genomic alterations (reviewed in [6]).

Since the HPV genome suffers irreversible physical disruption as a consequence of integration into a host chromosome, carcinogenic progression of a high-risk HPV-infected cell is a terminal event and not part of the normal viral life cycle. Rather, malignant progression ensues as a result of dysregulated expression of HPV E6/E7 genes, which normally play essential roles for the viral life cycle (reviewed in [22]).

Remarkably, the majority of human solid tumors have abnormalities in the pRB and p53 tumor suppressor pathways and maintain stable telomere length (reviewed in [27]), illustrating the notion that the signal transduction pathways targeted by high-risk HPV E6 and E7 oncoproteins are also rendered dysfunctional in many other human solid tumors that are not HPV associated.

HPVs, genomic instability and malignant progression

Genomic instability is a defining characteristic of human solid tumors, and human cancer has been described as a disease of genomic instability [28]. There is much debate as to whether genomic destabilization mechanistically contributes to malignant progression, or if it arises as a consequence of cell cycle checkpoint abnormalities and/or continued division of cells containing chromosomes with eroded telomeres. The mutation rate of normal human cells is exceedingly low, and thus may not suffice to permit accumulation of the multitude of genetic alterations that are necessary for multistep human carcinogenesis (reviewed in [29]). Moreover, human cancer-like cells can be generated *in vitro* by targeting a minimal set of critical regulatory pathways including the pRB and p53 tumor suppressors, protein phosphatase 2A, and telomerase, all of which are also commonly rendered abnormal in human cancers (reviewed in [30]). Such *in vitro* generated human tumor-like cells retain a high degree of genomic stability [31]. This result lends powerful support to the notion that genomic stability does not inevitably ensue in cells with abnormal patterns of proliferation. Genomic instability may require additional alterations or a combination of molecular changes to enable an emerging tumor cell to accumulate the necessary oncogenic mutations, and hence may represent a vital step for cancer formation *in vivo* [32].

Multiple cooperating mechanisms likely contribute to genomic destabilization in tumors (reviewed in [33]). Subversion of "quality control" functions such as cell cycle checkpoints and DNA repair functions allow cells that have accumulated mutations to remain in the proliferative pool. Other oncogenic hits, however, may directly affect genomic instability by generating a mutator phenotype, which enhances the mutation rate at every round of DNA synthesis and cell division of an emerging tumor cell [34, 35].

Cervical cancers exhibit both structural and numerical chromosomal aberrations and genomic instability is observed in early pre-malignant lesions (reviewed in [36]). In addition, ectopic expression of high-risk HPV E6 and E7 in primary human cells can each interfere with genomic stability. In experiments where cells were selected for acquiring resistance to the drug *N*-phosphonoacetyl-L-aspartate (PALA), HPV E6-expressing cells accumulated structural chromosomal abnormalities, whereas numerical chromosomal abnormalities and aneuploidy emerged in HPV E7-expressing cells [37].

Tetrasomy/multinucleation

Morphological examination of HPV-associated cervical lesions revealed the presence of distinct nuclear abnormalities, including enlarged nuclei as well as multinucleation (reviewed in [38]). Enlarged nuclei are a hallmark of increased ploidy and numerical chromosomal abnormalities. Lesions caused by high-risk HPV infections, but not those associated with low-risk HPV infections, showed an increased degree of tetrasomy [39, 40]. Ectopic high-risk HPV E6 or E7 expression can each independently induce tetraploidization both in actively dividing basal and intrinsically growth-arrested suprabasal cells [41, 42]. The mechanism of tetrasomy induction by HPV oncoproteins has not been delineated. Expression of high-risk HPV E6 may cause increased ploidy and multinucleation through inactivation of the p53 tumor suppressor [43]. Interestingly, however, the ability of HPV E7 to induce tetrasomy is unrelated to the capacity to inactivate the retinoblastoma tumor suppressor [42].

Tetrasomy is often regarded as a prelude to aneusomy, since, as discussed in more detail later, such cells are more prone to mitotic errors when they undergo additional rounds of cell division (reviewed in [44]). Tetraploidy arises in cells that undergo DNA synthesis without completing nuclear and cellular division, most frequently as a consequence of cytokinesis problems. It is important to point out that a tetraploid cell will have to be able to successfully complete a subsequent cell division to generate aneuploid progeny. Cells that re-encounter cytokinesis problems, however, will become polyploid and/or multinucleated (reviewed in [44]). The emergence of cells with severe nuclear abnormalities may be of relevance diagnostically, but since such cells were generated through persistent cytokinesis defects, and thus are incapable of undergoing full cell division, they represent abortive structures [48, 49] that do not contribute to malignant progression.

Centrosome duplication errors, multipolar mitoses and aneusomy

Aneuploidy arises as a consequence of chromosome segregation errors during mitosis. One of the typical mitotic abnormalities that pathologists have observed in high-risk HPV-associated pre-malignant lesions and cancers are tri-polar mitotic figures (Fig. 2) [45]. Such abnormalities can arise when cells contain supernumerary mitotic spindle pole bodies, centrosomes. Centrosomes consist of two centrioles that are surrounded by a pericentriolar protein matrix, which functions to anchor microtubules during mitosis. Immediately after cell division each daughter cell contains a single centrosome, which undergoes semi-conservative duplication in exact synchrony with the cell division cycle. Entry into S phase of the cell division cycle is believed to generate a "licensing signal" that renders the centrosome competent for duplication. Once a centrosome is licensed for duplication the two centrioles separate and each serves as a template for synthesis of a single daughter centriole. Daughter centriole synthesis is complete at the end of S phase, and the resulting two centrosomes form the mitotic spindle pole bodies that are critical for bipolar mitosis (Fig. 3). The mechanistic details of this unique and important cellular process remain enigmatic (reviewed in [46]).

Abnormal centrosome numbers can arise by two principal mechanisms. As described in the previous section, a cell that experiences cytokinesis problems

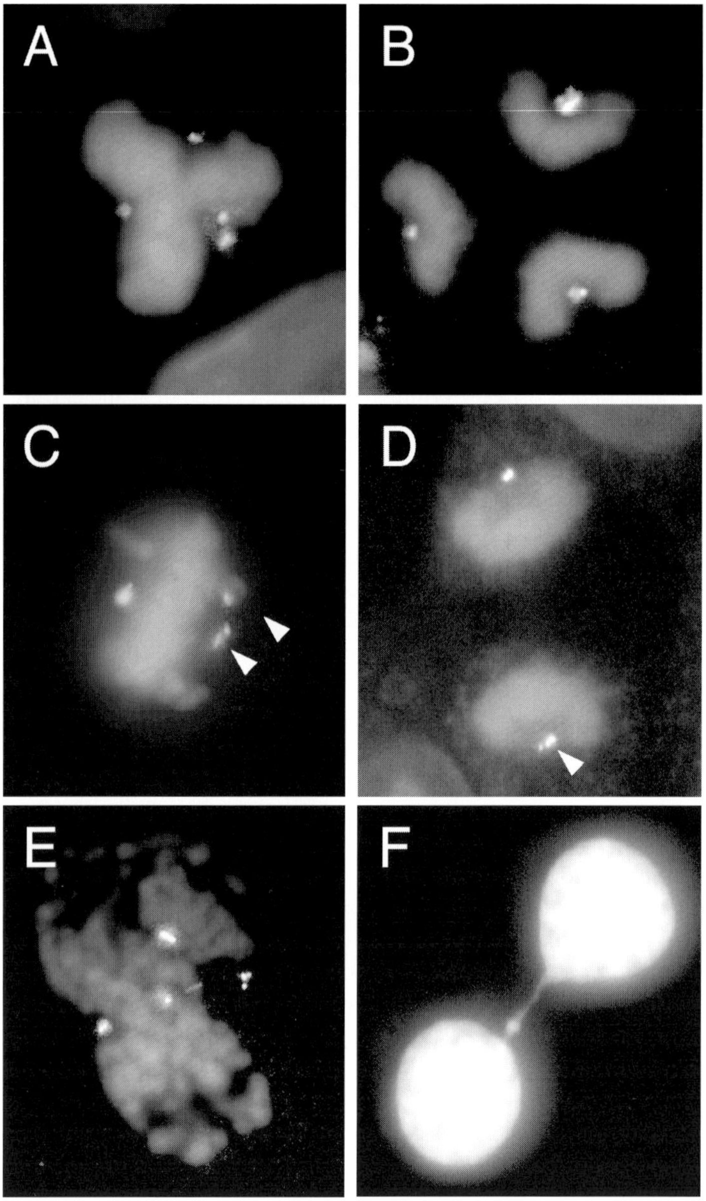

Figure 2. Examples of mitotic abnormalities in HPV-16 oncogene-expressing cells. HPV-16 E7 oncogene expression causes centrosome duplication errors, which can give rise to tripolar metaphases (A), which can undergo anaphase progression (B). Multiple centrosomes (indicated by arrowheads) in HPV-16 E7-expressing cells can undergo coalescence and form abnormal bipolar metaphases (C) and anaphases (D). Highly irregular multipolar metaphase in HPV-16 E7-expressing cells (E). HPV oncogene-expressing cells also contain centrosome-independent mitotic abnormalities including anaphase bridges (F). These may be caused by dicentric chromosomes that might have formed as a consequence of breakage fusion bridge cycles. See text for details.

progresses into a G_1-like phase without completing cellular and/or nuclear division (Fig. 3). Hence, the resulting tetraploid and/or binucleated cell contains two centrosomes. If such a cell re-enters the cell division cycle, it will also duplicate its centrosomes and enter mitosis with four centrosomes. Each individual centrosome may then act as a mitotic spindle pole body and a tetrapolar mitotic spindle can form. Some centrosomes may not fully separate and multiple centrosomes may form a single mitotic spindle pole leading to tripolar or bipolar mitosis (Fig. 2). If such cells can complete cell division and do not re-encounter cytokinesis problems (which caused the formation of the tetraploid cell during the previous round of the cell division cycle), there is an increased probability for chromosome missegregation, leading to formation of aneuploid daughter cells (reviewed in [47]). In this scenario, centrosome abnormalities are generated through normal duplication cycles and arise as a consequence of cytokinesis defects and not through aberrant synthesis. In many cases, however, including in cells lacking p53 tumor suppressor function, cytokinesis errors persist [48]. Such cells may be able to reduplicate their chromosomes and centrosomes but remain incapable of successfully undergoing cell division, and acquire progressive nuclear abnormalities including large multilobulated nuclei, multiple nuclei or micronuclei (Fig. 3). Such abnormal cells are ultimately removed form the proliferative pool by apoptosis, senescence or other abortive mechanisms, and hence are unlikely to contribute to carcinogenic progression [48, 49].

Centrosome abnormalities, however, may also emerge as a primary defect in cells through uncoupling of centrosome duplication from the cell division cycle. In such a scenario, a single maternal centriole may serve as a template for the synthesis of more than one daughter centrioles during S phase, or alternatively newly formed daughters may be immediately "licensed" to serve as templates for the synthesis of granddaughters (Fig. 3). In the late 19th century, the eminent German embryologist Theodor Boveri performed studies with polyspermic embryos and recognized that abnormal centrosome numbers gave rise to abnormal multipolar mitoses [50, 51], which severely compromised the viability of the resulting embryos. Based on earlier observations that human cancer cells frequently displayed abnormal multipolar mitotic figures [52], he first postulated that "a single multipolar mitosis going on in a healthy tissue, caused perhaps by the simultaneous multiple division of the centrosome, might produce the primordial cell of a malignant tumor" [53]. This attractive hypothesis, however, remains largely unproven experimentally.

Like many other human tumors, cervical cancers as well as high-risk HPV-associated pre-malignant lesions contain centrosome abnormalities [49], and their incidence appears to increase in parallel with malignant progression [54]. In addition to numerical aberrations, high-risk HPV-associated lesions also display structural centrosome abnormalities, including excess pericentriolar material [49], but the mechanistic basis of the structural defects has not been assessed in detail. Cervical cancer is unique among human solid tumors in that the fundamental carcinogenic insult that causes these tumors, infection with

sentinel response in aberrantly proliferating cells. Abrogation of cell cycle checkpoints, particularly those that ensure mitotic fidelity, may be necessary for perpetuation of genomic instability, and hence it may be argued that at later stages of carcinogenic progression the mutator activity provided by such infectious entities may no longer be necessary. Since excessive genomic instability ("error catastrophe", [101]) may ensue when the initiating mutator activity is retained there may in fact be a powerful evolutionary advantage for such a stimulus to be removed. Although attractive, such a model will be very difficult to prove in experimental models or through epidemiological studies.

Concluding remarks

Studies with DNA tumor virus oncogenes have been instrumental in the discovery of critical growth regulatory pathways that control proliferation, apoptosis and differentiation in normal human cells. Based in part on these discoveries it has been possible to define a minimal set of regulatory nodes that are rendered dysfunctional in almost any human solid tumor [27]. Tumorigenic human cell populations can be generated when these pathways are disrupted *in vitro* (reviewed in [30]). Unlike naturally occurring human tumors such artificially generated human tumor-like cells do not exhibit marked genomic instability [31]. Thus, it appears that genome destabilization may be a necessary step to set the stage for carcinogenic progression. It is an exciting possibility that the study of viral oncoproteins will once again be instrumental in discovering cellular pathways that control the genomic integrity, and that these studies will have important ramifications for our understanding of tumorigenic pathways. Cellular processes that control genomic instability may also be attractive targets for development of novel anticancer therapies. Inhibition of genomic instability in early pre-malignant lesions may restrain malignant progression, whereas therapeutic interventions that lead to increased genomic destabilization in later stage tumors may create genomic chaos [101] that could interfere with clonal expansion and the viability of the tumor [29].

Acknowledgements
The work on HPV-induced genomic destabilization in K.M.'s laboratory is supported by grants CA81135 and CA66980 from the National Institutes of Health, and a grant from CYTYC Corporation. C.L.N. is supported by NIH training grant T32 CA09031. S.D.'s and A.D.'s laboratory is supported by a Special Innovation Award from the PNC Foundation and a Pilot Project Developmental Grant from the University of Pittsburgh Cancer Institute.

References

1 Weinberg RA (1997) The cat and mouse games that genes, viruses, and cells play. *Cell* 88: 573–575
2 zur Hausen H (2001) Proliferation-inducing viruses in non-permissive systems as possible causes of human cancers. *Lancet* 357: 381–384

3 zur Hausen H (2001) Oncogenic DNA viruses. *Oncogene* 20: 7820–7823
4 Nevins JR (2001) Cell Transformation by Viruses. In: DM Knipe, PM Howley (eds): *Fields Virology*. Lippincott-Williams and Wikins, Philadelphia, 245–283
5 Parada LF, Tabin CJ, Shih C, Weinberg RA (1982) Human EJ bladder carcinoma oncogene is homologue of Harvey sarcoma virus ras gene. *Nature* 297: 474–478
6 Howley PM, Lowy DR (2001) Papillomaviruses and their replication. In: DM Knipe, PM Howley (eds): *Fields Virology*. Lippincott Williams and Wilkins, Philadelphia, 2197–2229
7 de Villiers EM, Fauquet C, Broker TR, Bernard HU, zur Hausen H (2004) Classification of papillomaviruses. *Virology* 324: 17–27
8 zur Hausen H (2002) Papillomaviruses and cancer: from basic studies to clinical application. *Nat Rev Cancer* 2: 342–350
9 Gillison ML, Koch WM, Capone RB, Spafford M, Westra WH, Wu L, Zahurak ML, Daniel RW, Viglione M, Symer DE et al. (2000) Evidence for a causal association between human papillomavirus and a subset of head and neck cancers. *J Natl Cancer Inst* 92: 709–720
10 Crum CP, McLachlin CM, Tate JE, Mutter GL (1997) Pathobiology of vulvar squamous neoplasia. *Curr Opin Obstet Gynecol* 9: 63–69
11 Klencke BJ, Palefsky JM (2003) Anal cancer: an HIV-associated cancer. *Hematol Oncol Clin North Am* 17: 859–872
12 Koutsky LA, Ault KA, Wheeler CM, Brown DR, Barr E, Alvarez FB, Chiacchierini LM, Jansen KU (2002) A controlled trial of a human papillomavirus type 16 vaccine. *N Engl J Med* 347: 1645–1651
13 Frazer IH (2004) Prevention of cervical cancer through papillomavirus vaccination. *Nat Rev Immunol* 4: 46–54
14 Dyson N, Howley PM, Münger K, Harlow E (1989) The human papillomavirus-16 E7 oncoprotein is able to bind to the retinoblastoma gene product. *Science* 243: 934–937
15 Werness BA, Levine AJ, Howley PM (1990) Association of human papillomavirus types 16 and 18 E6 proteins with p53. *Science* 248: 76–79
16 Klingelhutz AJ, Foster SA, McDougall JK (1996) Telomerase activation by the E6 gene product of human papillomavirus type 16. *Nature* 380: 79–82
17 Winer RL, Lee SK, Hughes JP, Adam DE, Kiviat NB, Koutsky LA (2003) Genital human papillomavirus infection: incidence and risk factors in a cohort of female university students. *Am J Epidemiol* 157: 218–226
18 Lowy DR, Howley PM (2001) Papillomaviruses. In: DM Knipe, PM Howley (eds): *Fields Virology*. Lippincott Williams and Wilkins, Philadelphia, 2231–2264
19 Thorland EC, Myers SL, Gostout BS, Smith DI (2003) Common fragile sites are preferential targets for HPV16 integrations in cervical tumors. *Oncogene* 22: 1225–1237
20 Wentzensen N, Vinokurova S, von Knebel Doeberitz M (2004) Systematic review of genomic integration sites of human papillomavirus genomes in epithelial dysplasia and invasive cancer of the female lower genital tract. *Cancer Res* 64: 3878–3884
21 Jeon S, Allen-Hoffmann BL, Lambert PF (1995) Integration of human papillomavirus type 16 into the human genome correlates with a selective growth advantage of cells. *J Virol* 69: 2989–2997
22 Munger K, Baldwin A, Edwards KM, Hayakawa H, Nguyen CL, Owens M, Grace M, Huh KW (2004) Mechanisms of human papillomavirus-induced oncogensis. *J Virol* 78: 11451–11460
23 Hawley-Nelson P, Vousden KH, Hubbert NL, Lowy DR, Schiller JT (1989) HPV16 E6 and E7 proteins cooperate to immortalize human foreskin keratinocytes. *EMBO J* 8: 3905–3910
24 Münger K, Phelps WC, Bubb V, Howley PM, Schlegel R (1989) The E6 and E7 genes of the human papillomavirus type 16 together are necessary and sufficient for transformation of primary human keratinocytes. *J Virol* 63: 4417–4421
25 McCance DJ, Kopan R, Fuchs E, Laimins LA (1988) Human papillomavirus type 16 alters human epithelial cell differentiation *in vitro*. *Proc Natl Acad Sci USA* 85: 7169–7173
26 Arbeit JM, Howley PM, Hanahan D (1996) Chronic estrogen-induced cervical and vaginal squamous carcinogenesis in human papillomavirus type 16 transgenic mice. *Proc Natl Acad Sci USA* 93: 2930–2935
27 Hanahan D, Weinberg RA (2000) The hallmarks of cancer. *Cell* 100: 57–70
28 Klausner RD (2002) The fabric of cancer cell biology-Weaving together the strands. *Cancer Cell* 1: 3–10
29 Cahill DP, Kinzler KW, Vogelstein B, Lengauer C (1999) Genetic instability and darwinian selection in tumours. *Trends Cell Biol* 9: M57–M60

30 Hahn WC, Weinberg RA (2002) Rules for making human tumor cells. *N Engl J Med* 347: 1593–1603
31 Zimonjic D, Brooks MW, Popescu N, Weinberg RA, Hahn WC (2001) Derivation of human tumor cells *in vitro* without widespread genomic instability. *Cancer Res* 61: 8838–8844
32 Nowak MA, Komarova NL, Sengupta A, Jallepalli PV, Shih Ie M, Vogelstein B, Lengauer C (2002) The role of chromosomal instability in tumor initiation. *Proc Natl Acad Sci USA* 99: 16226–16231
33 Kinzler KW, Vogelstein B (1997) Cancer-susceptibility genes. Gatekeepers and caretakers. *Nature* 386: 761, 763
34 Loeb LA (1991) Mutator phenotype may be required for multistage carcinogenesis. *Cancer Res* 51: 3075–3079
35 Tomlinson IP, Novelli MR, Bodmer WF (1996) The mutation rate and cancer. *Proc Natl Acad Sci USA* 93: 14800–14803
36 Duensing S, Munger K (2003) Centrosomes, genomic instability, and cervical carcinogenesis. *Crit Rev Eukaryot Gene Expr* 13: 9–23
37 White AE, Livanos EM, Tlsty TD (1994) Differential Disruption of Genomic Integrity and Cell Cycle Regulation in Normal Human Fibroblasts by the HPV Oncoproteins. *Genes and Develop* 8: 666–677
38 Duensing S, Munger K (2004) Mechanisms of genomic instability in human cancer: insights from studies with human papillomavirus oncoproteins. *Int J Cancer* 109: 157–162
39 Southern SA, Evans MF, Herrington CS (1997) Basal cell tetrasomy in low-grade cervical squamous intraepithelial lesions infected with high-risk human papillomaviruses. *Cancer Res* 57: 4210–4213
40 Giannoudis A, Evans MF, Southern SA, Herrington CS (2000) Basal keratinocyte tetrasomy in low-grade squamous intra-epithelial lesions of the cervix is restricted to high and intermediate risk HPV infection but is not type-specific. *Br J Cancer* 82: 424–428
41 Southern SA, Noya F, Meyers C, Broker TR, Chow LT, Herrington CS (2001) Tetrasomy is induced by human papillomavirus type 18 E7 gene expression in keratinocyte raft cultures. *Cancer Res* 61: 4858–4863
42 Southern SA, Lewis MH, Herrington CS (2004) Induction of tetrasomy by human papillomavirus type 16 E7 protein is independent of pRb binding and disruption of differentiation. *Br J Cancer* 90: 1949–1954
43 Meraldi P, Honda R, Nigg EA (2002) Aurora-A overexpression reveals tetraploidization as a major route to centrosome amplification in p53(–/–) cells. *Embo J* 21: 483–492
44 Storchova Z, Pellman D (2004) From polyploidy to aneuploidy, genome instability and cancer. *Nat Rev Mol Cell Biol* 5: 45–54
45 Winkler B, Crum CP, Fujii T, Ferenczy A, Boon M, Braun L, Lancaster WD, Richart RM (1984) Koilocytotic lesions of the cervix. The relationship of mitotic abnormalities to the presence of papillomavirus antigens and nuclear DNA content. *Cancer* 53: 1081–1087
46 Hinchcliffe EH, Sluder G (2001) "It Takes Two to Tango": understanding how centrosome duplication is regulated throughout the cell cycle. *Genes Dev* 15: 1167–1181
47 Nigg EA (2002) Centrosome aberrations: cause or consequence of cancer progression? *Nat Rev Cancer* 2: 815–825
48 Bunz F, Dutriaux A, Lengauer C, Waldman T, Zhou S, Brown JP, Sedivy JM, Kinzler KW, Vogelstein B (1998) Requirement for p53 and p21 to sustain G2 arrest after DNA damage. *Science* 282: 1497–1501
49 Duensing S, Lee LY, Duensing A, Basile J, Piboonniyom S, Gonzalez S, Crum CP, Munger K (2000) The human papillomavirus type 16 E6 and E7 oncoproteins cooperate to induce mitotic defects and genomic instability by uncoupling centrosome duplication from the cell division cycle. *Proc Natl Acad Sci USA* 97: 10002–10007
50 Boveri T (1900) *Zellenstudien IV. Über die Natur der Zentrosomen*. G. Fischer, Jena
51 Boveri T (1902) Über mehrpolige Mitosen als Mittel zur Analyse des Zellkerns. *Verh d phys-med Ges Würzburg NF* 35: 67–90
52 Hansemann D (1890) Ueber asymmetrische Zelltheilung in Epithelkrebsen und deren biologische Bedeutung. *Arch Pathol Anat Physiol Klin Med* 119: 299–326
53 Boveri T (1914) *Zur Frage der Entstehung Maligner Tumoren*. Fischer, Jena
54 Skyldberg B, Fujioka K, Hellstrom AC, Sylven L, Moberger B, Auer G (2001) Human papillomavirus infection, centrosome aberration, and genetic stability in cervical lesions. *Mod Pathol* 14:

279–284
55 Duensing S, Duensing A, Flores ER, Do A, Lambert PF, Munger K (2001) Centrosome abnormalities and genomic instability by episomal expression of human papillomavirus type 16 in raft cultures of human keratinocytes. *J Virol* 75: 7712–7716
56 Duensing S, Duensing A, Crum CP, Munger K (2001) Human papillomavirus type 16 E7 oncoprotein-induced abnormal centrosome synthesis is an early event in the evolving malignant phenotype. *Cancer Res* 61: 2356–2360
57 Brinkley BR (2001) Managing the centrosome numbers game: from chaos to stability in cancer cell division. *Trends Cell Biol* 11: 18–21
58 Schaeffer AJ, Nguyen M, Liem A, Lee D, Montagna C, Lambert PF, Ried T, Difilippantonio MJ (2004) E6 and E7 oncoproteins induce distinct patterns of chromosomal aneuploidy in skin tumors from transgenic mice. *Cancer Res* 64: 538–546
59 Riley RR, Duensing S, Brake T, Munger K, Lambert PF, Arbeit JM (2003) Dissection of human papillomavirus E6 and E7 function in transgenic mouse models of cervical carcinogenesis. *Cancer Res* 63: 4862–4871
60 Hernando E, Nahle Z, Juan G, Diaz-Rodriguez E, Alaminos M, Hemann M, Michel L, Mittal V, Gerald W, Benezra R et al. (2004) Rb inactivation promotes genomic instability by uncoupling cell cycle progression from mitotic control. *Nature* 430: 797–802
61 Duensing S, Munger K (2003) Human papillomavirus type 16 E7 oncoprotein can induce abnormal centrosome duplication through a mechanism independent of retinoblastoma protein family members. *J Virol* 77: 12331–12335
62 Duensing S, Duensing A, Lee DC, Edwards KM, Piboonniyom S, Manuel E, Skaltsounis L, Meijer L, Munger K (2004) The cyclin-dependent kinase inhibitor indirubin-3'-oxime selectively inhibits human papillomavirus type 16 E7-induced numerical centrosome anomalies. *Oncogene* 23: 8206–8215
63 Ortega S, Prieto I, Odajima J, Martin A, Dubus P, Sotillo R, Barbero JL, Malumbres M, Barbacid M (2003) Cyclin-dependent kinase 2 is essential for meiosis but not for mitotic cell division in mice. *Nat Genet* 35: 25–31
64 Tetsu O, McCormick F (2003) Proliferation of cancer cells despite CDK2 inhibition. *Cancer Cell* 3: 233–245
65 Duensing S, Münger K (2002) The human papillomavirus type 16 E6 and E7 oncoproteins independently induce numerical and structural chromosome instability. *Cancer Res* 62: 7075–7082
66 McClintock B (1940) The stability of broken ends of chromosomes in Zea mays. *Genetics* 26: 234–282
67 Hackett JA, Feldser DM, Greider CW (2001) Telomere dysfunction increases mutation rate and genomic instability. *Cell* 106: 275–286
68 Kessis TD, Connolly DC, Hedrick L, Cho KR (1996) Expression of HPV16 E6 or E7 increases integration of foreign DNA. *Oncogene* 13: 427–431
69 Iftner T, Elbel M, Schopp B, Hiller T, Loizou JI, Caldecott KW, Stubenrauch F (2002) Interference of papillomavirus E6 protein with single-strand break repair by interaction with XRCC1. *Embo J* 21: 4741–4748
70 Thompson DA, Belinsky G, Chang TH-T, Jones DL, Schlegel R, Münger K (1997) The human papillomavirus-16 E6 oncoprotein decreases the vigilance of mitotic checkpoints. *Oncogene* 15: 3025–3036
71 Thomas JT, Laimins LA (1998) Human papillomavirus oncoproteins E6 and E7 independently abrogate the mitotic spindle checkpoint. *J Virol* 72: 1131–1137
72 Patel D, Incassati A, Wang N, McCance DJ (2004) Human papillomavirus type 16 E6 and E7 cause polyploidy in human keratinocytes and up-regulation of G2-M-phase proteins. *Cancer Res* 64: 1299–1306
73 Thierry F, Benotmane MA, Demeret C, Mori M, Teissier S, Desaintes C (2004) A genomic approach reveals a novel mitotic pathway in papillomavirus carcinogenesis. *Cancer Res* 64: 895–903
74 Tsukasaki K (2002) Genetic instability of adult T-cell leukemia/lymphoma by comparative genomic hybridization analysis. *J Clin Immunol* 22: 57–63
75 Tsukasaki K, Krebs J, Nagai K, Tomonaga M, Koeffler HP, Bartram CR, Jauch A (2001) Comparative genomic hybridization analysis in adult T-cell leukemia/lymphoma: correlation with clinical course. *Blood* 97: 3875–3881
76 Jin DY, Spencer F, Jeang KT (1998) Human T cell leukemia virus type 1 oncoprotein Tax targets

the human mitotic checkpoint protein MAD1. *Cell* 93: 81–91
77 Haoudi A, Daniels RC, Wong E, Kupfer G, Semmes OJ (2003) Human T-cell leukemia virus-I tax oncoprotein functionally targets a subnuclear complex involved in cellular DNA damage-response. *J Biol Chem* 278: 37736–37744
78 Semmes OJ, Jeang KT (1996) Localization of human T-cell leukemia virus type 1 tax to subnuclear compartments that overlap with interchromatin speckles. *J Virol* 70: 6347–6357
79 Park HU, Jeong JH, Chung JH, Brady JN (2004) Human T-cell leukemia virus type 1 Tax interacts with Chk1 and attenuates DNA-damage induced G2 arrest mediated by Chk1. *Oncogene* 23: 4966–4974
80 Jeang KT, Giam CZ, Majone F, Aboud M (2004) Life, death, and tax: role of HTLV-I oncoprotein in genetic instability and cellular transformation. *J Biol Chem* 279: 31991–31994
81 Lemoine FJ, Marriott SJ (2002) Genomic instability driven by the human T-cell leukemia virus type I (HTLV-I) oncoprotein, Tax. *Oncogene* 21: 7230–7234
82 Shimura M, Onozuka Y, Yamaguchi T, Hatake K, Takaku F, Ishizaka Y (1999) Micronuclei formation with chromosome breaks and gene amplification caused by Vpr, an accessory gene of human immunodeficiency virus. *Cancer Res* 59: 2259–2264
83 Shimura M, Tanaka Y, Nakamura S, Minemoto Y, Yamashita K, Hatake K, Takaku F, Ishizaka Y (1999) Micronuclei formation and aneuploidy induced by Vpr, an accessory gene of human immunodeficiency virus type 1. *Faseb J* 13: 621–637
84 Minemoto Y, Shimura M, Ishizaka Y, Masamune Y, Yamashita K (1999) Multiple centrosome formation induced by the expression of Vpr gene of human immunodeficiency virus. *Biochem Biophys Res Commun* 258: 379–384
85 Rickinson AB, Kieff E (2001) Epstein-Barr Virus. In: DM Knipe, PM Howley (eds): *Fields Virology*. Lippincott Williams and Wilkins, Philadelphia, 2575–2627
86 Liu MT, Chen YR, Chen SC, Hu CY, Lin CS, Chang YT, Wang WB, Chen JY (2004) Epstein-Barr virus latent membrane protein 1 induces micronucleus formation, represses DNA repair and enhances sensitivity to DNA-damaging agents in human epithelial cells. *Oncogene* 23: 2531–2539
87 Chang Y, Cesarman E, Pessin MS, Lee F, Culpepper J, Knowles DM, Moore PS (1994) Identification of herpesvirus-like DNA sequences in AIDS-associated Kaposi's sarcoma. *Science* 266: 1865–1869
88 Moore PS, Chang YE (2001) Kaposi's Sarcoma-Associated Herpesvirus. In: DM Knipe, PM Howley (eds): *Fields Virology*. Lippincott Williams and Wilkins, Philadelphia, 2803–2833
89 Verschuren EW, Klefstrom J, Evan GI, Jones N (2002) The oncogenic potential of Kaposi's sarcoma-associated herpesvirus cyclin is exposed by p53 loss *in vitro* and *in vivo*. *Cancer Cell* 2: 229–241
90 Verschuren EW, Hodgson JG, Gray JW, Kogan S, Jones N, Evan GI (2004) The role of p53 in suppression of KSHV cyclin-induced lymphomagenesis. *Cancer Res* 64: 581–589
91 Pan H, Zhou F, Gao SJ (2004) Kaposi's sarcoma-associated herpesvirus induction of chromosome instability in primary human endothelial cells. *Cancer Res* 64: 4064–4068
92 Wu X, Avni D, Chiba T, Yan F, Zhao Q, Lin Y, Heng H, Livingston D (2004) SV40 T antigen interacts with Nbs1 to disrupt DNA replication control. *Genes Dev* 18: 1305–1316
93 Cotsiki M, Lock RL, Cheng Y, Williams GL, Zhao J, Perera D, Freire R, Entwistle A, Golemis EA, Roberts TM et al. (2004) Simian virus 40 large T antigen targets the spindle assembly checkpoint protein Bub1. *Proc Natl Acad Sci USA* 101: 947–952
94 Ricciardiello L, Baglioni M, Giovannini C, Pariali M, Cenacchi G, Ripalti A, Landini MP, Sawa H, Nagashima K, Frisque RJ et al. (2003) Induction of chromosomal instability in colonic cells by the human polyomavirus JC virus. *Cancer Res* 63: 7256–7262
95 Hollinger FB, Liang TJ (2001) Hepatitis B Virus. In: DM Knipe, PM Howley (eds): *Fields Virology*. Lippincott Williams and Wilkins, Philadelphia, 2971–3036
96 Forgues M, Difilippantonio MJ, Linke SP, Ried T, Nagashima K, Feden J, Valerie K, Fukasawa K, Wang XW (2003) Involvement of Crm1 in hepatitis B virus X protein-induced aberrant centriole replication and abnormal mitotic spindles. *Mol Cell Biol* 23: 5282–5292
97 Yun C, Cho H, Kim SJ, Lee JH, Park SY, Chan GK (2004) Mitotic aberration coupled with centrosome amplification is induced by hepatitis B virus X oncoprotein via the Ras-mitogen-activated protein/extracellular signal-regulated kinase-mitogen-activated protein pathway. *Mol Cancer Res* 2: 159–169
98 De Luca A, Mangiacasale R, Severino A, Malquori L, Baldi A, Palena A, Mileo AM, Lavia P,

Paggi MG (2003) E1A deregulates the centrosome cycle in a Ran GTPase-dependent manner. *Cancer Res* 63: 1430–1437

99 Carbone M, Klein G, Gruber J, Wong M (2004) Modern criteria to establish human cancer etiology. *Cancer Res* 64: 5518–5524

100 McDougall JK (2001) "Hit and run" transformation leading to carcinogenesis. *Dev Biol (Basel)* 106: 77–82

101 Eigen M (2002) Error catastrophe and antiviral strategy. *Proc Natl Acad Sci USA* 99: 13374–13376

102 Vidwans SJ, Wong ML, O'Farrell PH (2003) Anomalous centriole configurations are detected in Drosophila wing disc cells upon Cdk1 inactivation. *J Cell Sci* 116: 137–143

103 Duensing S, Lee BH, Dal Cin P, Münger K (2003) Excessive centrosome abnormalities without ongoing numerical chromosome instability in a Burkitt's lymphoma. *Molecular Cancer* 2: 30

102–292) region of p53 that is evolutionarily conserved between p53 and its homologues from *Drosophila* and *C. elegans*. In human tumors, amino acid residues that are essential for contact with DNA target sequence (two repeats of PuPuPuC(A/T)(A/T)GpyPyPy; in which Pu is a purine and Py is a pyrimidine) are frequently found to be mutated [69]. In addition, mutations of residues that do not contact DNA directly but are required for structural maintenance also cause disruption of the p53-DNA interaction. Frequently, mutations in one allele are sufficient to interfere with p53-dependent apoptosis by a dominant negative mechanism since in most cases mutant p53 negates wild-type p53 function through heteromerization.

Under normal conditions, p53 has a short half-life and is maintained at very low levels by Mdm2-mediated degradation [71]. However, in response to stress by DNA damage, hypoxia, oxidative stress and oncogene activation, p53 is stabilized and activated by post-translational modification [69]. In tumor cells, transcriptionally inactive mutant p53 is unable to induce the expression of the Mdm2 protein which would normally provide a feedback mechanism that downregulates p53 protein levels [72]. Moreover, some p53 mutants exhibit lower affinity for association with Mdm2 [73]. Hence, mutant p53 proteins that are impervious to these negative regulations accumulate to high levels in cancer cells and negate the functions of the wild-type protein.

Pathways through which p53 induces apoptosis may involve both transcriptional transactivation and transrepression of multiple p53-target genes, as well as transcription-independent mechanisms that engage the mitochondrial-apoptotic pathways [70]. In general, apoptotic target genes of p53 may be divided into two major categories: (1) proteins acting at the level of receptor signaling for apoptosis, and (2) proteins acting downstream by activating apoptotic effector proteins [74]. The former includes the insulin-like growth factor-1-binding protein 3 (IGF-BP3), which induces apoptosis by blocking the IGF-1 survival signal [75] and Fas/APO-1/CD95, which functions in the T cell killing triggered by anticancer drugs [76]. Essential downstream p53-targeted apoptotic effector proteins are primarily associated with mitochondrial changes, including caspase-9 and its cofactor Apaf-1 in myc oncogene-induced apoptosis [77], and Bax, necessary for p53-mediated cell death in brain tumors [78]. In addition to acting as a regulatory gene coordinating the expression of many proteins involved in apoptosis, recent research also suggests that p53 is involved in mediating apoptosis at the mitochondrial level by directly and physically interacting with the Bcl-2 member Bak, resulting in the release of cytochrome *c* from the mitochondria [79, 80].

Bax, Bak and Bcl-2

The pro-apoptotic proteins Bax and Bak are mediators of mitochondrial membrane damage that are mutated or downregulated in gastric and colorectal cancers [81–83]. In particular, combined mutation of p53 and Bax results in an extremely aggressive tumor progression and poor clinical prognosis [83]. Bax and Bak were found to be sufficient but not necessary for drug-induced apoptosis [84]. By contrast, increased copy numbers of the anti-apoptotic Bcl-X_L occurs in breast carcinoma, glioblastomas, and Hodgkin lymphoma and other specific tumor types [52]. Similarly, the anti-apoptotic Bcl-2 protein is frequently overexpressed in many tumors including acute lymphoblastic leukemia (ALL), precursor B-lymphoblastic leukemia/lymphoma and diffuse large B cell lymphoma [52]. Bcl-2 is an antagonist to Bax and Bak and inhibits mitochondrial membrane disruption, a mechanism that likely accounts for drug resistance in Bcl-2-overpressing lymphomas [85].

Akt, PI3K and PTEN

The pro-apoptotic Bad is a substrate for Akt/protein kinase B [86] and acts as a negative regulator for other anti-apoptotic family members. Upon phosphorylation by Akt, Bad dissociates from anti-apoptotic Bcl-X_L, allowing it to hinder the progression of apoptosis [87, 88]. Amplification of *akt* has been found in ovarian, pancreatic, breast and gastric malignancies [89, 90] and hyperactivation of Akt is known to induce resistance to a range of apoptotic stimuli including chemotherapeutic drugs [91, 92]. Akt activity can be induced indirectly by Ras in various growth factor receptor-initiated signaling cascades [93]. This Ras-mediated survival signal is connected to the effector phosphatidylinositol 3-kinase (PI3K), a lipid kinase responsible for the activation of Akt [94, 95]. Ras and PI3K are deregulated in many cancers, and the inhibition of PI3K enhances chemotherapeutic drug-induced apoptosis [94, 95]. The PI3K-Akt pathway is negatively regulated by PTEN (phosphatase and tensin homologue deleted on chromosome 10), a lipid phosphatase that inhibits PI3K-induced signaling by dephosphorylating PI3K-generated 3'-phosphorylated phosphatidylinositides [96, 97]. PTEN is frequently mutated in advanced stages of several human tumors, notably in glioblastoma, endometrial and prostate cancers [97], and some PTEN mutations are associated with a higher risk in the development of malignant breast tumors [98].

Death receptors

Fas/APO-1/CD95 and TRAIL-R1/R2 are sensors on the cell surface that, upon binding to their respective ligands, initiate the extrinsic apoptotic pathway (see above). Fas ligand and TRAIL are components of a tumor surveillance mechanism that partakes in the killing of cancer cells by cytotoxic lymphocytes [99, 100]. Tumorigenic disruptions, found in the intrinsic pathway, may also occur in the extrinsic pathway, albeit far less frequently. Fas is observed to be mutated and downregulated in lymphoid and solid tumors [101], whereas TRAIL-R1/R2 is mutated in metastatic breast cancers [102]. Suppression of the death receptor pathway could allow immune escape and provide a survival advantage to tumor cells. This loss of function is also associated with resistance to drug-induced cell death.

Caspases and non-IAP regulators

Caspase-8 is the initiator caspase for the extrinsic apoptotic pathway, and it is also the mediator for cross-talk between the extrinsic and the intrinsic pathways (see the 'Bcl-2 family' section above). Caspase-8 is silenced through DNA methylation as well as through gene deletion in childhood neuroblastomas, rendering these cancers resistant to apoptosis triggered by death-receptor ligation and by doxorubicin, a chemotherapeutic drug [103]. The expression level of c-FLIP, an endogenous inhibitor to caspase-8, is upregulated in some cancers, thus preventing caspase-8-mediated apoptosis induced by some chemotherapeutic drugs [104]. In the intrinsic pathway, Apaf-1 is necessary for activation of caspase-9 following cytochrome *c* release for the early amplification of apoptotic signals. In malignant melanoma and leukemia cell lines, Apaf-1 is mutated and transcriptionally silenced. Notably, Apaf-1-negative melanomas are chemoresistant, failing to execute typical apoptosis in response to p53 activation [105, 106].

Inhibitors of apoptosis and antagonists

As the only known endogenous proteins that function as direct, physiological inhibitors of both initiator and effector caspases, the IAPs occupy a central position in the apoptotic cascade, representing an important survival factor in resistant cancer [42, 107]. The IAPs, especially XIAP, are frequently overexpressed in the NCI 60 cell line panel of cancer cells as well as in cancer tissues compared to normal tissues [108–112]. Interestingly, at least in breast, colon and pancreatic cancers, a strong positive correlation was found between the levels of XIAP and caspase-3 [113, 114], suggesting that opportunities exist in which the downregulation of XIAP might release caspase-3 inhibition and promote the execution of apoptosis in cancer cells. Indeed, the inhibition of XIAP by antisense oligonucleotides, peptide inhibitors or small-molecule antagonists has been shown to sensitize cancer cells to apoptosis in chemoresistant tumors [108, 115, 116].

The potential therapeutic utility of IAP suppression for cancer treatment had sparked an explosion of research into finding and identifying endogenous IAP antagonists. As discussed in the section 'Inhibitors of apoptosis' above, to date, Smac/DIABLO, Omi/Htra2 and XAF1 are the three negative regulators of IAPs activity. The expression of Smac is decreased in various types of cancer, including lung, prostate and hepatocellular carcinomas [117, 118]. Recombinant adenovirus carrying Smac is able to sensitize ovarian carcinoma cells to chemotherapeutic drugs cisplatin and paclitaxel [119], whereas the combination of TRAIL and cell-permeable peptides that mimic Smac activity has been shown to eradicate established malignant glioma in mice [120]. Conversely, the elimination of endogenous Omi by RNA interference increases resistance to TRAIL-induced apoptosis [121]. While XAF1 is ubiquitously expressed in normal tissues and cells, it is found at less than 1% of control levels in the majority of the NCI 60 cell line panel of cancer cells [109]. Furthermore, overexpression of XAF1 by adenoviral vector transduction is capable of inducing apoptosis by unblocking caspase-3 and -9 inhibitions in certain pancreatic and colon cancer cell lines [114].

NF-κB

The nuclear factor of κB (NF-κB) family is composed of a number of heterodimeric transcription factors that regulate the expression of over 200 genes that are involved in the control of immune, inflammatory and stress responses, as well as growth and apoptosis [122, 123]. The activity of NF-κB is deregulated in many cancers, notably in B cell lymphomas [124]. Although NF-κB transcriptionally activates both anti- and pro-apoptotic genes, on a balance, NF-κB activation favors the suppression of apoptosis [123]. Key anti-apoptotic genes activated by NF-κB include the IAPs and the anti-apoptotic members of the Bcl-2 families [124, 125]. Since the IAPs and the anti-apoptotic members of the Bcl-2 families are crucial inhibitors to both the extrinsic and intrinsic death pathways, as expected, active NF-κB can inhibit these pathways and induce drug resistance in cancer cells [124].

Therapeutic opportunities

Given that apoptosis suppression is fundamental to cancer cell survival, it is not surprising that components of the apoptotic pathway have emerged as important therapeutic targets. A variety of antisense oligonucleotides, traditional small molecules, biologically active peptides, peptidomimetics, monoclonal antibodies and gene therapy pay loads have been incorporated into strategies that target apoptotic pathways in cancer cells [64, 126–130]. Although factors such as unexpected toxicities, poor pharmacokinetics, stability and oral bioavailability may limit the use of these compounds in anticancer treatment, these apoptosis-based antitumor agents might still serve as precursor molecules for the development of more effective therapies.

The importance of Bcl-2 in tumor cells resistant to most cytotoxic anticancer drugs has propelled this anti-apoptotic gene to the forefront as a candidate for antisense oligonucleotides (ASONs)-based therapies. ASONs are short pieces of DNA that hybridize to a specific target mRNA, thereby blocking its translation to a functional protein. *In vitro* experiments and xenograft models have demonstrated that Bcl-2 ASONs chemosensitizes human cancer cells [131, 132]. In a phase I clinical trial, a combination of Bcl-2 ASON and mitoxantrone has been shown to be well tolerated in combination [133]. In fact, ASONs targeting the Bcl-2 mRNA have advanced to phase II clinical trials for a variety of solid tumors, and phase III for melanoma, myeloma, chronic lymphocytic leukemia and acute myeloid leukemia [64]. One concern for targeting Bcl-2 alone is the ability of some tumor cells to switch expression from Bcl-2 to Bcl-X_L, thereby potentially retaining their apoptosis resistance [134]. Therefore, the simultaneous inhibition of Bcl-2 and Bcl-X_L expression in tumors by a single bi-specific ASON [135] or by small-molecule antagonists [136] may represent an appealing approach in certain cancers. Alternatively, inducing the expression of pro-apoptotic Bax with p53 adenovirus is a potentially useful gene therapy, particularly in human brain tumors [137]. In addition, short peptides that represent the BH3 domains of Bid or Bim have been shown to be capable of inducing oligomerization and activation of Bak and Bax, promoting killings of leukemic cells [138].

By virtue of their anti-caspase activity, the IAPs serve as pivotal regulators of the core apoptotic machinery, thereby representing another promising target for enhancing the re-activation of the death program. Numerous proof-of-principle studies have demonstrated that the downregulation of XIAP leads to enhanced chemotherapy sensitivity in various types of cancer cells [42, 64, 139]. For example, in both *in vitro* and *in vivo* xenograft human lung cancer models, ASONs targeting XIAP induce apoptosis and enhance chemotherapeutic activity [140]. These validations for XIAP as an important gate keeper to the apoptosis cascade have led to the launching of phase I clinical trials of an XIAP-specific ASON designed to stimulate apoptosis in cancer cells [141]. An alternative approach to suppress IAP function utilizes short peptides or small molecules that mimic IAP antagonists. In an intracranial malignant glioma xenograft model *in vivo*, synthetic peptides that mimic IAP antagonists Smac and HtrA2 are able to induce complete regression of the tumors caused by TRAIL-mediated apoptosis without detectable toxicity to normal brain tissue [120]. Similarly, non-peptidyl small-molecule XIAP antagonists screened from combinatorial chemical libraries have been shown to sensitize cancer cells to chemotherapeutic drugs and to suppress growth of established tumors in xenograft models in mice, while displaying little toxicity to normal tissues [116, 142]. Clearly, the effective tumor suppression activities of these IAP antagonists warrant further studies into their applicability in anticancer regimens.

Conclusions

Cancer is the consequence of parallel pathways that lead to both inappropriate cell proliferation and aberrant control of apoptosis. The inherent suppression of apoptosis in cancer cells has emerged to be a fundamental mechanism of tumor formation, progression and resistance to therapy. Advances made in elucidating the underlying mechanisms for the inhibition to apoptosis in tumor cells have identified important therapeutic targets and facilitated the development of novel strategies for resensitizing cancer cells to apoptosis. As evident in the leaps and bounds made in our understanding of apoptosis in cancer, clinical trials in progress are employing new approaches that are designed to directly modulate key apoptosis regulators. We anticipate that future advances will continually be made to these rational molecular approaches, such that apoptosis-based cancer therapies will match the diversity of the disease itself. Although much remains to be learned regarding apoptosis in cancers as well as other aspects of resistance and tumorigenesis, progress made to date indeed justifies our optimism that eradicating this disease will be a reality.

Acknowledgements
We apologize to those whose work was not cited or discussed due to space limitations. We thank the staff and students of the Apoptosis Research Centre and Ægera Oncology Inc. for fruitful discussions. H.H.C. is the recipient of a Michael Cuccione Foundation and CIHR Institute of Cancer Research postdoctoral fellowship. R.G.K. is a HHMI International Research Scholar, a CIHR Senior Investigator and a Fellow of the Royal Society of Canada (FRCS).

References

1 Lockshin RA, Williams CM (1964) Programmed cell death. II. Endocrine potentiation of the breakdown of the intersegmental muscles of silkmoths. *J Insect Physiol* 10: 643–649
2 Kerr JF, Wyllie AH, Currie AR (1972) Apoptosis: a basic biological phenomenon with wide-ranging implications in tissue kinetics. *Br J Cancer* 26: 239–257
3 Thompson CB (1995) Apoptosis in the pathogenesis and treatment of disease. *Science* 267: 1456–1462
4 Hanahan D, Weinberg RA (2000) The hallmarks of cancer. *Cell* 100: 57–70
5 Soini Y, Paakko P, Lehto VP (1998) Histopathological evaluation of apoptosis in cancer. *Am J Pathol* 153: 1041–1053
6 Tittel JN, Steller H (2000) A comparison of programmed cell death between species. *Genome Biol* 1: REVIEWS0003
7 Susin SA, Zamzami N, Castedo M, Daugas E, Wang HG, Geley S, Fassy F, Reed JC, Kroemer G (1997) The central executioner of apoptosis: multiple connections between protease activation and mitochondria in Fas/APO-1/CD95- and ceramide-induced apoptosis. *J Exp Med* 186: 25–37
8 Wyllie AH, Kerr JF, Currie AR (1980) Cell death: the significance of apoptosis. *Int Rev Cytol* 68: 251–306
9 Walker NI, Harmon BV, Gobe GC, Kerr JF (1988) Patterns of cell death. *Methods Achiev Exp Pathol* 13: 18–54
10 Cohen JJ (1991) Programmed cell death in the immune system. *Adv Immunol* 50: 55–85
11 Cummings MC, Winterford CM, Walker NI (1997) Apoptosis. *Am J Surg Pathol* 21: 88–101
12 Clarke PG (1990) Developmental cell death: morphological diversity and multiple mechanisms. *Anat Embryol (Berl)* 181: 195–213
13 Wyllie AH (1980) Glucocorticoid-induced thymocyte apoptosis is associated with endogenous endonuclease activation. *Nature* 284: 555–556
14 Huettenbrenner S, Maier S, Leisser C, Polgar D, Strasser S, Grusch M, Krupitza G (2003) The evolution of cell death programs as prerequisites of multicellularity. *Mutat Res* 543: 235–249
15 Thornberry NA, Lazebnik Y (1998) Caspases: enemies within. *Science* 281: 1312–1316
16 Brady HJ (2003) Apoptosis and leukaemia. *Br J Haematol* 123: 577–585
17 Talanian RV, Quinlan C, Trautz S, Hackett MC, Mankovich JA, Banach D, Ghayur T, Brady KD, Wong WW (1997) Substrate specificities of caspase family proteases. *J Biol Chem* 272: 9677–9682

18 Thornberry NA (1997) The caspase family of cysteine proteases. *Br Med Bull* 53: 478–490
19 Strasser A, O'Connor L, Dixit VM (2000) Apoptosis signaling. *Annu Rev Biochem* 69: 217–245
20 Nachmias B, Ashhab Y, Ben-Yehuda D (2004) The inhibitor of apoptosis protein family (IAPs): an emerging therapeutic target in cancer. *Semin Cancer Biol* 14: 231–243
21 Green DR, Kroemer G (2004) The pathophysiology of mitochondrial cell death. *Science* 305: 626–629
22 Goldstein JC, Waterhouse NJ, Juin P, Evan GI, Green DR (2000) The coordinate release of cytochrome c during apoptosis is rapid, complete and kinetically invariant. *Nat Cell Biol* 2: 156–162
23 Cain K, Bratton SB, Langlais C, Walker G, Brown DG, Sun XM, Cohen GM (2000) Apaf-1 oligomerizes into biologically active approximately 700-kDa and inactive approximately 1.4-MDa apoptosome complexes. *J Biol Chem* 275: 6067–6070
24 Saelens X, Festjens N, Vande Walle L, van Gurp M, van Loo G, Vandenabeele P (2004) Toxic proteins released from mitochondria in cell death. *Oncogene* 23: 2861–2874
25 Scorrano L, Oakes SA, Opferman JT, Cheng EH, Sorcinelli MD, Pozzan T, Korsmeyer SJ (2003) BAX and BAK regulation of endoplasmic reticulum Ca^{2+}: a control point for apoptosis. *Science* 300: 135–139
26 Nakagawa T, Yuan J (2000) Cross-talk between two cysteine protease families. Activation of caspase-12 by calpain in apoptosis. *J Cell Biol* 150: 887–894
27 Morishima N, Nakanishi K, Takenouchi H, Shibata T, Yasuhiko Y (2002) An endoplasmic reticulum stress-specific caspase cascade in apoptosis. Cytochrome c-independent activation of caspase-9 by caspase-12. *J Biol Chem* 277: 34287–34294
28 Rao RV, Hermel E, Castro-Obregon S, del Rio G, Ellerby LM, Ellerby HM, Bredesen DE (2001) Coupling endoplasmic reticulum stress to the cell death program. Mechanism of caspase activation. *J Biol Chem* 276: 33869–33874
29 Rao RV, Castro-Obregon S, Frankowski H, Schuler M, Stoka V, del Rio G, Bredesen DE, Ellerby HM (2002) Coupling endoplasmic reticulum stress to the cell death program. An Apaf-1-independent intrinsic pathway. *J Biol Chem* 277: 21836–21842
30 Zong WX, Li C, Hatzivassiliou G, Lindsten T, Yu QC, Yuan J, Thompson CB (2003) Bax and Bak can localize to the endoplasmic reticulum to initiate apoptosis. *J Cell Biol* 162: 59–69
31 Itoh N, Nagata S (1993) A novel protein domain required for apoptosis. Mutational analysis of human Fas antigen. *J Biol Chem* 268: 10932–10937
32 Tartaglia LA, Ayres TM, Wong GH, Goeddel DV (1993) A novel domain within the 55 kd TNF receptor signals cell death. *Cell* 74: 845–853
33 Boldin MP, Varfolomeev EE, Pancer Z, Mett IL, Camonis JH, Wallach D (1995) A novel protein that interacts with the death domain of Fas/APO1 contains a sequence motif related to the death domain. *J Biol Chem* 270: 7795–7798
34 Chinnaiyan AM, O'Rourke K, Tewari M, Dixit VM (1995) FADD, a novel death domain-containing protein, interacts with the death domain of Fas and initiates apoptosis. *Cell* 81: 505–512
35 Kischkel FC, Hellbardt S, Behrmann I, Germer M, Pawlita M, Krammer PH, Peter ME (1995) Cytotoxicity-dependent APO-1 (Fas/CD95)-associated proteins form a death-inducing signaling complex (DISC) with the receptor. *EMBO J* 14: 5579–5588
36 Boldin MP, Goncharov TM, Goltsev YV, Wallach D (1996) Involvement of MACH, a novel MORT1/FADD-interacting protease, in Fas/APO-1- and TNF receptor-induced cell death. *Cell* 85: 803–815
37 Muzio M, Chinnaiyan AM, Kischkel FC, O'Rourke K, Shevchenko A, Ni J, Scaffidi C, Bretz JD, Zhang M, Gentz R et al. (1996) FLICE, a novel FADD-homologous ICE/CED-3-like protease, is recruited to the CD95 (Fas/APO-1) death—inducing signaling complex. *Cell* 85: 817–827
38 Muzio M, Stockwell BR, Stennicke HR, Salvesen GS, Dixit VM (1998) An induced proximity model for caspase-8 activation. *J Biol Chem* 273: 2926–2930
39 Stennicke HR, Jurgensmeier JM, Shin H, Deveraux Q, Wolf BB, Yang X, Zhou Q, Ellerby HM, Ellerby LM, Bredesen D et al. (1998) Pro-caspase-3 is a major physiologic target of caspase-8. *J Biol Chem* 273: 27084–27090
40 Marsden VS, Strasser A (2003) Control of apoptosis in the immune system: Bcl-2, BH3-only proteins and more. *Annu Rev Immunol* 21: 71–105
41 Holcik M, Gibson H, Korneluk RG (2001) XIAP: apoptotic brake and promising therapeutic target. *Apoptosis* 6: 253–261
42 Liston P, Fong WG, Korneluk RG (2003) The inhibitors of apoptosis: there is more to life than

Bcl2. *Oncogene* 22: 8568–8580
43 Deveraux QL, Takahashi R, Salvesen GS, Reed JC (1997) X-linked IAP is a direct inhibitor of cell-death proteases. *Nature* 388: 300–304
44 Takahashi R, Deveraux Q, Tamm I, Welsh K, Assa-Munt N, Salvesen GS, Reed JC (1998) A single BIR domain of XIAP sufficient for inhibiting caspases. *J Biol Chem* 273: 7787–7790
45 Pickart CM (2001) Ubiquitin enters the new millennium. *Mol Cell* 8: 499–504
46 Huang H, Joazeiro CA, Bonfoco E, Kamada S, Leverson JD, Hunter T (2000) The inhibitor of apoptosis, cIAP2, functions as a ubiquitin-protein ligase and promotes *in vitro* monoubiquitination of caspases 3 and 7. *J Biol Chem* 275: 26661–26664
47 Suzuki Y, Nakabayashi Y, Takahashi R (2001) Ubiquitin-protein ligase activity of X-linked inhibitor of apoptosis protein promotes proteasomal degradation of caspase-3 and enhances its anti-apoptotic effect in Fas-induced cell death. *Proc Natl Acad Sci USA* 98: 8662–8667
48 Salvesen GS, Duckett CS (2002) IAP proteins: blocking the road to death's door. *Nat Rev Mol Cell Biol* 3: 401–410
49 Liston P, Fong WG, Kelly NL, Toji S, Miyazaki T, Conte D, Tamai K, Craig CG, McBurney MW, Korneluk RG (2001) Identification of XAF1 as an antagonist of XIAP anti-Caspase activity. *Nat Cell Biol* 3: 128–133
50 Distelhorst CW, Shore GC (2004) Bcl-2 and calcium: controversy beneath the surface. *Oncogene* 23: 2875–2880
51 Droin NM, Green DR (2004) Role of Bcl-2 family members in immunity and disease. *Biochim Biophys Acta* 1644: 179–188
52 Kirkin V, Joos S, Zornig M (2004) The role of Bcl-2 family members in tumorigenesis. *Biochim Biophys Acta* 1644: 229–249
53 Zha J, Weiler S, Oh KJ, Wei MC, Korsmeyer SJ (2000) Posttranslational N-myristoylation of BID as a molecular switch for targeting mitochondria and apoptosis. *Science* 290: 1761–1765
54 Wei MC, Lindsten T, Mootha VK, Weiler S, Gross A, Ashiya M, Thompson CB, Korsmeyer SJ (2000) tBID, a membrane-targeted death ligand, oligomerizes BAK to release cytochrome c. *Genes Dev* 14: 2060–2071
55 Eskes R, Desagher S, Antonsson B, Martinou JC (2000) Bid induces the oligomerization and insertion of Bax into the outer mitochondrial membrane. *Mol Cell Biol* 20: 929–935
56 Kuwana T, Smith JJ, Muzio M, Dixit V, Newmeyer DD, Kornbluth S (1998) Apoptosis induction by caspase-8 is amplified through the mitochondrial release of cytochrome c. *J Biol Chem* 273: 16589–16594
57 Li H., Zhu H, Xu CJ, Yuan J (1998) Cleavage of BID by caspase 8 mediates the mitochondrial damage in the Fas pathway of apoptosis. *Cell* 94: 491–501
58 Luo X, Budihardjo I, Zou H, Slaughter C, Wang X (1998) Bid, a Bcl2 interacting protein, mediates cytochrome c release from mitochondria in response to activation of cell surface death receptors. *Cell* 94: 481–490
59 Borner C (2003) The Bcl-2 protein family: sensors and checkpoints for life-or-death decisions. *Mol Immunol* 39: 615–647
60 Pinton P, Ferrari D, Rapizzi E, Di Virgilio F, Pozzan T, Rizzuto R (2001) The Ca^{2+} concentration of the endoplasmic reticulum is a key determinant of ceramide-induced apoptosis: significance for the molecular mechanism of Bcl-2 action. *EMBO J* 20: 2690–2701
61 Darios F, Lambeng N, Troadec JD, Michel PP, Ruberg M (2003) Ceramide increases mitochondrial free calcium levels via caspase 8 and Bid: role in initiation of cell death. *J Neurochem* 84: 643–654
62 Boehning D, Patterson RL, Sedaghat L, Glebova NO, Kurosaki T, Snyder SH (2003) Cytochrome c binds to inositol (1,4,5) trisphosphate receptors, amplifying calcium-dependent apoptosis. *Nat Cell Biol* 5: 1051–1061
63 Ionov Y, Yamamoto H, Krajewski S, Reed JC, Perucho M (2000) Mutational inactivation of the proapoptotic gene BAX confers selective advantage during tumor clonal evolution. *Proc Natl Acad Sci USA* 97: 10872–10877
64 Reed JC (2003) Apoptosis-targeted therapies for cancer. *Cancer Cell* 3: 17–22
65 Sjostrom J, Bergh J (2001) How apoptosis is regulated, and what goes wrong in cancer. *BMJ* 322: 1538–1539
66 Sigal A, Rotter V (2000) Oncogenic mutations of the p53 tumor suppressor: the demons of the guardian of the genome. *Cancer Res* 60: 6788–6793
67 Jimenez GS, Nister M, Stommel JM, Beeche M, Barcarse EA, Zhang XQ, O'Gorman S, Wahl GM

(2000) A transactivation-deficient mouse model provides insights into Trp53 regulation and function. *Nat Genet* 26: 37–43
68 Vogelstein B, Kinzler KW (1992) p53 function and dysfunction. *Cell* 70: 523–526
69 Haupt S, Haupt Y (2004) Improving Cancer Therapy Through p53 Management. *Cell Cycle* 3: 912–916
70 Slee EA, O'Connor DJ, Lu X (2004) To die or not to die: how does p53 decide? *Oncogene* 23: 2809–2818
71 Haupt Y, Maya R, Kazaz A, Oren M (1997) Mdm2 promotes the rapid degradation of p53. *Nature* 387: 296–299
72 Midgley CA, Lane DP (1997) p53 protein stability in tumour cells is not determined by mutation but is dependent on Mdm2 binding. *Oncogene* 15: 1179–1189
73 Buschmann T, Minamoto T, Wagle N, Fuchs SY, Adler V, Mai M, Ronai Z (2000) Analysis of JNK, Mdm2 and p14(ARF) contribution to the regulation of mutant p53 stability. *J Mol Biol* 295:1009–10021
74 Sionov RV, Haupt Y (1999) The cellular response to p53: the decision between life and death. *Oncogene* 18: 6145–6157
75 Buckbinder L, Talbott R, Velasco-Miguel S, Takenaka I, Faha B, Seizinger BR, Kley N (1995) Induction of the growth inhibitor IGF-binding protein 3 by p53. *Nature* 377: 646–649
76 Friesen C, Herr I, Krammer PH, Debatin KM (1996) Involvement of the CD95 (APO-1/FAS) receptor/ligand system in drug-induced apoptosis in leukemia cells. *Nat Med* 2: 574–577
77 Soengas MS, Alarcon RM, Yoshida H, Giaccia AJ, Hakem R, Mak TW, Lowe SW (1999) Apaf-1 and caspase-9 in p53-dependent apoptosis and tumor inhibition. *Science* 284: 156–159
78 Yin C, Knudson CM, Korsmeyer SJ, Van Dyke T (1997) Bax suppresses tumorigenesis and stimulates apoptosis *in vivo*. *Nature* 385: 637–640
79 Erster S, Mihara M, Kim RH, Petrenko O, Moll UM (2004) *In vivo* mitochondrial p53 translocation triggers a rapid first wave of cell death in response to DNA damage that can precede p53 target gene activation. *Mol Cell Biol* 24: 6728–6741
80 Leu JI, Dumont P, Hafey M, Murphy ME, George DL (2004) Mitochondrial p53 activates Bak and causes disruption of a Bak-Mcl1 complex. *Nat Cell Biol* 6: 443–450
81 Rampino N, Yamamoto H, Ionov Y, Li Y, Sawai H, Reed JC, Perucho M (1997) Somatic frameshift mutations in the BAX gene in colon cancers of the microsatellite mutator phenotype. *Science* 275: 967–969
82 Kondo S, Shinomura Y, Miyazaki Y, Kiyohara T, Tsutsui S, Kitamura S, Nagasawa Y, Nakahara M, Kanayama S, Matsuzawa Y (2000) Mutations of the bak gene in human gastric and colorectal cancers. *Cancer Res* 60: 4328–4330
83 Mrozek A, Petrowsky H, Sturm I, Kraus J, Hermann S, Hauptmann S, Lorenz M, Dorken B, Daniel PT (2003) Combined p53/Bax mutation results in extremely poor prognosis in gastric carcinoma with low microsatellite instability. *Cell Death Differ* 10: 461–467
84 Johnstone RW, Ruefli AA, Lowe SW (2002) Apoptosis: a link between cancer genetics and chemotherapy. *Cell* 108: 153–164
85 Reed JC (1999) Dysregulation of apoptosis in cancer. *J Clin Oncol* 17: 2941–2953
86 Datta SR, Brunet A, Greenberg ME (1999) Cellular survival: a play in three Akts. *Genes Dev* 13: 2905–2927
87 Datta SR, Dudek H, Tao X, Masters S, Fu H, Gotoh Y, Greenberg ME (1997) Akt phosphorylation of BAD couples survival signals to the cell-intrinsic death machinery. *Cell* 91: 231–241
88 del Peso L, Gonzalez-Garcia M, Page C, Herrera R, Nunez G (1997) Interleukin-3-induced phosphorylation of BAD through the protein kinase Akt. *Science* 278: 687–689
89 Bellacosa A, de Feo D, Godwin AK, Bell DW, Cheng JQ, Altomare DA, Wan M, Dubeau L, Scambia G, Masciullo V et al. (1995) Molecular alterations of the AKT2 oncogene in ovarian and breast carcinomas. *Int J Cancer* 64: 280–285
90 Cheng JQ, Ruggeri B, Klein WM, Sonoda G, Altomare DA, Watson DK, Testa JR (1996) Amplification of AKT2 in human pancreatic cells and inhibition of AKT2 expression and tumorigenicity by antisense RNA. *Proc Natl Acad Sci USA* 93: 3636–3641
91 Mayo LD, Dixon JE, Durden DL, Tonks NK, Donner DB (2002) PTEN protects p53 from Mdm2 and sensitizes cancer cells to chemotherapy. *J Biol Chem* 277: 5484–5489
92 Wendel HG, De Stanchina E, Fridman JS, Malina A, Ray S, Kogan S, Cordon-Cardo C, Pelletier J, Lowe SW (2004) Survival signaling by Akt and eIF4E in oncogenesis and cancer therapy. *Nature* 428: 332–337

93 Franke TF, Hornik CP, Segev L, Shostak GA, Sugimoto C (2003) PI3K/Akt and apoptosis: size matters. *Oncogene* 22: 8983–8998
94 el-Deiry WS (1997) Role of oncogenes in resistance and killing by cancer therapeutic agents. *Curr Opin Oncol* 9: 79–87
95 Roymans D, Slegers H (2001) Phosphatidylinositol 3-kinases in tumor progression. *Eur J Biochem* 268: 487–498
96 Di Cristofano A, Pandolfi PP (2000) The multiple roles of PTEN in tumor suppression. *Cell* 100: 387–390
97 Fresno Vara JA, Casado E, de Castro J, Cejas P, Belda-Iniesta C, Gonzalez-Baron M (2004) PI3K/Akt signaling pathway and cancer. *Cancer Treat Rev* 30: 193–204
98 Simpson L, Parsons R (2001) PTEN: life as a tumor suppressor. *Exp Cell Res* 264: 29–41
99 Rosen D, Li JH, Keidar S, Markon I, Orda R, Berke G (2000) Tumor immunity in perforin-deficient mice: a role for CD95 (Fas/APO-1). *J Immunol* 164: 3229–3235
100 Takeda K, Hayakawa Y, Smyth MJ, Kayagaki N, Yamaguchi N, Kakuta S, Iwakura Y, Yagita H, Okumura K (2001) Involvement of tumor necrosis factor-related apoptosis-inducing ligand in surveillance of tumor metastasis by liver natural killer cells. *Nat Med* 7: 94–100
101 Muschen M, Warskulat U, Beckmann MW (2000) Defining CD95 as a tumor suppressor gene. *J Mol Med* 78: 312–325
102 Shin MS, Kim HS, Lee SH, Park WS, Kim SY, Park JY, Lee JH, Lee SK, Lee SN, Jung SS et al. (2001) Mutations of tumor necrosis factor-related apoptosis-inducing ligand receptor 1 (TRAIL-R1) and receptor 2 (TRAIL-R2) genes in metastatic breast cancers. *Cancer Res* 61: 4942–4946
103 Teitz T, Wei T, Valentine MB, Vanin EF, Grenet J, Valentine VA, Behm FG, Look AT, Lahti JM, Kidd VJ (2000) Caspase 8 is deleted or silenced preferentially in childhood neuroblastomas with amplification of MYCN. *Nat Med* 6: 529–535
104 Tepper CG, Seldin MF (1999) Modulation of caspase-8 and FLICE-inhibitory protein expression as a potential mechanism of Epstein-Barr virus tumorigenesis in Burkitt's lymphoma. *Blood* 94: 1727–1737
105 Soengas MS, Capodieci P, Polsky D, Mora J, Esteller M, Opitz-Araya X, McCombie R, Herman JG, Gerald WL, Lazebnik YAet al. (2001) Inactivation of the apoptosis effector Apaf-1 in malignant melanoma. *Nature* 409: 207–211
106 Dai DL, Martinka M, Bush JA, Li G (2004) Reduced Apaf-1 expression in human cutaneous melanomas. *Br J Cancer* 91: 1089–1095
107 Hersey P, Zhang XD (2003) Overcoming resistance of cancer cells to apoptosis. *J Cell Physiol* 196: 9–18
108 Bilim V, Kasahara T, Hara N, Takahashi K, Tomita Y (2003) Role of XIAP in the malignant phenotype of transitional cell cancer (TCC) and therapeutic activity of XIAP antisense oligonucleotides against multidrug-resistant TCC *in vitro*. *Int J Cancer* 103: 29–37
109 Fong WG, Liston P, Rajcan-Separovic E, St Jean M, Craig C, Korneluk RG (2000) Expression and genetic analysis of XIAP-associated factor 1 (XAF1) in cancer cell lines. *Genomics* 70: 113–122
110 Ross DT, Scherf U, Eisen MB, Perou CM, Rees C, Spellman P, Iyer V, Jeffrey SS, Van de Rijn M, Waltham M et al. (2000) Systematic variation in gene expression patterns in human cancer cell lines. *Nat Genet* 24: 227–235
111 Tamm I, Kornblau SM, Segall H, Krajewski S, Welsh K, Kitada S, Scudiero DA, Tudor G, Qui YH, Monks A et al. (2000) Expression and prognostic significance of IAP-family genes in human cancers and myeloid leukemias. *Clin Cancer Res* 6: 1796–1803
112 Yang L, Mashima T, Sato S, Mochizuki M, Sakamoto H, Yamori T, Oh-Hara T, Tsuruo T (2003) Predominant suppression of apoptosome by inhibitor of apoptosis protein in non-small cell lung cancer H460 cells: therapeutic effect of a novel polyarginine-conjugated Smac peptide. *Cancer Res* 63: 831–837
113 Parton M, Krajewski S, Smith I, Krajewska M, Archer C, Naito M, Ahern R, Reed J, Dowsett M (2002) Coordinate expression of apoptosis-associated proteins in human breast cancer before and during chemotherapy. *Clin Cancer Res* 8: 2100–2108
114 Yang L, Cao Z, Yan H, Wood WC (2003) Coexistence of high levels of apoptotic signaling and inhibitor of apoptosis proteins in human tumor cells: implication for cancer specific therapy. *Cancer Res* 63: 6815–6824
115 Sasaki H, Sheng Y, Kotsuji F, Tsang BK (2000) Down-regulation of X-linked inhibitor of apoptosis protein induces apoptosis in chemoresistant human ovarian cancer cells. *Cancer Res* 60:

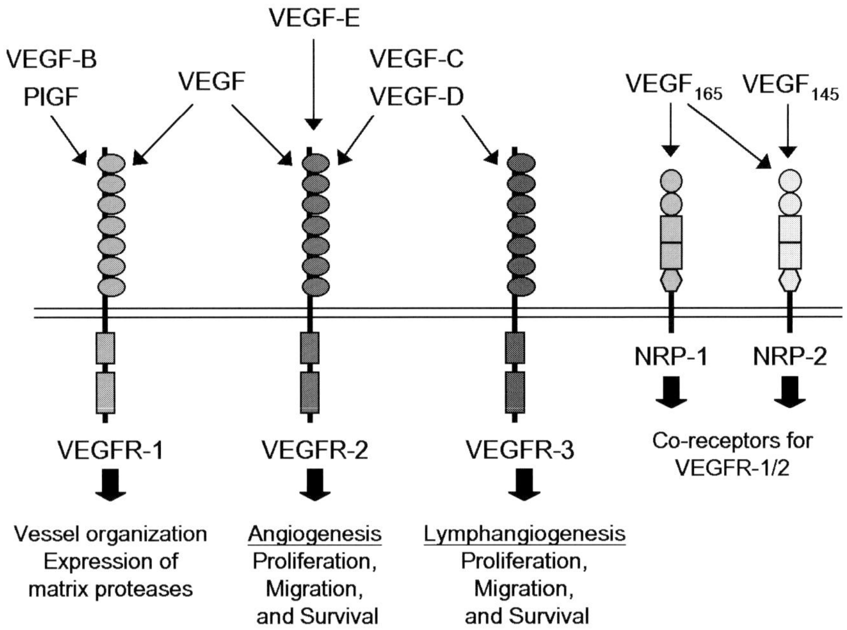

Figure 1. Interactions between VEGF family members and VEGF receptors and downstream consequences. Preferential binding has been reported between the various members of the VEGF family and the known VEGF receptors. Each specific ligand/receptor interaction mediates a specific downstream effect on endothelial cells.

mation of the corpus luteum [37–40], as well as endochondral ossification and bone formation [41–44]. Genetic ablation of the VEGF gene, even of only one allele, results in embryonic lethality primarily due to severe defects in the vascularization of multiple organ systems in the embryo [45, 46].

VEGF stimulation of endothelial cells leads to morphological changes that contribute to an angiogenic phenotype. For example, VEGF is capable of stimulating proliferation and migration of endothelial cells [20, 28] and is an endothelial cell survival factor *in vitro* and *in vivo* [47–49]. As its original name implies, VEGF stimulation leads to increased vascular permeability [50–52], which permits greater perfusion of the tumor and could also facilitate entry of metastatic cells into the circulation.

More recently, a role for VEGF in the recruitment of EPCs [53] has been described. The mobilization of EPCs from the bone marrow and incorporation of these cells into angiogenic tumor vessels has been well documented [9, 10]. EPCs express functional VEGF receptors (VEGFRs) [54–56], and recruitment of EPCs to the angiogenic tumor vasculature appears to be dependent on upregulated expression of VEGF [57, 58] or PlGF [59].

Many molecules classified as angiogenic factors actually induce angiogenesis by upregulating expression of VEGF (Fig. 2). Many growth factors and

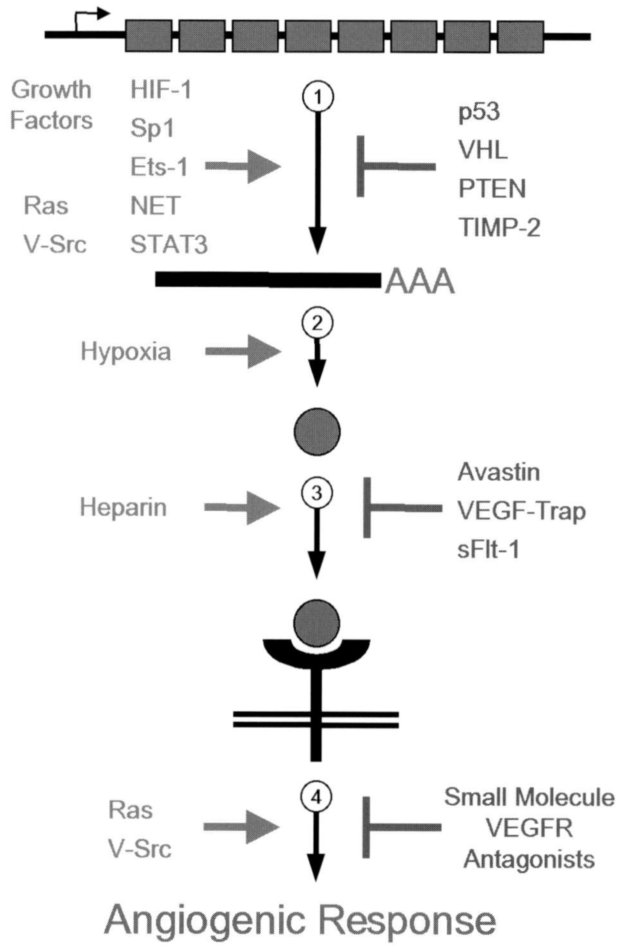

Figure 2. Positive and negative regulation of VEGF expression and activity. Several positive and negative regulators of VEGF have been identified. Regulation can occur at the level of transcription (1), mRNA stability/translation (2), ligand binding (3), or signaling pathways downstream of ligand/receptor interactions (4).

cytokines have been reported to stimulate VEGF expression including, but not limited to, transforming growth factor-β (TGF-β) [60], PDGF [61], EGF [62], insulin-like growth factor-1 (IGF-I) [63], interleukin-1β (IL-1β) [64], IL-6 [65], tumor necrosis factor α (TNF-α) [66], and FGF4 [67]. Nitric oxide (NO) is also a potent inducer of VEGF expression [68, 69] and could potentially form an autocrine loop with VEGF since VEGF upregulates NO synthesis [70, 71].

Transcription of VEGF is regulated by a variety of different mechanisms (Fig. 2). Most notably, hypoxia is a strong stimulator of VEGF expression *in vitro* and *in vivo* [72, 73]. The hypoxia-induced upregulation of VEGF is primarily mediated through activation of the hypoxia inducible factor-1 (HIF-1) transcription factor complex [74–76]. VEGF transcription is also positively regulated through the transcription factors Sp1 [77–79], AP-1 [80, 81], and NF-κB [82]. A number of oncogenic mutations are associated with increased expression of VEGF, for example mutations in the genes for Ras [83–85], and V-Src [86–88]. The transcriptional regulation of VEGF and how it relates to the initiation of the angiogenic phenotype are discussed in greater detail in the third section of this chapter.

While activation of VEGF transcription has been well characterized, little is known regarding negative regulation of VEGF expression. The tumor suppressors p53 and the von Hippel-Lindau gene product are both capable of downregulating VEGF via a number of mechanisms. It has been reported that each protein can bind to and sequester the VEGF transcriptional activators HIF-1α [89, 90] and Sp1 [91, 92], thereby preventing them from activating VEGF transcription. p53 can also act as a functional transcriptional repressor of VEGF, although it is not yet clear if it is through direct binding of p53 to the VEGF promoter [93].

VEGF-induced activities are mediated through preferential binding of VEGF homodimers to a number of receptor tyrosine kinases (RTKs) and other cell-surface receptors (Fig. 1). The VEGF-RTKs include VEGFR-1/Flt-1 [94, 95], VEGFR-2/KDR/Flk-1 [96–98], and VEGFR-3/Flt-4 [99]. These VEGFRs are transmembrane proteins, where the extracellular domain is composed of seven immunoglobulin-like domains responsible for ligand binding, and on the intracellular portion of the molecule, an interrupted tyrosine kinase domain, which mediates downstream signaling upon VEGF/VEGFR interactions [100–105]. Binding of VEGF to these receptors induces homodimerization or heterodimerization of the receptors, followed by transactivation of the tyrosine kinase domains, which initiates VEGF-induced signal transduction pathways [101, 103].

The second class of VEGF receptors are members of the neuropilin (NRP) family: NRP1 [106] and NRP2 [107]. NRPs are transmembrane proteins that were originally described as neuronal receptors for the semaphorins, a family of proteins that can act as chemoattractants or chemorepellents [108, 109]. More recently, NRP1 was shown to be an isoform-specific receptor for $VEGF_{165}$ but not $VEGF_{121}$ [110–112], and NRP2 is an isoform-specific receptor for $VEGF_{145}$ and $VEGF_{165}$ [113]. The cytoplasmic domains of NRP1 and NRP2 lack any kinase domains, so it is hypothesized that these NRPs act as co-receptors for VEGFR-1 and VEGFR-2 [114–116].

Elucidation of the role of these receptors through production of either knockout or transgenic animals has verified the importance of VEGF signaling through its receptors during development and disease. Targeted disruption of either the VEGFR-1, VEGFR-2 or VEGFR-3 genes results in embryonic lethality due to various defects in the vasculature. Increased hemangioblast commitment and disorganization of blood vessels are responsible for lethality in VEGFR-$1^{-/-}$ embryos [117, 118], while lethality was due to a complete failure of blood vessel development due to inhibition of endothelial cell differentiation in VEGFR-2-deficient mice [119]. Endothelial cell differentiation and vascular channel development are normal in VEGFR-$3^{-/-}$ mice, but these animals die *in utero* due to lack of lumen formation in, and proper maturation of, vessels [120]. NRP-1-deficient mice and zebra fish also exhibit an embryonic lethal phenotype due to cardiovascular defects [121, 122], as do NRP-1 transgenic mice [123].

VEGF receptor expression was originally thought to be endothelial cell specific [124]; however, various cell types have been shown to express VEGF receptors including smooth muscle cells [125, 126], hematopoietic cells [127, 128], EPCs [54–56, 129], osteoblasts [130, 131], and islet cells in the pancreas [132].

Interestingly, several types of cancer cells also express one or several of the VEGF receptors *in vitro* and *in vivo* [133]. Breast cancer cells express VEGFR-1 [134], VEGFR-2 [134, 135] and NRP-1 [136]; colon cancer cells express VEGFR-2 [135] and NRP-1 [137]; esophageal cancer cells express VEGFR-1, VEGFR-2 [138] and NRP-1 [137]; gallbladder cancer cells express NRP-1 [137]; gastric cancer cells express NRP-1 [137, 139]; pancreatic cancer cells express VEGFR-1, VEGFR-2 [140, 141] and NRP-1 [137, 142] and NRP-2 [142]; and prostate cancer cells express VEGFR-1 [143–145] and VEGFR-2 [143, 145] and NRP-1 [146, 147]. Even cells from oncological blood disorders express VEGF receptors. Non-Hodgkin's lymphoma cells express VEGFR-2 [135] as do leukemia cells [148].

The observation that tumor cells express VEGF receptors suggests that an autocrine loop of VEGF signaling could occur in these cells since many of the cell types that express VEGF receptors also express VEGF [136, 141, 143–145, 148, 149]. Thus, it is likely that anti-angiogenic therapies that target VEGF and its receptors may function to target both the endothelium as well as the tumor cells themselves.

Angiogenesis inhibitors

Angiogenesis inhibitors have been divided into two classes: direct and indirect inhibitors. Direct inhibitors, as their name suggests, function directly on endothelial cells either by arresting the proliferation of these cells, or by inducing apoptosis. Indirect angiogenesis inhibitors target the signaling induced by angiogenic stimuli either by sequestering the angiogenic factors secreted by tumor cells or by blocking the downstream signal transduction pathways that are activated upon the binding of these factors to their cognate receptors on endothelial cells. Several selected endogenous angiogenesis inhibitors with which we are particularly familiar are discussed below. A more-detailed discussion of selected indirect angiogenesis inhibitors that specifically target VEGF is presented in the second section of this chapter.

Thrombospondin

The first endogenous angiogenesis inhibitor to be identified was thrombospondin-1 (TSP-1) [150–152]. TSP-1 is a secreted, homotrimeric glycoprotein that is associated with either the cell surface or the extracellular matrix [153]. High levels of TSP-1 can be detected in the embryonic and developing heart, brain, lung, liver, kidney, skeletal muscle and bone [154–159]. It is abundantly expressed in cartilage and bone with much higher expression in osteoblasts than in chondrocytes [158, 159], and is among a group of cartilage-derived inhibitors of angiogenesis purified from articular cartilage [160].

TSP-1 is a potent inhibitor of angiogenesis *in vitro* as well as *in vivo*. Proliferation, migration and adhesion of endothelial cells *in vitro* are all blocked by intact TSP-1 or fragments of TSP-1 [161–164], and either TSP-1 or its fragments are capable of inducing apoptosis in endothelial cells [165, 166]. These effects of TSP-1 on endothelial cells are reportedly mediated through TSP-1 binding to either CD36 [167, 168] or the $\alpha_v\beta_3$ and $\alpha_3\beta_1$ integrins independently [169, 170] or complexed with integrin-associated protein (IAP) [171, 172].

Further evidence for the role of TSP-1 in cancer comes from data that demonstrate that transcription of TSP-1 is activated by tumor suppressor genes such as p53 [173–175] and PTEN [176]. Conversely, TSP-1 expression is downregulated by several oncogenes including *c-myc* [177], *v-src* [178, 179], *c-jun* [180] and *ras* [177, 181, 182], hypoxia [182], as well as by the Id1 transcription factor [183]. Verification of the role of Id1 in negatively regulating TSP-1 expression comes from *Id1* knockout mice [183]. TSP-1 expression is significantly increased in these mice, and tumor growth is markedly decreased due to potent inhibition of angiogenesis.

Immunohistochemical data from several tumor models demonstrates that TSP-1 protein levels are markedly reduced or non-existent in tumor tissues, while TSP-1 protein is abundant in surrounding, adjacent normal tissue [182, 184], suggesting that expression of TSP-1 in the tissues adjacent to the tumor may form a type of anti-angiogenic barrier [184].

Data from *in vivo* tumor models provides further evidence that TSP-1 inhibits angiogenesis and subsequent tumor growth. Genetic studies in which TSP-1-null mice were crossed with mice deficient in p53 demonstrated that melanoma tumors in these double knockout mice exhibited twice the growth rate of tumors grown in mice deficient in p53 alone [185]. Mammary-specific overexpression in mice prone to mammary tumors resulted in significantly inhibited or no tumor growth due to decreased vascularity [186]. Conversely, mammary-specific knockout of the TSP-1 gene in these same mice led to increased incidence of tumors, coupled with increased growth rates and hypervascularity of resultant tumors due to increased VEGF/VEGFR-2 interaction. Interestingly, elevated levels of active matrix metalloproteinases (MMPs), zinc-dependent proteases that are involved in endothelial cell migration during angiogenesis and in tumor invasivity (see below) were also observed in these animals, while overexpression of TSP-1 in mammary epithelial cells negatively regulated the activity of MMPs. Further studies with cells from these mice demonstrated that the interaction between TSP-1 and the proform of MMP-9 prevents the proteolytic cleavage of MMP-9 required to activate the enzyme from its latent state [186]. MMP-9 has been shown to release heparin-binding growth factors, such as VEGF from the extracellular matrix [187].

Taken together, these data suggest that the anti-angiogenic properties of TSP-1 include inhibition of endothelial cell migration, proliferation and adhesion, and prevention of MMP-9 activation to, in turn, prevent the release of angiogenic factors stored in the extracellular matrix.

Troponin I

Our laboratory first described the anti-angiogenic properties of troponin I (TnI), a protein first characterized as an inhibitor of actomyosin ATPase during contractility of cardiac and skeletal muscle [188–190]. TnI was purified to homogeneity from bovine scapular cartilage using an *in vitro* angiogenesis assay that measured inhibition of endothelial cell proliferation to screen the purification process, and the identity of TnI was verified by microsequencing. Since this was the first demonstration that TnI could be anti-angiogenic, the human TnI gene was cloned and expressed. Human TnI was found to inhibit both bFGF- and VEGF-stimulated proliferation of endothelial cells *in vitro*. TnI is also a potent inhibitor of *in vivo* angiogenesis, as evidenced by inhibition of embryonic angiogenesis in the chick chorioallantoic membrane assay and of bFGF-induced angiogenesis in the mouse corneal pocket assay [189].

More recent studies have focused on elucidating the mechanisms by which TnI inhibits angiogenesis. TnI has been shown to bind the bFGF receptor and, thus, may inhibit bFGF-induced angiogenesis by competing with the angiogenic factor bFGF for its receptor expressed on endothelial cells [191]. A 30-amino acid peptide of TnI (pTnI) was recently shown to inhibit endothelial cell proliferation and tube formation *in vitro* [192]. This peptide also inhibited *in vitro* VEGF expression by the pancreatic cancer cell line CAPAN-1, and treatment of mice injected with CAPAN-1 cells with pTnI significantly decreased the number of liver metastases compared to control animals [192]. Based on these reports, TnI is currently in clinical development for use as an inhibitor of solid tumor growth and metastasis.

TIMP-2/Loop 6

The MMPs, a multigene family of metal-dependent endoproteinases, are required for extracellular matrix remodeling *in vivo*, and are therefore critically important enzymes in the processes of tumor angiogenesis, progression and metastasis [187, 193–195]. Our laboratory was the first to demonstrate that inhibition of angiogenesis *in vivo* could be accomplished by inhibiting MMP activity with an endogenous tissue inhibitor of MMPs (TIMP) [196, 197]. This work was subsequently confirmed by a number of other groups, and demonstrated the key role that MMP activity played in successful neovascularization.

The TIMPs are important, endogenous negative regulators of MMPs. Currently, four TIMP family members have been identified: TIMP-1, TIMP-2, TIMP-3 and TIMP-4. Each of the TIMPs is capable of inhibiting the activity of MMPs, and thus could negatively regulate angiogenesis by preventing MMP-mediated migration of endothelial cells and tumor cells [188]. However, of the known TIMPs, TIMP-2 is unique in that it has been shown to inhibit endothelial cell proliferation as well [198], in contrast to, for example, TIMP-1, which has been shown to modestly stimulate the proliferation of many types of cells including endothelial cells and tumor cells [199, 200]. Therefore, it is possible that TIMP-2 could inhibit angiogenesis by inhibiting both the proliferation and migration of endothelial cells in addition to its MMP-inhibitory activity.

Little is known regarding the mechanism by which TIMP-2 inhibits endothelial cell proliferation. One possibility is that TIMP-2 inhibits the expression of VEGF. Overexpression of TIMP-2 in breast carcinoma cells was associated with downregulation of VEGF expression both *in vitro* and *in vivo*, resulting in decreased angiogenesis and tumor growth [201].

The ability of TIMP-2 to inhibit endothelial cell proliferation may be a receptor-mediated event and studies have recently focused on the identification of endothelial cell surface receptors for TIMP-2. It has been demonstrated that the anti-proliferative activity of TIMP-2 was dependent on the cellular expression of β_1 integrins [202], and further data showed that the $\alpha_3\beta_1$ integrin is a functional TIMP-2 receptor on endothelial cells.

To characterize the MMP-inhibitory and anti-proliferative activities of TIMP-2 vis-à-vis its anti-angiogenic activity, our laboratory performed a series of structure-function studies of TIMP-2 to determine the regions of the molecule responsible for its anti-angiogenic activity [203]. We first confirmed previous reports that the MMP-inhibitory activity was housed in the N-terminal domain (designated T2N) [204, 205]. Using *in vitro* endothelial cell proliferation assays, MMP radiometric enzyme assays, and the *in vivo* angiogenesis assays, the CAM assay and the mouse corneal pocket assay, we found that the anti-proliferative activity was housed in the C-terminal domain (designated T2C), a domain that lacks MMP-inhibitory activity [203]. Further structure-function mapping of the C-terminal domain determined that the anti-proliferative, anti-angiogenic region of the T2C domain resided in Loop 6 of the TIMP-2 molecule. Loop 6 was a potent inhibitor of endothelial cell proliferation *in vitro*, and inhibited *in vivo* angiogenesis in the CAM and corneal pocket assays [203]. To summarize, TIMP-2 contains two anti-angiogenic activities that are independent of one another: MMP-inhibitory activity in the T2N and anti-proliferative activity in the T2C. Furthermore, Loop 6, which is responsible for the anti-proliferative activity of T2C, represents a novel, small molecular weight inhibitor of angiogenesis [203].

Angiostatin

The phenomenon of spontaneous growth of dormant metastases after surgical resection of certain types of primary tumors [206] led to the hypothesis that perhaps the primary tumor was producing a circulating inhibitor that suppressed the growth of these dormant metastases. Murine tumor models using Lewis Lung carcinoma were developed that could mimic this phenomenon. Research to identify these tumor-derived inhibitors using this tumor system led to the discovery of the angiogenesis inhibitor angiostatin [207].

Using *in vitro* angiogenesis assays to monitor the purification process, angiostatin was purified from the urine and serum of Lewis lung carcinoma-bearing mice [207]. Sequencing of purified angiostatin, a 38-kDa protein, revealed that it is an internal fragment of plasminogen encompassing the first four of the five kringle domains of the molecule. Angiostatin, but not intact plasminogen, inhibited endothelial cell proliferation *in vitro* in a cell-specific manner, and the anti-angiogenic activity of angiostatin could be blocked through immunodepletion with angiostatin-specific antibodies. Angiostatin was also effective in inhibiting embryonic angiogenesis in the CAM assay. Furthermore, angiostatin was shown to inhibit the growth of primary tumors and metastases in a variety of *in vivo* human tumor models. In a separate report, it was demonstrated that angiostatin restricts the growth of micrometastases by increasing the rate of apoptosis in these tumors [208].

Reports in the literature suggested that tumor cells do not secrete angiostatin. Rather, it is hypothesized that tumors secrete a number of enzymes that generate angiostatin through proteolysis of plasminogen [209]. A number of molecules that release angiostatin from its parent molecule have been identified. Several members of the MMP family are able to cleave plasminogen in such a manner as to produce angiostatin: MMP-2 [209, 210]; MMP-3/stromelysin [211], MMP-7 [212], MMP-9 [212, 213], and MMP-12/human macrophage metalloelastase [214, 215]. In addition, both tissue-type plasminogen activator (tPA) and urokinase plasminogen activator (uPA) can generate angiostatin from plasminogen [216].

The full complement of anti-angiogenic mechanisms utilized by angiostatin is still being elucidated. Significant effort has been focused on the identification of receptors on the surface of endothelial cells that could mediate angiostatin's anti-angiogenic activities. To date, several molecules have been identified as putative endothelial cell surface receptors for angiostatin, including the ATP synthase F1 complex [217, 218], the $\alpha_v\beta_3$ integrin [219], angiomotin [220], and annexin II [221]. Angiostatin is currently being tested in a number of human clinical trials for its antitumor effects.

Endostatin

Like angiostatin, endostatin is an internal fragment of a larger, parent molecule. Endostatin, a 20-kDa C-terminal fragment of collagen XVIII, is a potent inhibitor of angiogenesis, while its parent molecule does not possess anti-angiogenic activities [222]. Unlike the simple cleavage of angiostatin from plasminogen via one enzyme, the generation of endostatin from collagen XVIII appears to be a two-step process. An as-of-yet unidentified MMP activity is responsible for cleavage of the non-collagenous-1 (NC1) domain from the parent molecule, which is then cleaved by an elastase activity that processes the NC1 domain into functional endostatin [223]. Other enzymes can also perform this second digestion producing endostatin from the NC1 domain, such as cathepsins L, K and B, MMP-3, MMP-9, MMP-12, MMP-13 and MMP-20, as well as MMP-2 and MMP-14, although to a lesser degree [224, 225].

Endostatin was originally purified from the conditioned medium of hemangioendothelioma (EOMA) cells, and was shown to specifically inhibit endothelial cell proliferation [222] and migration [226], and to induce apoptosis in endothelial cells [227].

Systemic treatment with recombinant endostatin was shown to drastically inhibit the growth of multiple tumor types through decreased angiogenesis coupled with increased apoptosis of tumor cells [222, 228]. Most remarkably, however, cyclic rounds of endostatin treatment not only inhibited tumor growth during the treatment stages, but after several rounds of treatment, the tumors entered remission [229]. This suggested that endostatin might be an effective anti-angiogenic cancer therapy without the induction of drug resistance, which is common to many chemotherapeutic agents.

A considerable amount of data suggests that at least one of the major activities of endostatin appears to be its inhibition of endothelial cell migration [230]. Many of the observed activities of endostatin are consistent with this premise. Endostatin can block the activation of MMP-2, an MMP shown to be important in endothelial cell motility, by binding to the catalytic domain of MMP-2 [231, 232]. Recently, *in vitro* endothelial cell assays coupled with transcriptional profiling experiments demonstrated that endostatin inhibits migration of endothelial cells via suppression of *c-myc* and other genes associated with cell migration [226, 233].

Endostatin may also inhibit endothelial cell migration by negatively regulating VEGF expression and/or activity. Endostatin treatment in the *in vitro* mouse aortic ring assay and in *in vivo* murine tumor models, resulted in significant downregulation of VEGF mRNA and protein [234]. A direct interaction between endostatin and VEGFR-2 has been reported [235], and this interaction blocks VEGF-stimulated chemotaxis [236]. These negative effects of endostatin on VEGF and VEGF signaling may explain the ability of endostatin to block the VEGF-stimulated mobilization of circulating endothelial cells (CECs), which express VEGFR-2 [237].

Endostatin is a heparin-binding protein, and this ability to bind heparin may mediate its anti-angiogenic activities. Recently, it was reported that a heparin-binding motif within endostatin is responsible for the ability of endostatin to inhibit endothelial cell migration stimulated by VEGF or bFGF *in vitro* and to block VEGF- or bFGF-stimulated angiogenesis *in vivo* in the CAM assay [238].

In addition to binding VEGFR-2 on the cell surface, a number of potential endothelial cell-surface receptors for endostatin have been identified, including the $\alpha_5\beta_1$ integrin [239] and heparan sulfate proteoglycans glypican-1 and -4 [240], and heparin sulfate may act as a co-receptor for endostatin [230].

A number of inhibitors of angiogenesis have been discussed in this section and synthetic inhibitors that target VEGF are discussed in the next. For detailed discussions of other inhibitors of angiogenesis not discussed here, we recommend the review "Endogenous inhibitors of angiogenesis" by Judah Folkman [241].

MMPs: a dichotomy of positive and negative regulation of angiogenesis

It has been well documented that expression and activity of MMPs are required during angiogenesis, and that MMP activity is one of the earliest and most sustained events during the course of the angiogenic process [187, 193–195]. Suppression of angiogenesis *in vivo* via inhibition of MMP activity provides evidence of the importance of these proteins in tumor angiogenesis [196, 242]. MMPs are utilized by the primary tumor to facilitate invasion into the surrounding stroma, and by metastatic tumor cells to intravasate into and extravasate out of the vasculature, and to degrade extracellular matrix and stroma as it migrates to establish a metastatic foci [188, 243]. MMPs are also expressed by endothelial cells [244–249], and expression of MMPs by endothelial cells is regulated by several angiogenesis-related molecules including VEGF [250–252] and bFGF [253]. MMPs secreted by stimulated endothelial cells are initially utilized to degrade the vascular basement membrane, and then to digest extracellular matrix components enabling the cells to migrate towards the tumor.

In addition to their role in the migration of tumor and endothelial cells, MMPs can also promote angiogenesis through release of angiogenic factors stored in the extracellular matrix, such as VEGF [32, 187] and bFGF [254, 255]. Matrix-bound growth factors represent a potential pool of stored growth factors whose release is mediated through proteolysis of the extracellular matrix. This results in an increase in the bioavailability of angiogenic stimulators without a concomitant increase in the expression of these stimulators.

Thus far, this section has focused on the pro-angiogenic properties of MMPs. However, the activities of several MMPs can serve protective functions against tumor growth and progression. The fact that MMPs can be both pro- and anti-angiogenic may help to explain why disappointing results have been obtained from clinical trials with non-specific MMP inhibitors [188, 195, 209, 256, 257].

results from this dose-escalation study indicate that like Avastin, VEGF-Trap does not induce toxicity and the maximum tolerated dose has not yet been reached. The clinical results from this study should be released later this year.

Potential pitfalls of anti-VEGF strategies

The clinical success of VEGF-targeting molecules validates this modality for the treatment of human cancers. However, there is some concern regarding the general applicability of this strategy to the broad spectrum of human cancers. As observed in the phase III clinical trial studying the efficacy of Avastin treatment of refractory breast cancer [289], and in clinical trials of SU5416, a VEGFR tyrosine kinase inhibitor [290, 291], targeting VEGF may simply not be sufficient for full therapeutic efficacy in all cancer types.

VEGF was found to be expressed or upregulated in approximately 60% of human cancers [241], with approximately 40% that do not express VEGF, and would therefore not be susceptible to anti-VEGF strategies. Cancers may express other angiogenic factors, such as bFGF, PDGF, or EGF to promote neovascularization. In addition, most types of cancer express more than one type of angiogenic stimulator or may change the expression of these angiogenic factors during the course of tumor progression [292]. While molecules that target VEGF have been successful, the phenotype of the cancer must be taken into consideration before employing anti-VEGF strategies alone as with any anti-tumor strategy.

Anti-angiogenic/low-dose/metronomic chemotherapy

One of the most interesting applications of anti-angiogenic therapy has come from the discovery that low-dose, or metronomic scheduling of conventional chemotherapeutic agents has the potential to be anti-angiogenic. Cytotoxic chemotherapy is designed to indiscriminately induce apoptosis of proliferating cells, i.e., tumor cells, and is administered at the clinically determined maximum-tolerated dose (MTD). Treatment regimens generally include short courses of treatment followed by breaks to allow recovery, usually in a periodic or cyclical fashion. The indiscriminate death of normal proliferating cells throughout the body (hematopoietic cells, hair follicles, intestinal lining etc.) results in many side effects, which necessitate breaks between treatments.

The main targets of conventional chemotherapy are the tumor cells themselves. However, since these agents target dividing cells, in theory they should also possess anti-angiogenic activity by killing endothelial cells stimulated to proliferate by angiogenic factors. Another important facet of endothelial cells that makes them ideal targets for chemotherapeutics is that they are essentially genetically stable, and, therefore, would not acquire drug resistance commonly observed in many types of cancer due to the genetic instability of cancer cells [229]. Much work has been done recently to enhance the efficacy of traditional chemotherapeutics by modifying the treatment regimens to attack not only the tumor cells but also the endothelial cells of the tumor's vasculature in other words, anti-angiogenic chemotherapy.

The seminal research that literally opened this new field of cancer therapy was first reported by Browder et al. [293]. Rather than treat tumors explanted into mice with the conventional MTD-determined regimen, tumor-bearing mice were treated with an anti-angiogenic schedule of chemotherapy delivery: more continuous, lower doses of the chemotherapeutic agent, cyclophosphamide. The anti-angiogenic schedule of cyclophosphamide resulted in the stabilization of cyclophosphamide-resistant Lewis lung carcinomas and EMT-6/CTX breast carcinomas. This suppression of tumor growth in drug-resistant tumors was attributed to increased apoptotic rates in the endothelial cells of the tumor vasculature coupled with subsequently increased apoptosis of the tumor cells themselves. This anti-angiogenic schedule also eradicated cyclophosphamide-sensitive Lewis Lung carcinomas and L1210 leukemias without drug-resistance, which is not possible with the MTD-based schedule. Finally, the anti-angiogenic schedule of cyclophosphamide coupled with the angiogenesis inhibitor TNP-470 not only suppressed tumor growth, but also led to complete regression of these tumors. The importance of this work is that it demonstrated that by altering the treatment schedule, treatment of cyclophosphamide-resistant tumors with cyclophosphamide could still be effective.

The question arises: Since most chemotherapeutic agents are capable of inducing apoptosis of proliferating endothelial cells [294], why is it that MTD-based chemotherapy is not effective in attacking the tumor vasculature, while anti-angiogenic or "metronomic" [295] chemotherapy is? It now appears that the break period between doses in MTD-based scheduling allows for the recovery of the vasculature [293]. The metronomic schedule arose as a means to prevent the repair of the tumor vasculature by administering lowered doses of chemotherapy more frequently. The effectiveness of metronomic scheduling of chemotherapy either alone or coupled with other anti-angiogenic agents has been confirmed by several other reports from preclinical and clinical studies [296–309].

While the most effective anti-angiogenic chemotherapy regimens must still be evaluated on a case by case basis, these reports give some hope as to the effectiveness of this anti-angiogenic strategy for treating cancer. Currently several clinical trials are underway to measure the effectiveness of metronomic chemotherapy against many types of tumors [310–314]. Recently, preclinical studies using the RIP1-Tag2 mouse model of pancreatic tumorigenesis demonstrated that a combinatorial strategy involving MTD scheduling of chemotherapeutics, followed by a metronomic chemotherapy schedule coupled with anti-VEGF was most effective in treating intractable end-stage pancreatic tumors [315]. These data suggest that similar combinatorial strategies coupling anti-angiogenic therapies with anti-angiogenic scheduling of chemotherapy may be most successful in treating human cancers in clinical settings.

Molecular determinants of the angiogenic switch

In the previous section, several cancer therapies based on the inhibition of angiogenesis were discussed. In each case, these treatments were designed to treat an established tumor that already possesses its own vascular supply. Based on the widely held paradigm in cancer therapy that treating a tumor early in its progression yields a greater chance of disease-free survival, it is obviously more advantageous to treat a tumor shortly after the tumor has initially become vascularized, rather than treating a tumor with an extensive vascular network. In fact, the best-case scenario would be to prevent the tumor from acquiring an angiogenic phenotype in the first place. By identifying and characterizing the molecular mechanisms that regulate the angiogenic switch, new diagnostic and therapeutic targets can be discovered for detecting and treating cancer, respectively, at one of the earliest stages in tumor progression.

The angiogenic switch is formally defined as the transition from an avascular, or pre-angiogenic, dormant tumor to a vascularized, or angiogenic, active tumor that has the capacity to progress to an invasive phenotype [1]. This acquisition of the angiogenic phenotype is one of the earliest events during the course of tumor progression and represents a critical checkpoint in tumor progression [1, 316, 317]. The angiogenic switch has been correlated with tumor progression in breast cancer [1, 318], such that, based on quantitative assessment of microvessel density in progressive stages of breast cancer from ductal carcinoma *in situ* to metastatic disease, the degree of vascularity correlates directly with an increased occurrence of metastasis. Therapeutic interventions aimed at blocking the angiogenic switch have the potential to prevent further tumor progression to an invasive, metastatic, and subsequently lethal phenotype [319].

This transition to the angiogenic phenotype occurs when changes in the genetic make-up and subsequent molecular expression, which have recently begun to be characterized, lead to a shift from a state that favors inhibition of angiogenesis to a phenotype that stimulates angiogenesis. For example, expression of the angiogenic growth factors VEGF and bFGF are upregulated and expression of negative regulators of angiogenesis, such as thrombospondin and TIMP-2, are downregulated leading to tumor neovascularization [84, 184, 187, 320–322].

While many of the molecules involved in the positive and negative regulation of angiogenesis have been identified, little is known about the molecular mechanisms that regulate the initiation of the angiogenic phenotype in a tumor. Which genetic or molecular differences exist that can determine if a tumor remains avascular or progresses to a vascular phenotype? Only recently have some of the genetic changes and subsequent alterations in protein expression and activity that promote the onset of neovascularization in a tumor been identified.

The Role of MMPs

Our laboratory was among the first to identify some of the molecules involved in the initiation of the angiogenic phenotype. Using an *in vivo* tumor model that reliably recapitulates the angiogenic switch (Fig. 3), we characterized a crucial role for MMP-2 in the transition from an avascular to a vascular tumor [193]. MMP-2 expression and activity was significantly upregulated in angiogenic tumors compared to avascular tumors, and treatment of tumors with antisense oligonucleotides directed against MMP-2 inhibited angiogenesis *in vitro* and *in vivo*, and significantly inhibited the growth of the treated tumors compared to controls. In a separate study utilizing a different model of the angiogenic switch, the RIP1-Tag2 mouse tumor model in which pancreatic tumors arise spontaneously, MMP-9 was required for the onset of neovascularization of the tumors that arose [187]. MMPs have been implicated in the invasivity of primary tumors and metastases [243] and, taken together, these data and data from other reports suggest that MMP expression/activity, and in particular gelatinase activity, is also necessary to initiate the earliest stages of the angiogenic program.

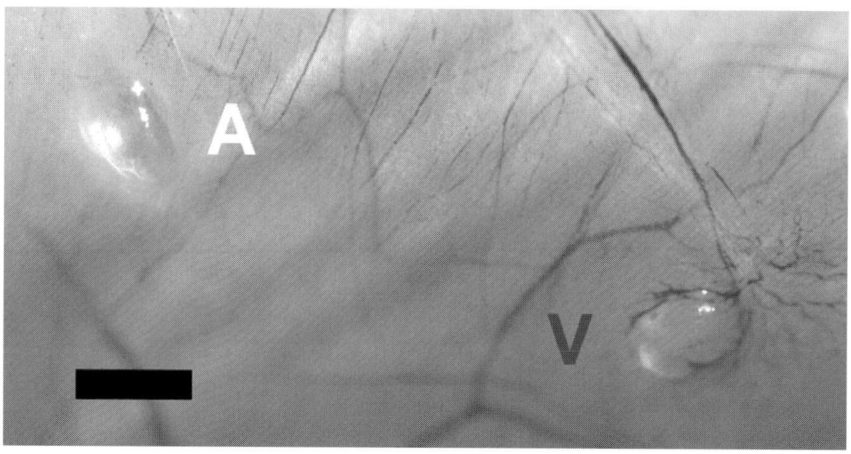

Figure 3. *In vivo* model of the angiogenic switch. A tumor model that reliably recapitulates the angiogenic switch has been developed by our laboratory in which avascular tumor nodules (A) and vascular tumor nodules (V) can be harvested simultaneously for biochemical and molecular analyses to characterize the molecular determinants of the angiogenic switch. Scale bar = 1 mm.

Hypoxia and the angiogenic switch

The role of VEGF and its receptors in both normal and pathophysiological angiogenesis has been discussed previously in this chapter. Given its importance in tumor angiogenesis, numerous studies have focused on identifying molecules that modulate the expression or activity of VEGF (Fig. 2). Since VEGF plays such a crucial role in initiating an angiogenic response, much recent work has focused on identifying transcription factors that regulate expression of VEGF, and determining how these molecules are involved in the angiogenic switch. One of these is the transcription factor HIF-1 [73].

In a pre-angiogenic tumor, hypoxic conditions can arise within the tumor as the tumor's mass increases to a point where it is no longer supported by the oxygen and nutrients that are able to diffuse from a nearby vessel. It has been well established that hypoxia induces upregulation of VEGF expression [72, 73]. This hypoxia-induced VEGF would then recruit angiogenic vessels to vascularize the tumor and relieve the hypoxic stress. Given this relationship between hypoxia and VEGF expression, it is likely that hypoxia and hypoxia-inducible elements are key modulators of the angiogenic switch [73].

HIF-1 is one of the major mediators of hypoxia-inducible gene expression [75, 76, 323]. It is a heterodimeric transcription factor composed of two subunits: HIF-1α and HIF-1β [74, 324]. HIF-1β is constitutively expressed, while expression of HIF-1α can be stimulated by hypoxia and certain growth factors, as well as by various genetic mutations. Activity of the HIF-1 transcription factor complex is regulated by both the expression of HIF-1α and the cellular localization of this subunit [73, 320]. Under normoxic conditions, HIF-1α is sequestered in the cytoplasm of the cell, while HIF-1β is localized to the nucleus. Upon the onset of hypoxia, HIF-1α is translocated into the nucleus where it can bind to its partner HIF-1β, resulting in the activation of transcription of HIF-1-inducible genes, such as VEGF.

There is a growing body of evidence supporting a role for HIF-1 in the induction of the angiogenic phenotype. Interestingly, in a variety of studies, VEGF has been shown to be upregulated in avascular tumors [187, 320, 325], which at the onset may appear to be counterintuitive. Based on the observed activity of the HIF-1 transcription factor in these models, it is hypothesized that hypoxia within these pre-angiogenic lesions may be at least partly responsible for this upregulation of VEGF expression. In avascular tumors, which express high levels of VEGF, HIF-1α was localized to the nucleus where it can activate HIF-1-mediated gene expression by binding to HIF-1β [320]. In the angiogenic tumors, which are relieved of hypoxic stress, HIF-1α is sequestered in the cytoplasm, rendering the HIF-1 transcription factor inactive, and resulting in the significantly decreased VEGF expression observed in these tumors. It should be noted that as VEGF levels drop after the initial vascularization of the avascular tumors, expression of the angiogenic factor bFGF is significantly upregulated to sustain the angiogenic response initiated by VEGF [320, 325]. Furthermore, during later stages of tumor growth and development, the VEGF expression is again increased as expected.

Bi-mechanistic regulation of the angiogenic switch

The initiation of angiogenesis is characterized by an overall increase in angiogenic stimulus through increased expression of angiogenic stimulators and/or decreased expression of inhibitors of angiogenesis. There are several molecules that represent excellent regulatory candidates of the angiogenic switch because they modulate angiogenesis by affecting the expression or activity of both positive and negative regulators of angiogenesis. It also comes as no surprise that the many of the molecules that promote angiogenesis in this bi-mechanistic fashion are oncogenes, and that many of the molecules that impede the initiation of the angiogenic phenotype in this manner are tumor suppressor genes. Since genetic instability is one of the hallmarks of cancer and is associated with both early and late stages in tumor progression [326], it is likely that activating mutations of oncogenes and/or deletion or inactivating mutations of tumor suppressor genes could be genetic determinants of the angiogenic switch.

Ras

Ras is perhaps one of the best characterized oncogenes. The Ras proteins, H-Ras, N-Ras, and two forms of K-Ras (K-Ras4A and K-Ras 4B) are GTP-binding proteins whose activity is regulated in a positive manner by guanine nucleotide exchange factors (GEFs) and negatively through hydrolysis of GTP by GTPase-activating proteins (GAPs) [327, 328]. Ras is in an inactive state when bound to GDP and becomes activated when GEFs, such as Ras-GAP [329], dissociate GDP from Ras allowing GTP to bind. Ras returns to an inactive state when GTPases, for example p120/GAP, hydrolyze the bound GTP to form GDP [328]. GTP-bound, activated Ras is farnesylated and then localizes to the inner surface of the cell membrane [330, 331], where it can act as an effector of a number of activated receptor tyrosine kinases including the VEGF and EGF receptors (Fig. 2) [332–335]. The downstream activity of Ras is mediated through a series of effectors including, but not limited to, Raf serine/threonine kinases, MAP kinases and phosphoinositide 3-kinases (PI3Ks) [327].

Expression of mutant forms of Ras is one of the most common genetic changes detected in human cancers, present in approximately 30% of all human cancers [83, 336]. These mutations interfere with the ability of GAPs to hydrolyze GTP bound to Ras, resulting in a constitutively active protein. Oncogenic Ras mutations are associated with virtually every aspect of malignant transformation including proliferation, transformation, invasion and metastasis [327, 328, 337]. In addition to Ras conferring a survival advantage directly upon the cancer cells themselves, Ras and mutated Ras proteins are also involved in promoting angiogenesis via a number of different mechanisms.

A link between Ras signaling and subsequent upregulation of VEGF expression has been clearly established [78, 83–85, 177, 181, 338–342]. Overexpression of *K-ras* and *H-ras* oncogenes in normal epithelial cell lines or expression of mutant Ras in immortalized epithelial cells results in significant upregulation of VEGF [83, 181], and expression of mutant Ras results in significantly increased VEGF expression in fibroblasts [181, 338] and endothelial cells [84]. Inhibition of the Ras pathway via transfection and expression of a dominant negative form of H-Ras, H-Ras(N17) or via treatment with farnesyl transferase inhibitors, which remove Ras from the inner cell surface, each results in significant reduction of VEGF expression and synthesis [83, 341]. Inactivation of the mutant K-Ras utilizing a Ras-specific hammerhead ribozyme decreased VEGF expression in colorectal cancer cells *in vitro*, and *in vivo* angiogenesis and subsequent tumor growth were markedly inhibited when these cells were injected into mice [342]. Ras is also capable of upregulating other members of the VEGF family. PlGF expression is upregulated in a Ras-dependent manner in carcinogen-induced skin cancers in mice [343].

The downstream effectors of Ras-mediated VEGF expression include, but are not limited to the transcription factors Sp1, Net, and HIF-1. Sp1 is a well-characterized transcriptional activator of VEGF [79]. It was recently demonstrated that PKC zeta, acting downstream of Ras, promotes the binding of Sp1 to the VEGF promoter [78]. Activated Ras also phosphorylates and activates the transcription factor Net, which is capable of binding to and activating the VEGF promoter [344]. Additionally, downregulation of Net results in inhibition of angiogenesis *in vitro* and *in vivo* due to significantly decreased VEGF expression [344]. Ras-mediated upregulation of VEGF may also be involved with the HIF-1 transcription factor, as downstream effectors of Ras have been implicated in simulating or synergizing the effects of hypoxia upon VEGF expression [339, 341, 345].

Ras can also promote the angiogenic phenotype through upregulation of both the expression and bioavailability/activity of other angiogenic factors, in particular bFGF. Upregulated Ras activity results in increased expression of bFGF [345, 346], and activation of bFGF localized to the extracellular matrix is mediated by oncogenic Ras in skeletal muscle cells [347]. This latter activity may be through Ras-mediated expression of FGF binding protein (FGF-BP). FGF-BP is itself a potential candidate of the angiogenic switch that binds to, mobilizes and activates FGF stored in the extracellular matrix [348]. Message levels of FGF-BP are significantly increased as a function of Ras activation *in vitro* [349].

The MMPs, discussed previously, are important regulators of angiogenesis and are implicated in the angiogenic switch [187, 193]. Activation of Ras through overexpression or via mutant forms of Ras are implicated in upregulation of several members of the MMP family, and in particular MMP-2 [84, 350–354] and MMP-9 [84, 350, 351, 355–358]. TIMP-1 expression was concomitantly downregulated with upregulation of MMP-9 as a function of Ras activity [359]. MMP-1 expression is also increased upon Ras activation [360]. Taken together, these data suggest that Ras is involved in the positive regulation of several members of the MMP family.

The mechanisms Ras utilizes to initiate the initiation of the angiogenic phenotype are multi-faceted, and thus far the discussion has focused on the ability of Ras to stimulate pro-angiogenic factors. However, Ras is capable of stimulating a pro-angiogenic environment by downregulating key inhibitors of angiogenesis as well, in particular TSP-1.

As discussed previously, TSP-1 is a potent and well-characterized inhibitor of angiogenesis. Several published reports demonstrate that activation of Ras leads to downregulation of TSP-1 [177, 183, 342, 361, 362]. An inverse correlation has been reported between transfection with either N-Ras or oncogenic H-Ras or K-Ras and TSP-1 expression [361]. Transformation of VEGF-deficient fibroblasts with Ras significantly downregulated expression of TSP-1 [362]. Injection of these cells gave rise to highly angiogenic fibrosarcomas, which in turn could be inhibited with treatment of a fragment of TSP-1 [362]. Conversely, inactivation of mutant Ras using an anti-Ras hammerhead ribozyme resulted in significantly increased TSP-1 expression [342].

The pathway linking Ras activation and repression of TSP-1 was elegantly described by Watnick et al., 2003 [177], using human mammary epithelial and embryonic kidney cell lines that are transformed with the SV40 early region, the catalytic subunit of human telomerase, and the oncogenic allele of H-Ras, H-Ras(G12V). Ras induces the activation of PI3K, which in turn activates Rho, which in turn activates ROCK, which phosphorylates and activates the transcription factor Myc. Activated Myc then represses transcription of TSP-1. PI3K and ROCK are thus upstream of TSP-1 repression, a fact that was verified using breast carcinoma cell lines, which express no endogenous TSP-1. Treatment of these cell lines with inhibitors to either PI3K or ROCK resulted in dramatic upregulation of TSP-1.

V-Src

Another oncogene that may exert its oncogenic effects through activation of the angiogenic switch via upregulation of VEGF and downregulation of TSP-1 is V-Src. *V-src*, the first described oncogene, encodes a tyrosine kinase that is a downstream effector of several RTKs [363]. Treatment of tumors with a Src tyrosine kinase inhibitor (AZM475271; AstraZeneca) results in significant reduction in tumor volume and number of metastases due to decreased microvessel density [364], suggesting that Src may be playing a role in the regulation of angiogenesis. Numerous reports indicate that expression of oncogenic V-Src upregulates expression of VEGF [86–88, 346, 365] and suggest that oncogenic mutations of V-Src might contribute to initiating the angiogenic phenotype.

Overexpression of V-Src results in the activation of the VEGF promoter in promoter reporter assays, indicating that downstream effectors of V-Src upregulate VEGF expression through increased transcription of the VEGF gene [365]. Two transcription factors that activate VEGF transcription are reported to be downstream of V-Src.

Earlier in this section the role of HIF-1-mediated VEGF expression in the angiogenic switch was discussed. Cells that overexpress oncogenic V-Src display increased expression of HIF-1α with subsequent upregulation of VEGF [87]. Conversely, inhibition of Src activity through either expression of a dom-

A, Nusgens BV et al. (2001) Down-regulation of vascular endothelial growth factor by tissue inhibitor of metalloproteinase-2: effect on *in vivo* mammary tumor growth and angiogenesis. *Cancer Res* 61: 3450–3457
202 Stetler-Stevenson WG, Seo DW (2005) TIMP-2: an endogenous inhibitor of angiogenesis. *Trends Mol Med* 11: 97–103
203 Fernandez CA, Butterfield C, Jackson G, Moses MA (2003) Structural and functional uncoupling of the enzymatic and angiogenic inhibitory activities of tissue inhibitor of metalloproteinase-2 (TIMP-2): loop 6 is a novel angiogenesis inhibitor. *J Biol Chem* 278: 40989–40995
204 Murphy G, Houbrechts A, Cockett MI, Williamson RA, O'Shea M, Docherty AJ (1991) The N-terminal domain of tissue inhibitor of metalloproteinases retains metalloproteinase inhibitory activity, *Biochemistry* 30: 8097–8102
205 Muskett FW, Frenkiel TA, Feeney J, Freedman RB, Carr MD, Williamson RA (1998) High resolution structure of the N-terminal domain of tissue inhibitor of metalloproteinases-2 and characterization of its interaction site with matrix metalloproteinase-3. *J Biol Chem* 273: 21736–21743
206 Sugarbaker EV, Ketcham AS (1977) Mechanisms and prevention of cancer dissemination: an overview. *Semin Oncol* 4: 19–32
207 O'Reilly MS, Holmgren L, Shing Y, Chen C, Rosenthal RA, Moses M, Lane WS, Cao Y, Sage EH, Folkman J (1994) Angiostatin: a novel angiogenesis inhibitor that mediates the suppression of metastases by a Lewis lung carcinoma. *Cell* 79: 315–328
208 Holmgren L, O'Reilly MS, Folkman J (1995) Dormancy of micrometastases: balanced proliferation and apoptosis in the presence of angiogenesis suppression. *Nat Med* 1: 149–153
209 O'Reilly MS, Wiederschain D, Stetler-Stevenson WG, Folkman J, Moses MA (1999) Regulation of angiostatin production by matrix metalloproteinase-2 in a model of concomitant resistance. *J Biol Chem* 274: 29568–29571
210 Moses MA, O'Reilly MS (2003) Regulation of angiostatin mobilization by tumor-derived matrix metalloproteinase-2. *Methods Mol Med* 74: 375–390
211 Lijnen HR, Ugwu F, Bini A, Collen D (1998) Generation of an angiostatin-like fragment from plasminogen by stromelysin-1 (MMP-3). *Biochemistry* 37: 4699–4702
212 Patterson BC, Sang QA (1997) Angiostatin-converting enzyme activities of human matrilysin (MMP-7) and gelatinase B/type IV collagenase (MMP-9). *J Biol Chem* 272: 28823–28825
213 Dong Z, Kumar R, Fidler IJ (1996) Generation of the angiogenesis inhibitor, angiostatin, by Lewis lung carcinoma is mediated by macrophage elastase. *Proc Am Assoc Cancer Research* 37: 403a
214 Dong Z, Kumar R, Yang X, Fidler IJ (1997) Macrophage-derived metalloelastase is responsible for the generation of angiostatin in Lewis lung carcinoma. *Cell* 88: 801–810
215 Gorrin Rivas MJ, Arii S, Furutani M, Harada T, Mizumoto M, Nishiyama H, Fujita J, Imamura M (1998) Expression of human macrophage metalloelastase gene in hepatocellular carcinoma: correlation with angiostatin generation and its clinical significance. *Hepatology* 28: 986–993
216 Ramsden JD, Yarram S, Mathews E, Watkinson JC, Eggo MC (2002) Thyroid follicular cells secrete plasminogen activators and can form angiostatin from plasminogen. *J Endocrinol* 173: 475–481
217 Moser TL, Stack MS, Asplin I, Enghild JJ, Hojrup P, Everitt L, Hubchak S, Schnaper HW, Pizzo SV (1999) Angiostatin binds ATP synthase on the surface of human endothelial cells. *Proc Natl Acad Sci USA* 96: 2811–2816
218 Moser TL, Kenan DJ, Ashley TA, Roy JA, Goodman MD, Misra UK, Cheek DJ, Pizzo SV (2001) Endothelial cell surface F1–F0 ATP synthase is active in ATP synthesis and is inhibited by angiostatin. *Proc Natl Acad Sci USA* 98: 6656–6661
219 Tarui T, Miles LA, Takada Y (2001) Specific interaction of angiostatin with integrin alpha(v)beta(3) in endothelial cells. *J Biol Chem* 276: 39562–39568
220 Troyanovsky B, Levchenko T, Mansson G, Matvijenko O, Holmgren L (2001) Angiomotin: an angiostatin binding protein that regulates endothelial cell migration and tube formation. *J Cell Biol* 152: 1247–1254
221 Tuszynski, G. P., Sharma, M. R., Rothman, V. L. and Sharma, M. C. (2002) Angiostatin binds to tyrosine kinase substrate annexin II through the lysine-binding domain in endothelial cells. *Microvasc Res* 64: 448–462
222 O'Reilly MS, Boehm T, Shing Y, Fukai N, Vasios G, Lane WS, Flynn E, Birkhead JR, Olsen BR, Folkman J (1997) Endostatin: an endogenous inhibitor of angiogenesis and tumor growth. *Cell*

88: 277–285
223 Wen W, Moses MA, Wiederschain D, Arbiser JL, Folkman J (1999) The generation of endostatin is mediated by elastase. *Cancer Res* 59: 6052–6056
224 Felbor U, Dreier L, Bryant RA, Ploegh HL, Olsen BR, Mothes W (2000) Secreted cathepsin L generates endostatin from collagen XVIII. *Embo J* 19: 1187–1194
225 Ferreras M, Felbor U, Lenhard T, Olsen BR, Delaisse J (2000) Generation and degradation of human endostatin proteins by various proteinases. *FEBS Lett* 486: 247–251
226 Shichiri M, Hirata Y (2001) Antiangiogenesis signals by endostatin. *Faseb J* 15: 1044–1053
227 Dhanabal M, Ramchandran R, Waterman MJ, Lu H, Knebelmann B, Segal M, Sukhatme VP (1999) Endostatin induces endothelial cell apoptosis. *J Biol Chem* 274: 11721–11726
228 Bergers G, Javaherian K, Lo KM, Folkman J, Hanahan D (1999) Effects of angiogenesis inhibitors on multistage carcinogenesis in mice. *Science* 284: 808–812
229 Boehm T, Folkman J, Browder T, O'Reilly MS (1997) Anti-angiogenic therapy of experimental cancer does not induce acquired drug resistance. *Nature* 390: 404–407
230 Dixelius J, Cross MJ, Matsumoto T, Claesson-Welsh L (2003) Endostatin action and intracellular signaling: beta-catenin as a potential target? *Cancer Lett* 196: 1–12
231 Kim YM, Jang JW, Lee OH, Yeon J, Choi EY, Kim KW, Lee ST, Kwon YG (2000) Endostatin inhibits endothelial and tumor cellular invasion by blocking the activation and catalytic activity of matrix metalloproteinase. *Cancer Res* 60: 5410–5413
232 Lee SJ, Jang JW, Kim YM, Lee HI, Jeon JY, Kwon YG, Lee ST (2002) Endostatin binds to the catalytic domain of matrix metalloproteinase-2. *FEBS Lett* 519: 147–152
233 Abdollahi A, Hahnfeldt P, Maercker C, Grone HJ, Debus J, Ansorge W, Folkman J, Hlatky L, Huber PE (2004) Endostatin's anti-angiogenic signaling network. *Mol Cell* 13: 649–663
234 Hajitou A, Grignet C, Devy L, Berndt S, Blacher S, Deroanne CF, Bajou K, Fong T, Chiang Y, Foidart JM et al. (2002) The antitumoral effect of endostatin and angiostatin is associated with a down-regulation of vascular endothelial growth factor expression in tumor cells. *Faseb J* 16: 1802–1804
235 Kim YM, Hwang S, Kim YM, Pyun BJ, Kim TY, Lee ST, Gho YS, Kwon YG (2002) Endostatin blocks vascular endothelial growth factor-mediated signaling via direct interaction with KDR/Flk-1. *J Biol Chem* 277: 27872–27879
236 Eriksson K, Magnusson P, Dixelius J, Claesson-Welsh L, Cross MJ (2003) Angiostatin and endostatin inhibit endothelial cell migration in response to FGF and VEGF without interfering with specific intracellular signal transduction pathways. *FEBS Lett* 536: 19–24
237 Schuch G, Heymach JV, Nomi M, Machluf M, Force J, Atala A, Eder JP Jr, Folkman J, Soker S (2003) Endostatin inhibits the vascular endothelial growth factor-induced mobilization of endothelial progenitor cells. *Cancer Res* 63: 8345–8350
238 Olsson AK, Johansson I, Akerud H, Einarsson B, Christofferson R, Sasaki T, Timpl R, Claesson-Welsh L (2004) The minimal active domain of endostatin is a heparin-binding motif that mediates inhibition of tumor vascularization. *Cancer Res* 64: 9012–9017
239 Wickstrom SA, Alitalo K, Keski-Oja J (2002) Endostatin associates with integrin alpha5beta1 and caveolin-1, and activates Src via a tyrosyl phosphatase-dependent pathway in human endothelial cells. *Cancer Res* 62: 5580–5589
240 Karumanchi SA, Jha V, Ramchandran R, Karihaloo A, Tsiokas L, Chan B, Dhanabal M, Hanai JI, Venkataraman G, Shriver Z et al. (2001) Cell surface glypicans are low-affinity endostatin receptors. *Mol Cell* 7: 811–822
241 Folkman J (2004) Endogenous angiogenesis inhibitors. *Apmis* 112: 496–507
242 Johnson MD, Kim HR, Chesler L, Tsao-Wu G, Bouck N, Polverini PJ (1994) Inhibition of angiogenesis by tissue inhibitor of metalloproteinase. *J Cell Physiol* 160: 194–202
243 Kleiner DE, Stetler-Stevenson WG (1999) Matrix metalloproteinases and metastasis. *Cancer Chemother Pharmacol* 43, Suppl: S42–S51
244 Moscatelli D, Jaffe E, Rifkin DB (1980) Tetradecanoyl phorbol acetate stimulates latent collagenase production by cultured human endothelial cells. *Cell* 20: 343–351
245 Gross JL, Moscatelli D, Jaffe EA, Rifkin DB (1982) Plasminogen activator and collagenase production by cultured capillary endothelial cells. *J Cell Biol* 95: 974–981
246 Herron GS, Banda MJ, Clark EJ, Gavrilovic J, Werb Z (1986) Secretion of metalloproteinases by stimulated capillary endothelial cells. II. Expression of collagenase and stromelysin activities is regulated by endogenous inhibitors. *J Biol Chem* 261: 2814–2818
247 Herron GS, Werb Z, Dwyer K, Banda MJ (1986) Secretion of metalloproteinases by stimulated

capillary endothelial cells. I. Production of procollagenase and prostromelysin exceeds expression of proteolytic activity. *J Biol Chem* 261: 2810–2813
248 Unemori EN, Bouhana KS, Werb Z (1990) Vectorial secretion of extracellular matrix proteins, matrix-degrading proteinases, and tissue inhibitor of metalloproteinases by endothelial cells. *J Biol Chem* 265: 445–451
249 Hanemaaijer R, Koolwijk P, le Clercq L, de Vree WJ, van Hinsbergh VW (1993) Regulation of matrix metalloproteinase expression in human vein and microvascular endothelial cells. Effects of tumour necrosis factor alpha, interleukin 1 and phorbol ester. *Biochem J* 296 (Pt 3): 803–809
250 Unemori EN, Ferrara N, Bauer EA, Amento EP (1992) Vascular endothelial growth factor induces interstitial collagenase expression in human endothelial cells. *J Cell Physiol* 153: 557–562
251 Lamoreaux WJ, Fitzgerald ME, Reiner A, Hasty KA, Charles ST (1998) Vascular endothelial growth factor increases release of gelatinase A and decreases release of tissue inhibitor of metalloproteinases by microvascular endothelial cells *in vitro*. *Microvasc Res* 55: 29–42
252 Zucker S, Mirza H, Conner CE, Lorenz AF, Drews MH, Bahou WF, Jesty J (1998) Vascular endothelial growth factor induces tissue factor and matrix metalloproteinase production in endothelial cells: conversion of prothrombin to thrombin results in progelatinase A activation and cell proliferation. *Int J Cancer* 75: 780–786
253 Mignatti P, Tsuboi R, Robbins E, Rifkin DB (1989) *In vitro* angiogenesis on the human amniotic membrane: requirement for basic fibroblast growth factor-induced proteinases. *J Cell Biol* 108: 671–682
254 Emoto N, Onose H, Sugihara H, Minami S, Shimizu K, Wakabayashi I (1998) Fibroblast growth factor-2 free from extracellular matrix is increased in papillary thyroid carcinomas and Graves' thyroids. *Thyroid* 8: 491–497
255 Chai YC, Binion DG, Macklis R, Chisolm GM 3rd. (2002) Smooth muscle cell proliferation induced by oxidized LDL-borne lysophosphatidylcholine. Evidence for FGF-2 release from cells not extracellular matrix. *Vascul Pharmacol* 38: 229–237
256 Zucker S, Cao J, Chen WT (2000) Critical appraisal of the use of matrix metalloproteinase inhibitors in cancer treatment. *Oncogene* 19: 6642–6650
257 Coussens LM, Fingleton B, Matrisian LM (2002) Matrix metalloproteinase inhibitors and cancer: trials and tribulations. *Science* 295: 2387–2392
258 Levi E, Fridman R, Miao HQ, Ma YS, Yayon A, Vlodavsky I (1996) Matrix metalloproteinase 2 releases active soluble ectodomain of fibroblast growth factor receptor 1. *Proc Natl Acad Sci USA* 93: 7069–7074
259 Frater-Schroder M, Risau W, Hallmann R, Gautschi P, Bohlen P (1987) Tumor necrosis factor type alpha, a potent inhibitor of endothelial cell growth *in vitro*, is angiogenic *in vivo*. *Proc Natl Acad Sci USA* 84: 5277–5281
260 Gearing AJ, Beckett P, Christodoulou M, Churchill M, Clements J, Davidson AH, Drummond AH, Galloway WA, Gilbert R, Gordon JL et al. (1994) Processing of tumour necrosis factor-alpha precursor by metalloproteinases. *Nature* 370: 555–557
261 Brooks PC, Silletti S, von Schalscha TL, Friedlander M, Cheresh DA (1998) Disruption of angiogenesis by PEX, a noncatalytic metalloproteinase fragment with integrin binding activity. *Cell* 92: 391–400
262 Silletti S, Kessler T, Goldberg J, Boger DL, Cheresh DA (2001) Disruption of matrix metalloproteinase 2 binding to integrin alpha vbeta 3 by an organic molecule inhibits angiogenesis and tumor growth *in vivo*. *Proc Natl Acad Sci USA* 98: 119–124
263 Folkman J (1972) Anti-angiogenesis: new concept for therapy of solid tumors. *Ann Surg* 175: 409–416
264 McCarthy M (2003) Antiangiogenesis drug promising for metastatic colorectal cancer. *Lancet* 361: 1959
265 [No authors listed] (2003) Adding a humanized antibody to vascular endothelial growth factor (Bevacizumab, Avastin) to chemotherapy improves survival in metastatic colorectal cancer. *Clin Colorectal Cancer* 3: 85–88
266 Hurwitz H, Fehrenbacher L, Novotny W, Cartwright T, Hainsworth J, Heim W, Berlin J, Baron A, Griffing S, Holmgren E et al. (2004) Bevacizumab plus irinotecan, fluorouracil, and leucovorin for metastatic colorectal cancer. *N Engl J Med* 350: 2335–2342
267 Bergsland EK (2004) Update on clinical trials targeting vascular endothelial growth factor in cancer. *Am J Health Syst Pharm* 61: S12–S20

268 Kim KJ, Li B, Winer J, Armanini M, Gillett N, Phillips HS, Ferrara N (1993) Inhibition of vascular endothelial growth factor-induced angiogenesis suppresses tumour growth *in vivo*. *Nature* 362: 841–844
269 Ignoffo RJ (2004) Overview of bevacizumab: a new cancer therapeutic strategy targeting vascular endothelial growth factor. *Am J Health Syst Pharm* 61: S21–S26
270 Warren RS, Yuan H, Matli MR, Gillett NA, Ferrara N (1995) Regulation by vascular endothelial growth factor of human colon cancer tumorigenesis in a mouse model of experimental liver metastasis. *J Clin Invest* 95: 1789–1797
271 Borgstrom P, Gold DP, Hillan KJ, Ferrara N (1999) Importance of VEGF for breast cancer angiogenesis *in vivo*: implications from intravital microscopy of combination treatments with an anti-VEGF neutralizing monoclonal antibody and doxorubicin. *Anticancer Res* 19: 4203–4214
272 Fernando NH, Hurwitz HI (2003) Inhibition of vascular endothelial growth factor in the treatment of colorectal cancer. *Semin Oncol* 30: 39–50
273 Gordon MS, Margolin K, Talpaz M, Sledge GW Jr, Holmgren E, Benjamin R, Stalter S, Shak S, Adelman D (2001) Phase I safety and pharmacokinetic study of recombinant human anti-vascular endothelial growth factor in patients with advanced cancer. *J Clin Oncol* 19: 843–850
274 Margolin K, Gordon MS, Holmgren E, Gaudreault J, Novotny W, Fyfe G, Adelman D, Stalter S, Breed J (2001) Phase Ib trial of intravenous recombinant humanized monoclonal antibody to vascular endothelial growth factor in combination with chemotherapy in patients with advanced cancer: pharmacologic and long-term safety data. *J Clin Oncol* 19: 851–856
275 Kabbinavar F, Hurwitz HI, Fehrenbacher L, Meropol NJ, Novotny WF, Lieberman G, Griffing S, Bergsland E (2003) Phase II, randomized trial comparing bevacizumab plus fluorouracil (FU)/leucovorin (LV) with FU/LV alone in patients with metastatic colorectal cancer. *J Clin Oncol* 21: 60–65
276 Yang JC, Haworth L, Sherry RM, Hwu P, Schwartzentruber DJ, Topalian SL, Steinberg SM, Chen HX, Rosenberg SA (2003) A randomized trial of bevacizumab, an anti-vascular endothelial growth factor antibody, for metastatic renal cancer. *N Engl J Med* 349: 427–434
277 Cobleigh MA, Langmuir VK, Sledge GW, Miller KD, Haney L, Novotny WF, Reimann JD, Vassel A (2003) A phase I/II dose-escalation trial of bevacizumab in previously treated metastatic breast cancer. *Semin Oncol* 30: 117–124
278 Johnson DH, Fehrenbacher L, Novotny WF, Herbst RS, Nemunaitis JJ, Jablons DM, Langer CJ, DeVore RF 3rd, Gaudreault J, Damico LA et al. (2004) Randomized phase II trial comparing bevacizumab plus carboplatin and paclitaxel with carboplatin and paclitaxel alone in previously untreated locally advanced or metastatic non-small-cell lung cancer. *J Clin Oncol* 22: 2184–2191
279 Miller KD, Rugo HS, Cobleigh MA et al. (2002) Phase III trial of capecitabine (Xeloda) plus bevacizumab (Avastin) *versus* capecitabine alone in women with metastatic breast cancer previously treated with an anthracycline and a taxane. *Breast Cancer Res Treat* 76: S37
280 Holash J, Davis S, Papadopoulos N, Croll SD, Ho L, Russell M, Boland P, Leidich R, Hylton D, Burova E et al. (2002) VEGF-Trap: a VEGF blocker with potent antitumor effects. *Proc Natl Acad Sci USA* 99: 11393–11398
281 Konner J, Dupont J (2004) Use of soluble recombinant decoy receptor vascular endothelial growth factor trap (VEGF Trap) to inhibit vascular endothelial growth factor activity. *Clin Colorectal Cancer*, 4 Suppl 2: S81–S85
282 Kim ES, Serur A, Huang J, Manley CA, McCrudden KW, Frischer JS, Soffer SZ, Ring L, New T, Zabski S et al. (2002) Potent VEGF blockade causes regression of coopted vessels in a model of neuroblastoma. *Proc Natl Acad Sci USA* 99: 11399–11404
283 Huang J, Frischer JS, Serur A, Kadenhe A, Yokoi A, McCrudden KW, New T, O'Toole K, Zabski S, Rudge JS et al. (2003) Regression of established tumors and metastases by potent vascular endothelial growth factor blockade. *Proc Natl Acad Sci USA* 100: 7785–7790
284 Byrne AT, Ross L, Holash J, Nakanishi M, Hu L, Hofmann JI, Yancopoulos GD, Jaffe RB (2003) Vascular endothelial growth factor-trap decreases tumor burden, inhibits ascites, and causes dramatic vascular remodeling in an ovarian cancer model. *Clin Cancer Res* 9: 5721–5728
285 Fukasawa M, Korc M (2004) Vascular endothelial growth factor-trap suppresses tumorigenicity of multiple pancreatic cancer cell lines. *Clin Cancer Res* 10: 3327–3332
286 Inai T, Mancuso M, Hashizume H, Baffert F, Haskell A, Baluk P, Hu-Lowe DD, Shalinsky DR, Thurston G, Yancopoulos GD et al. (2004) Inhibition of vascular endothelial growth factor (VEGF) signaling in cancer causes loss of endothelial fenestrations, regression of tumor vessels, and appearance of basement membrane ghosts. *Am J Pathol* 165: 35–52

287 Frischer JS, Huang J, Serur A, Kadenhe-Chiweshe A, McCrudden KW, O'Toole K, Holash J, Yancopoulos GD, Yamashiro DJ, Kandel JJ (2004) Effects of potent VEGF blockade on experimental Wilms tumor and its persisting vasculature, *Int J Oncol* 25: 549–553
288 Dupont J, Camastra D, Gordon MS, Mendelson D, Murren J, Hsu A, Lucarelli A, Cedarbaum JM, Spriggs DR (1998) Phase I study of VEGF Trap in patients with solid tumors and lymphoma, *Proc Am Soc Clin Oncol* 22: A776
289 Miller KD (2002) Issues and challenges for anti-angiogenic therapies. *Breast Cancer Res Treat* 75, Suppl 1, S45–S50; discussion S57–S58
290 Garber K (2002) Angiogenesis inhibitors suffer new setback. *Nat Biotechnol* 20: 1067–1068
291 Stadler WM, Cao D, Vogelzang NJ, Ryan CW, Hoving K, Wright R, Karrison T, Vokes EE (2004) A randomized Phase II trial of the anti-angiogenic agent SU5416 in hormone-refractory prostate cancer. *Clin Cancer Res* 10: 3365–3370
292 Relf M, LeJeune S, Scott PA, Fox S, Smith K, Leek R, Moghaddam A, Whitehouse R, Bicknell R, Harris AL (1997) Expression of the angiogenic factors vascular endothelial cell growth factor, acidic and basic fibroblast growth factor, tumor growth factor beta-1, platelet-derived endothelial cell growth factor, placenta growth factor, and pleiotrophin in human primary breast cancer and its relation to angiogenesis. *Cancer Res* 57: 963–969
293 Browder T, Butterfield CE, Kraling BM, Shi B, Marshall B, O'Reilly MS, Folkman J (2000) Anti-angiogenic scheduling of chemotherapy improves efficacy against experimental drug-resistant cancer. *Cancer Res* 60: 1878–1886
294 Kerbel RS, Kamen BA (2004) The anti-angiogenic basis of metronomic chemotherapy. *Nat Rev Cancer* 4: 423–436
295 Hanahan D, Bergers G, Bergsland E (2000) Less is more, regularly: metronomic dosing of cytotoxic drugs can target tumor angiogenesis in mice. *J Clin Invest* 105: 1045–1047
296 Fennelly D, Aghajanian C, Shapiro F, O'Flaherty C, McKenzie M, O'Connor C, Tong W, Norton L, Spriggs D (1997) Phase I and pharmacologic study of paclitaxel administered weekly in patients with relapsed ovarian cancer. *J Clin Oncol* 15: 187–192
297 Link MP, Shuster JJ, Donaldson SS, Berard CW, Murphy SB (1997) Treatment of children and young adults with early-stage non-Hodgkin's lymphoma. *N Engl J Med* 337: 1259–1266
298 Kakolyris S, Samonis G, Koukourakis M, Vlachonicolis I, Chalkiadakis G, Kalbakis K, Souglakos I, Agelaki S, Toloudis P, Georgoulias V (1998) Treatment of non-small-cell lung cancer with prolonged oral etoposide. *Am J Clin Oncol* 21: 505–508
299 Alvarez A, Mickiewicz E, Brosio C, Giglio R, Cinat G, Suarez A (1998) Weekly taxol (T) in patients who had relapsed or remained stable with T in a 21 day schedule. P*roc Am Soc Clin Oncol* 17: A726
300 Greco FA (1999) Docetaxel (Taxotere) administered in weekly schedules. *Semin Oncol* 26: 28–31
301 Klement G, Baruchel S, Rak J, Man S, Clark K, Hickli, DJ, Bohlen P, Kerbel RS (2000) Continuous low-dose therapy with vinblastine and VEGF receptor-2 antibody induces sustained tumor regression without overt toxicity. *J Clin Invest* 105: R15–R24
302 Bello L, Carrabba G, Giussani C, Lucini V, Cerutti F, Scaglione F, Landre J, Pluderi M, Tomei G, Villani R et al. (2001) Low-dose chemotherapy combined with an anti-angiogenic drug reduces human glioma growth *in vivo*. *Cancer Res* 61: 7501–7506
303 Soffer SZ, Moore JT, Kim E, Huang J, Yokoi A, Manley C, O'Toole K, Stolar C, Middlesworth W, Yamashiro DJ et al. (2001) Combination anti-angiogenic therapy: increased efficacy in a murine model of Wilms tumor. *J Pediatr Surg* 36: 1177–1181
304 Crist WM, Anderson JR, Meza JL, Fryer C, Raney RB, Ruymann FB, Breneman J, Qualman SJ, Wiener E, Wharam M et al. (2001) Intergroup rhabdomyosarcoma study-IV: results for patients with nonmetastatic disease. *J Clin Oncol* 19: 3091–3102
305 Zhang L, Yu D, Hicklin DJ, Hannay JA, Ellis LM, Pollock RE (2002) Combined anti-fetal liver kinase 1 monoclonal antibody and continuous low-dose doxorubicin inhibits angiogenesis and growth of human soft tissue sarcoma xenografts by induction of endothelial cell apoptosis. *Cancer Res* 62: 2034–2042
306 Bergers G, Hanahan D (2002) Combining anti-angiogenic agents with metronomic chemotherapy enhances efficacy against late-stage pancreatic islet carcinomas in mice. *Cold Spring Harb Symp Quant Biol* 67: 293–300
307 Man S, Bocci G, Francia G, Green SK, Jothy S, Hanahan D, Bohlen P, Hicklin DJ, Bergers G, Kerbel RS (2002) Antitumor effects in mice of low-dose (metronomic) cyclophosphamide

administered continuously through the drinking water. *Cancer Res* 62: 2731–2735

308 Klement G, Huang P, Mayer B, Green SK, Man S, Bohlen P, Hicklin D, Kerbel RS (2002) Differences in therapeutic indexes of combination metronomic chemotherapy and an anti-VEGFR-2 antibody in multidrug-resistant human breast cancer xenografts. *Clin Cancer Res* 8: 221–232

309 Abraham D, Abri S, Hofmann M, Holtl W, Aharinejad S (2003) Low dose carboplatin combined with angiostatic agents prevents metastasis in human testicular germ cell tumor xenografts. *J Urol* 170: 1388–1393

310 Colleoni M, Rocca A, Sandri MT, Zorzino L, Masci G, Nole F, Peruzzotti G, Robertson C, Orlando L, Cinieri S et al. (2002) Low-dose oral methotrexate and cyclophosphamide in metastatic breast cancer: antitumor activity and correlation with vascular endothelial growth factor levels. *Ann Oncol* 13: 73–80

311 Spieth K, Kaufmann R, Gille J (2003) Metronomic oral low-dose treosulfan chemotherapy combined with cyclooxygenase-2 inhibitor in pretreated advanced melanoma: a pilot study *Cancer Chemother Pharmacol* 52: 377–382

312 Gluck S, Mackinnon J, Lau H, Syme R, Dort J, Gluck D (2003) Metronomic therapy in recurrent and metastatic chemo-resistant SCCHN: Data from a pilot study. *Proc Am Soc Clin Oncol* 22: A2066

313 Glode LM, Barqawi A, Crighton F, Crawford ED, Kerbel R (2003) Metronomic therapy with cyclophosphamide and dexamethasone for prostate carcinoma. *Cancer* 98: 1643–1648

314 Kieran MW (2004) Anti-angiogenic chemotherapy in central nervous system tumors. *Cancer Treat Res* 117: 337–349

315 Pietras K, Hanahan D (2004) A Multitargeted, Metronomic, and Maximum-Tolerated Dose "Chemo-Switch" Regimen is Anti-angiogenic, Producing Objective Responses and Survival Benefit in a Mouse Model of Cancer. *J Clin Oncol* 23: 939–952

316 Shpitz B, Gochberg S, Neufeld D, Grankin M, Buklan G, Klein E, Bernheim J (2003) Angiogenic switch in earliest stages of human colonic tumorigenesis. *Anticancer Res* 23: 5153–5157

317 Nicholson B, Theodorescu D (2004) Angiogenesis and prostate cancer tumor growth. *J Cell Biochem* 91: 125–150

318 Weidner N, Semple JP, Welch WR, Folkman J (1991) Tumor angiogenesis and metastasis—correlation in invasive breast carcinoma. *N Engl J Med* 324: 1–8

319 Moses MA, Harper J, Fernandez CA (2004) A role for anti-angiogenic therapy in breast cancer. *Curr Oncol Rep* 6: 42–48

320 Fang J, Yan L, Shing Y, Moses MA (2001) HIF-1alpha-mediated up-regulation of vascular endothelial growth factor, independent of basic fibroblast growth factor, is important in the switch to the angiogenic phenotype during early tumorigenesis. *Cancer Res* 61: 5731–5735

321 Yano K, Kadoya K, Kajiya K, Hong YK, Detmar M (2005) Ultraviolet B irradiation of human skin induces an angiogenic switch that is mediated by upregulation of vascular endothelial growth factor and by downregulation of thrombospondin-1. *Br J Dermatol* 152: 115–121

322 de Fraipont F, El Atifi M, Gicquel C, Bertagna X, Chambaz EM, Feige JJ (2000) Expression of the angiogenesis markers vascular endothelial growth factor-A, thrombospondin-1, and platelet-derived endothelial cell growth factor in human sporadic adrenocortical tumors: correlation with genotypic alterations. *J Clin Endocrinol MeTab.* 85: 4734–4741

323 Wang GL, Semenza GL (1993) General involvement of hypoxia-inducible factor 1 in transcriptional response to hypoxia. *Proc Natl Acad Sci USA* 90: 4304–4308

324 Wang GL, Jiang BH, Rue EA, Semenza GL (1995) Hypoxia-inducible factor 1 is a basic-helix-loop-helix-PAS heterodimer regulated by cellular O_2 tension. *Proc Natl Acad Sci USA* 92: 5510–5514

325 Yoshiji H, Harris SR, Thorgeirsson UP (1997) Vascular endothelial growth factor is essential for initial but not continued *in vivo* growth of human breast carcinoma cells. *Cancer Res* 57: 3924–3928

326 Hanahan D, Weinberg RA (2000) The hallmarks of cancer. *Cell* 100: 57–70

327 Shields JM, Pruitt K, McFall A, Shaub A, Der CJ (2000) Understanding Ras: 'it ain't over 'til it's over'. *Trends Cell Biol* 10: 147–154

328 Campbell PM, Der CJ (2004) Oncogenic Ras and its role in tumor cell invasion and metastasis. *Semin Cancer Biol* 14: 105–114

329 Quilliam LA, Rebhun JF, Castro AF (2002) A growing family of guanine nucleotide exchange factors is responsible for activation of Ras-family GTPases. *Prog Nucleic Acid Res Mol Biol* 71:

391–444
330 Magee T, Marshall C (1999) New insights into the interaction of Ras with the plasma membrane. *Cell* 98: 9–12
331 Cox AD, Der CJ (1997) Farnesyltransferase inhibitors and cancer treatment: targeting simply Ras? *Biochim Biophys Acta* 1333: F51–F71
332 Cho CH, Lee CS, Chang M, Jang IH, Kim SJ, Hwang I, Ryu SH, Lee CO, Koh GY (2004) Localization of VEGFR-2 and PLD2 in endothelial caveolae is involved in VEGF-induced phosphorylation of MEK and ERK. *Am J Physiol Heart Circ Physiol* 286: H1881–H1888
333 Wilhelm SM, Carter C, Tang L, Wilkie D, McNabola A, Rong H, Chen C, Zhang X, Vincent P, McHugh M et al. (2004) BAY 43-9006 exhibits broad spectrum oral antitumor activity and targets the RAF/MEK/ERK pathway and receptor tyrosine kinases involved in tumor progression and angiogenesis. *Cancer Res* 64: 7099–7109
334 Ciardiello F, Caputo R, Bianco R, Damiano V, Fontanini G, Cuccato S, De Placido S, Bianco AR, Tortora G (2001) Inhibition of growth factor production and angiogenesis in human cancer cells by ZD1839 (Iressa), a selective epidermal growth factor receptor tyrosine kinase inhibitor. *Clin Cancer Res* 7: 1459–1465
335 Wu W, Shu X, Hovsepyan H, Mosteller RD, Broek D (2003) VEGF receptor expression and signaling in human bladder tumors. *Oncogene* 22: 3361–3370
336 Bos JL (1989) ras oncogenes in human cancer: a review. *Cancer Res* 49: 4682–4689
337 Malumbres M, Barbacid M (2001) To cycle or not to cycle: a critical decision in cancer. *Nat Rev Cancer* 1: 222–231
338 Grugel S, Finkenzeller G, Weindel K, Barleon B, Marme D (1995) Both v-Ha-Ras and v-Raf stimulate expression of the vascular endothelial growth factor in NIH 3 T3 cells. *J Biol Chem* 270: 25915–25919
339 Mazure NM, Chen EY, Yeh P, Laderoute KR, Giaccia AJ (1996) Oncogenic transformation and hypoxia synergistically act to modulate vascular endothelial growth factor expression. *Cancer Res 56*: 3436–3440
340 Al-Mulla F, Going JJ, Sowden ET, Winter A, Pickford IR, Birnie GD (1998) Heterogeneity of mutant *versus* wild-type Ki-ras in primary and metastatic colorectal carcinomas, and association of codon-12 valine with early mortality. *J Pathol* 185: 130–138
341 Feldkamp MM, Lau N, Rak J, Kerbel RS, Guha A (1999) Normoxic and hypoxic regulation of vascular endothelial growth factor (VEGF) by astrocytoma cells is mediated by Ras. *Int J Cancer* 81: 118–124
342 Tokunaga T, Tsuchida T, Kijima H, Okamoto K, Oshika Y, Sawa N, Ohnishi Y, Yamazaki H, Miura S, Ueyama Y et al. (2000) Ribozyme-mediated inactivation of mutant K-ras oncogene in a colon cancer cell line. *Br J Cancer* 83: 833–839
343 Larcher F, Franco M, Bolontrade M, Rodriguez-Puebla M, Casanova L, Navarro M, Yancopoulos G, Jorcano JL, Conti CJ (2003) Modulation of the angiogenesis response through Ha-ras control, placenta growth factor, and angiopoietin expression in mouse skin carcinogenesis. *Mol Carcinog* 37: 83–90
344 Zheng H, Wasylyk C, Ayadi A, Abecassis J, Schalken JA, Rogatsch H, Wernert N, Maira SM, Multon MC, Wasylyk B (2003) The transcription factor Net regulates the angiogenic switch. *Genes Dev* 17: 2283–2297
345 Rak J, Yu JL (2004) Oncogenes and tumor angiogenesis: the question of vascular "supply" and vascular "demand". *Semin Cancer Biol* 14: 93–104
346 Rak J, Filmus J, Finkenzeller G, Grugel S, Marme D, Kerbel RS (1995) Oncogenes as inducers of tumor angiogenesis. *Cancer Metastasis Rev* 14: 263–277
347 Fedorov YV, Rosenthal RS, Olwin BB (2001) Oncogenic Ras-induced proliferation requires autocrine fibroblast growth factor 2 signaling in skeletal muscle cells. *J Cell Biol* 152: 1301–1305
348 Czubayko F, Liaudet-Coopman ED, Aigner A, Tuveson AT, Berchem GJ, Wellstein A (1997) A secreted FGF-binding protein can serve as the angiogenic switch in human cancer. *Nat Med* 3: 1137–1140
349 Kurtz A, Wang HL, Darwiche N, Harris V, Wellstein A (1997) Expression of a binding protein for FGF is associated with epithelial development and skin carcinogenesis. *Oncogene* 14: 2671–2681
350 Aguirre-Ghiso JA, Frankel P, Farias EF, Lu Z, Jiang H, Olsen A, Feig LA, de Kier Joffe EB, Foster DA (1999) RalA requirement for v-Src- and v-Ras-induced tumorigenicity and overpro-

duction of urokinase-type plasminogen activator: involvement of metalloproteases. *Oncogene* 18: 4718–4725

351 Westermarck J, Kahari VM (1999) Regulation of matrix metalloproteinase expression in tumor invasion. *Faseb J* 13: 781–792

352 Liao J, Wolfman JC, Wolfman A (2003) K-ras regulates the steady-state expression of matrix metalloproteinase 2 in fibroblasts. *J Biol Chem* 278: 31871–31878

353 Kranenburg O, Gebbink MF, Voest EE (2004) Stimulation of angiogenesis by Ras proteins. *Biochim Biophys Acta* 1654: 23–37

354 Shin I, Kim S, Song H, Kim HR, Moon A (2005) H-ras-specific activation of Rac-MKK3/6-p38 pathway: Its critical role in invasion and migration of breast epithelial cells. J Biol Chem 280: 14675–14683

355 Thorgeirsson UP, Turpeenniemi-Hujanen T, Williams JE, Westin EH, Heilman C A, Talmadge JE, Liotta LA (1985) NIH/3 T3 cells transfected with human tumor DNA containing activated ras oncogenes express the metastatic phenotype in nude mice. *Mol Cell Biol* 5: 259–262

356 Ballin M, Gomez DE, Sinha CC, Thorgeirsson UP (1988) Ras oncogene mediated induction of a 92 kDa metalloproteinase; strong correlation with the malignant phenotype. *Biochem Biophys Res Commun* 154: 832–838

357 Bernhard EJ, Gruber SB, Muschel RJ (1994) Direct evidence linking expression of matrix metalloproteinase 9 (92-kDa gelatinase/collagenase) to the metastatic phenotype in transformed rat embryo cells. *Proc Natl Acad Sci USA* 91: 4293–4297

358 Simon C, Goepfert H, Boyd D (1998) Inhibition of the p38 mitogen-activated protein kinase by SB 203580 blocks PMA-induced Mr 92,000 type IV collagenase secretion and *in vitro* invasion. *Cancer Res* 58: 1135–1139

359 Yang JQ, Zhao W, Duan H, Robbins ME, Buettner GR, Oberley LW, Domann FE (2001) v-Ha-RaS oncogene upregulates the 92-kDa type IV collagenase (MMP-9) gene by increasing cellular superoxide production and activating NF-kappaB. *Free Radic Biol Med* 31: 520–529

360 Westermarck J, Li SP, Kallunki T, Han J, Kahari VM (2001) p38 mitogen-activated protein kinase-dependent activation of protein phosphatases 1 and 2A inhibits MEK1 and MEK2 activity and collagenase 1 (MMP-1) gene expression. *Mol Cell Biol* 21: 2373–2383

361 Zabrenetzky V, Harris CC, Steeg PS, Roberts DD (1994) Expression of the extracellular matrix molecule thrombospondin inversely correlates with malignant progression in melanoma, lung and breast carcinoma cell lines. *Int J Cancer* 59: 191–195

362 Viloria-Petit A, Miquerol L, Yu JL, Gertsenstein M, Sheehan C, May L, Henkin J, Lobe C, Nagy A, Kerbel RS et al. (2003) Contrasting effects of VEGF gene disruption in embryonic stem cell-derived *versus* oncogene-induced tumors. *Embo J* 22: 4091–4102

363 Bromann PA, Korkaya H, Courtneidge SA (2004) The interplay between Src family kinases and receptor tyrosine kinases. *Oncogene* 23: 7957–7968

364 Yezhelyev MV, Koehl G, Guba M, Brabletz T, Jauch KW, Ryan A, Barge A, Green T, Fennell M, Bruns CJ (2004) Inhibition of SRC tyrosine kinase as treatment for human pancreatic cancer growing orthotopically in nude mice. *Clin Cancer Res* 10: 8028–8036

365 Mukhopadhyay D, Tsiokas L, Sukhatme VP (1995) Wild-type p53 and v-Src exert opposing influences on human vascular endothelial growth factor gene expression. *Cancer Res* 55: 6161–6165

366 Fukuda R, Kelly B, Semenza GL (2003) Vascular endothelial growth factor gene expression in colon cancer cells exposed to prostaglandin E2 is mediated by hypoxia-inducible factor 1. *Cancer Res* 63: 2330–2334

367 Niu G, Bowman T, Huang M, Shivers S, Reintgen D, Daud A, Chang A, Kraker A, Jove R, Yu H (2002) Roles of activated Src and Stat3 signaling in melanoma tumor cell growth. *Oncogene* 21: 7001–7010

368 Niu G, Wright KL, Huang M, Song L, Haura E, Turkson J, Zhang S, Wang T, Sinibaldi D, Coppola D et al. (2002) Constitutive Stat3 activity up-regulates VEGF expression and tumor angiogenesis. *Oncogene* 21: 2000–2008

369 Wei D, Le X, Zheng L, Wang L, Frey JA, Gao AC, Peng Z, Huang S, Xiong HQ, Abbruzzese JL et al. (2003) Stat3 activation regulates the expression of vascular endothelial growth factor and human pancreatic cancer angiogenesis and metastasis. *Oncogene* 22: 319–329

370 Bromberg JF, Horvath CM, Besser D, Lathem WW, Darnell JE Jr, (1998) Stat3 activation is required for cellular transformation by v-src. *Mol Cell Biol* 18: 2553–2558

371 Chou MT, Wang J, Fujita DJ (2002) Src kinase becomes preferentially associated with the

VEGFR, KDR/Flk-1, following VEGF stimulation of vascular endothelial cells. *BMC Biochem* 3: 32

372 Lambeng N, Wallez Y, Rampon C, Cand F, Christe G, Gulino-Debrac D, Vilgrain I, Huber P (2005) Vascular Endothelial-Cadherin Tyrosine Phosphorylation in Angiogenic and Quiescent Adult Tissues. *Circ Res* 96: 384–391

373 Pedram A, Razandi M, Levin ER (2002) Deciphering vascular endothelial cell growth factor/vascular permeability factor signaling to vascular permeability. Inhibition by atrial natriuretic peptide. *J Biol Chem* 277: 44385–44398

374 Criscuoli ML, Nguyen M, Eliceiri BP (2005) Tumor metastasis but not tumor growth is dependent on Src-mediated vascular permeability. *Blood* 105: 1508–1514

375 Oren M, Levine AJ (1983) Molecular cloning of a cDNA specific for the murine p53 cellular tumor antigen. *Proc Natl Acad Sci USA* 80: 56–59

376 Hofseth LJ, Hussain SP, Harris CC (2004) p53: 25 years after its discovery. *Trends Pharmacol Sci* 25: 177–181

377 Slee EA, O'Connor DJ, Lu X (2004) To die or not to die: how does p53 decide? *Oncogene* 23: 2809–2818

378 de Fraipont F, Nicholson AC, Feige JJ, Van Meir EG (2001) Thrombospondins and tumor angiogenesis. *Trends Mol Med* 7: 401–407

379 Grossfeld GD, Ginsberg DA, Stein JP, Bochner BH, Esrig D, Groshen S, Dunn M, Nichols PW, Taylor CR, Skinner DG et al. (1997) Thrombospondin-1 expression in bladder cancer: association with p53 alterations, tumor angiogenesis, and tumor progression. *J Natl Cancer Inst* 89: 219–227

380 Mehta R, Kyshtoobayeva A, Kurosaki T, Small EJ, Kim H, Stroup R, McLaren CE, Li KT, Fruehauf JP (2001) Independent association of angiogenesis index with outcome in prostate cancer. *Clin Cancer Res* 7: 81–88

381 Fontanini G, Boldrini L, Calcinai A, Chine S, Lucchi M, Mussi A, Angeletti CA, Basolo F, Bevilacqua G (1999) Thrombospondins I and II messenger RNA expression in lung carcinoma: relationship with p53 alterations, angiogenic growth factors, and vascular density. *Clin Cancer Res* 5: 155–161

382 Linderholm BK, Lindahl T, Holmberg L, Klaar S, Lennerstrand J, Henriksson R, Bergh J (2001) The expression of vascular endothelial growth factor correlates with mutant p53 and poor prognosis in human breast cancer. *Cancer Res* 61: 2256–2260

383 Volpert OV, Dameron KM, Bouck N (1997) Sequential development of an angiogenic phenotype by human fibroblasts progressing to tumorigenicity. *Oncogene* 14: 1495–1502

384 Kieser A, Weich HA, Brandner G, Marme D, Kolch W (1994) Mutant p53 potentiates protein kinase C induction of vascular endothelial growth factor expression. *Oncogene* 9: 963–969

385 Moses MA, Wiederschain D, Loughlin KR, Zurakowski D, Lamb CC, Freeman MR (1998) Increased incidence of matrix metalloproteinases in urine of cancer patients. *Cancer Res* 58: 1395–1399

386 Roy R, Wewer UM, Zurakowski D, Pories SE, Moses MA (2004) ADAM 12 cleaves extracellular matrix proteins and correlates with cancer status and stage. *J Biol Chem* 279: 51323–51330

387 Chan LW, Moses MA, Goley E, Sproull M, Muanza T, Coleman CN, Figg WD, Albert PS, Menard C, Camphausen K (2004) Urinary VEGF and MMP levels as predictive markers of 1-year progression-free survival in cancer patients treated with radiation therapy: a longitudinal study of protein kinetics throughout tumor progression and therapy. *J Clin Oncol* 22: 499–506

Cancer: Cell Structures, Carcinogens and Genomic Instability
Edited by Leon P. Bignold

Genetic and environmental factors in hereditary predisposition to tumors: a conceptual overview

Okio Hino[1], Toshiyuki Kobayashi[1] and Kazuo Okimoto[2]

[1] *Department of Pathology, Juntendo University School of Medicine, 2-1-1 Hongo, Bunkyo-ku, Tokyo 113-8421, Japan*
[2] *Toxicology Group, Safety Research Laboratories, Dainippon Pharmaceutical Co., Ltd., 33-94, Enoki-cho, Suita, Osaka 564-0053, Japan*

Summary. Cancer is a heritable disorder of somatic cells. Carcinogenesis at the cellular level is like an opened Japanese fan, because initiated cells grow in several directions and tumors suggest the edge of the fan by having many gene abnormalities. We discuss here the primal force and gene networks (federal headship) in renal carcinogenesis. The Eker (*Tsc2* mutant) rat model of hereditary renal carcinoma (RC) is an example of a Mendelian dominantly inherited predisposition to a specific cancer in an experimental animal. Recently, we discovered a new hereditary renal carcinoma in the rat in Japan, and the rat was named the "Nihon" rat. We suggest that its predisposing (*Bhd*) gene is a novel renal tumor suppressor gene. We present these unique models as part of the study of problems in carcinogenesis; e.g., multistep carcinogenesis, cancer prevention and the development of the therapeutic treatments that can be translated to human patients, as well as how environmental factors interact with cancer susceptibility gene(s).

Key words: *Bhd* gene (Nihon) mutant, *Erc* gene, hereditary cancer, Knudson's two-hit, *Tsc 1* gene, *Tsc2* gene (Eker) mutant, *Niban* gene.

Introduction

Environment and heredity both operate in the origin of human cancer. These environment and genetic determinants of cancer can be classified into four groups designated "oncodems" by Knudson in 1985 [1]. Oncodem 1 is the irreducible "background" level of cancer due to spontaneous mutagenesis. Oncodem 2 is "environmentally induced" cancer, the causative agents among which are chemical carcinogens, radiation, or viruses. Oncodem 3 is basically "environmentally induced" cancer, but there are genetically determined differences among persons, e.g., in relation to tendencies for activation or inactivation of carcinogens. Most human cancers are believed to belong to oncodem 2 and/or 3 (about 80%), for which the probability of the occurrence of the carcinogenic steps is increased, although the number of steps is not decreased. Oncodem 1 would contain the 20% that would remain if "environmental induced" cancers (oncodem 2 and/or 3) were prevented. Oncodem 4 is all "hereditary" cancer. Hereditary cancers have been important in the understanding of carcinogenesis, and pointed to the first identification of a "tumor suppressor gene" and a "two-hit" model of carcinogenesis in the work of Knudson on human retinoblastoma in 1971 [2].

Hereditary cancers should continue to prove valuable in elucidating mechanisms of carcinogenesis, even though only a small proportion of cancers belong to this oncodem 4 group.

Human inherited predispositions to tumors: genes and functions of gene products and animal models

Over the last two decades, the genes responsible for familial cancer syndromes were cloned and identified using the positional cloning of candidate gene approach. A current list of such human familial syndromes is shown in Tables 1–3 [3]. Animal models of inherited tumor predisposition associated with tumor suppressor genes have also been discovered, and a current list of such models is also shown in Tables 1–3.

General history of the Eker rat model of inherited renal cell carcinoma

After 1971, Knudson, at the Fox Chase Cancer Center, Philadelphia, was looking for an animal model of "two-hit carcinogenesis" that might parallel hereditary retinoblastoma in man, and found it in the Eker rat. The naturally occurring hereditary cancer in the rat had been described by Reidar Eker in Oslo in 1954 [4]. In 1983, Reidar Eker kindly sent some Eker rats to Knudson in Philadelphia. Progress toward genetic linkage analysis was slow because few genetic markers were available in the rat. The author (O.H.) went to the Knudson's laboratory on an American Cancer Society–Eleanor Roosevelt International Fellowship (UICC) from 1989 to 1991 and began to isolate rat genetic markers [5]. Unfortunately, these initial linkage studies were unproductive. When the author (O.H.) returned to the Cancer Institute in Tokyo, Knudson sent him some Eker rats, and they have been bred on a normal Long–Evans strain background at the Animal Facility of the Cancer Institute since 1991. The author (O.H.) has since continued investigating them in Tokyo, independently of Knudson's group in Philadelphia. Thus, Eker rat research has been continued by a third generation of investigators (Oslo→ Philadelphia→Tokyo). Finally, we and Knudson's group independently identified a germline retrotransposon insertion in the rat homolog of the human tuberous sclerosis (*TSC2*) gene, resulting in aberrant RNA expression from the mutant allele, 40 years after the original discovery of the Eker rat [6–8]. To the best of our knowledge, this was the first isolation of a Mendelian dominantly predisposing cancer gene in a naturally occurring animal model. We then constructed transgenic Eker rats with a wild-type *Tsc2* gene and ascertained that germline suppression of the Eker phenotype is possible for both embryonic lethality of the homozygote and tumor predisposition in the heterozygote, and

Table 1. Tumor suppressor genes responsible for autosomal dominantly inherited tumor predisposing diseases

Gene name	Tumors syndrome	Chromosome localization	Function	Phenotypes of knockout mouse
APC	Colon cancer (Familial adenomatous polyposis coli)	5q22.2	Negative regulator of beta-catenin	E: intestinal tumor, O: embryonic lethality
BHD (*FLCN*)	Renal cell carcinoma, fibrofolliculoma (Birt-Hogg-Dubé syndrome)	17p11.2	Unknown	Unknown (renal cell carcinomas in the rat and dog)
BMPR1A	Colon cancer (Juvenile polyposis)	10q23.2	BMP receptor, Ser/Thr protein kinase	O: embryonic lethality
BRCA1	Breast carcinoma, Ovarian carcinoma	17q21.31	DNA damage repair	O: embryonic lethality
BRCA2	Breast carcinoma, Ovarian carcinoma	13q13.1	DNA damage repair	O: embryonic lethality
CDH1	Gastric carcinoma	16q22.1	Cell adhesion	O: embryonic lethality
CDKN2A	Melanoma	9p21.3	Cyclin-dependent protein kinase inhibitor	O: lymphoma, sarcoma
CHEK2	Osteosarcoma, brain tumor, breast carcinoma (Li-Fraumeni syndrome)	22q12.1	Regulation of cell cycle checkpoint	O: Lymphoma
CYLD	cylindroma	16q12.1	Negative regulator of NF-κB	Unknown
EXT1	Chondrosarcoma (Multiple exostoses)	8q24.1	Heparan sulfate synthetase	O: embryonic lethality
EXT2	Chondrosarcoma (Multiple exostoses)	11p11.2	Heparan sulfate synthetase	Unknown
FH	Skin leiomyoma, Uterus leiomyoma, Papillary renal cell carcinoma	1q43	Fumarate hydratase	Unknown
HRPT2	Parathyroid cancer	1q31.2	unknown	unknown
INI1 (*SNF5*)	Renal rhabdoid tumor	22q11.23	Chromatin remodeling	E: Rhabdoid tumor, O: embryonic lethality

(Continued on next page)

Table 1. (Continued)

Gene name	Tumors syndrome	Chromosome localization	Function	Phenotypes of knockout mouse
LKB1	Colon cancer (Peutz-Jeghers syndrome)	19p13.3	Protein kinase	E: Gastric tumor hapatocelluar carcinome, O: embryonic lethality
MEN1	Pituitary tumor, parathyroid tumor (Multiple endocrine neoplasia type I)	11q13.1	Unknown	E: Pancreatic islet tumor, parathyroid tumor, O: embryonic lethality
MLH1	Colon cancer (Hereditary non-polyposis colon cancer)	3p22.3	DNA mismatch repair	O: lymphoma, sterility
MSH2	Colon cancer (Hereditary non-polyposis colon cancer)	2p21	DNA mismatch repair	O: lymphoma
MSH6 (*GTBP*)	Colon cancer (Hereditary non-polyposis colon cancer)	2p16	DNA mismatch repair	O: lymphoma, skin tumor, uterin tumor
NF1	Neurofibroma (Neurofibromatosis type I)	17q11.2	Ras-GTPase activating protein (Ras-GAP)	E: learning defet, leukemia, O: embryonic lethality
NF2	Schwannoma, meningioma (Neurofibromatosis type II)	22q12.2	Cytoskeletal regulation	E: osteosarcoma, hepatocellular carcinoma, O: embryonic lethality
PRKAR1A	Cardiac myxoma, adrenal tumor (Carney complex)	17q24.2	cAMP-dependent protein kinase regulatory subunit	E: sarcoma, hepatocellular carcinoma, O: embryonic lethality
PTCH	Medulloblastoma, basal cell carcinoma (Gorlin syndrome)	9q22.32	Sonic hedgehog receotor, negative regulator of Smoothened	E: medulloblastoma, O: embryonic lethality
PTEN	Thyroid cancer, endometrial cancer (Cowden disease)	10q23.31	Protein/lipid phosphatase	E: endometrial cancer, thyroid cancer, O: embryonic lethality
RB	Retinoblastoma, osteosarcoma	13q14.2	Negative regulator of E2F	E: pituitary tumor, O: embryonic lethality

(Continued on next page)

Table 1. (Continued)

Gene name	Tumors syndrome	Chromosome localization	Function	Phenotypes of knockout mouse
SDHB	Pheochromocytoma, paraganglioma	1p36.13	Succinate dehydrogenase subunit	Unknown
SDHC	Pheochromocytoma, paraganglioma	1q23.3	Succinate dehydrogenase subunit	Unknown
SDHD	Pheochromocytoma, paraganglioma	11q23.1	Succinate dehydrogenase subunit	E: carotid body glomus cell hyperplasia, O: embryonic lethality
SMAD4	Colon cancer (Juvenile polyposis)	18q21.1	Transcription factor	E: Gastric hyperplasia, O: embryonic lethality
SUFU	Medulloblastoma	10q24.32	Negative regulator of Gli	Unknown
TCF1	Hepatoma	12q24.31	Transcription factor	O: Non-insulin dependent diabetes, sterility
TP53	Osteosarcoma, brain tumor, breast carcinoma (Li-Fraumeni syndrome)	17p13.1	Cell cycle checkpoint regulation	E O: lymphoma, osteosarcoma, hemangiosarcoma
TSC1	Renal angiomyolipoma, skin angiofibroma, renal cancer (tuberous sclerosis)	9q34.13	Tsc2 binding	E: renal tumor, liver hemangioma, O: embryonic lethality
TSC2	Renal angiomyolipoma, skin angiofibroma, renal cancer (tuberous sclerosis)	16p13.3	Rheb-GAP	E: renal tumor, liver hemangioma, O: embryonic lethality
VHL	Clear cell renal carcinoma, pheochromocytoma (von Hippel-Lindau disease)	3p25.3	Negative regulator of HIF1-alpha, Ubiquitin ligase subunit	O: embryonic lethality
WT1	Nephroblastoma (Wilms' tumor)	11p13	Transcription factor	O: embryonic lethality

Table 2. DNA repair-related genes responsible for recessively inherited tumor predisposing diseases

Gene name	Tumors syndrome	Chromosome localization	Function	Phenotypes of knockout mouse
ATM	Leukemia, lymphoma (Ataxia telangiectasia)	11q23	Protein kinase, cell cycle checkpoint regulation	O: growth defect, lymphoma, immunodeficiency, sterility
BLM	Leukemia, lymphoma (Bloom syndrome)	15q26.1	DNA helicase	O: lymphoma
FANCA	Leukemia, squamous cell carcinoma (Fanconi anemia)	16q24.3	FANCG binding, BRCA1 binding, FANCD2 mono-ubiqutitinnation regulation	O: genital defect, mitomycin C-hypersensitvity
FANCC	Leukemia, squamous cell carcinoma (Fanconi anemia)	9q22.32	FANCE binding, regulation of FANCD2 mono-ubiqutitinnation	O: hematopoietic defect, mitomycin C-hypersensitvity
FANCD2	Leukemia, squamous cell carcinoma (Fanconi anemia)	3p25.3	unknown	O: microphtalmia, various eptithelial cancer, mitomycin C-hypersensitvity
FANCE	Leukemia, squamous cell carcinoma (Fanconi anemia)	6p21.31	FANCC binding, regulation of FANCD2 monoubiquitination	unknown
FANCF	Leukemia, squamous cell carcinoma (Fanconi anemia)	11p14.3	FANCG binding, regulation of FANCD2 monoubiquitination	unknown
FANCG	Leukemia, squamous cell carcinoma (Fanconi anemia)	9p13.3	FANCA/FANCF binding, BRCA2 binding, regulation of FANCD2 monoubiquitination	O: genital defect, mitomycin C-hypersensitvity
FANCL	Leukemia, squamous cell carcinoma (Fanconi anemia)	2p16.1	E3 ubiquitin ligase	unknown
NBS1	Lymphoma, glioma (Nijmegen breakages syndrome)	8q21.3	RAD51/MRE11 complex DNA damage repair	E: various epithelial cancer, O: embryonic lethality
POLH	Skin cancer, melanoma (Xeroderma pigmentosum)	6p21.1	Trans-lesion DNA polymerase	Unknown

(Continued on next page)

Table 2. (Continued)

Gene name	Tumors syndrome	Chromosome localization	Function	Phenotypes of knockout mouse
RECQL4	Basal cell carcinoma, squamous cell carcinoma	8q24.3	DNA helicase	O: growth defect, skin abnormality
SBDS	Leukemia (Schwachman-Diamond syndrome)	7q11	RNA metabolism	unknown
WRN	Osteosarcoma (Werner syndrome) (Xeroderma pigmentosum)	8p12	DNA helicase	O: no obvious phenotype
XPA	Skin cancer, melanoma (Xeroderma pigmentosum)	19p13.2	DNA damage recognition	O: UV-induced skin tumor
XPB	Skin cancer, melanoma (Xeroderma pigmentosum)	2q14.3	DNA helicase	unknown
XPC	Skin cancer, melanoma (Xeroderma pigmentosum)	3p25.1	DNA binding	O: UV-induced skin tumor
XPD	Skin cancer, melanoma (Xeroderma pigmentosum)	19q13.32	DNA helicase	O: embryonic lethality
XPE (*DDB2*)	Skin cancer, melanoma (Xeroderma pigmentosum)	11p11.2	Damaged DNA binding	O: UV-induced skin tumor
XPF	Skin cancer, melanoma	16p13.12	DNA endonuclease	O: growth defect
XPG	Skin cancer, melanoma (Xeroderma pigmentosum)	13q33.1	DNA endonuclease	O: growth defect

Table 3. Oncogene responsible for autosomal dominantly inherited tumor predisposing disease

Gene name	Tumors syndrome	Chromosome localization	Function	Phenotypes of model mouse
CDK4	Melanoma	12q14.1	Cyclin-dependent protein kinase	caKI: melanoma, hemangiosarcoma
FAS	Lymphoma (Cannale-Smith syndrome)	10q23.31	FAS	dnTg: Hepatosplenomegary Lymphocyte overproduction KO lpr: Lymphoma autoimmune disease
KIT	Gastrointestinal stromal tumor	4q12	Receptor protein tyrosine kinase	caKI: gastrointestinal tumor
MET	Papillary renal cell carcinoma	7q31.2	Receptor protein tyrosine kinase (HGF receptor)	caKI: hamagiosarcoma, squamous cell carcinoma
RET	Thyroid cancer, pheochromocytoma (Multiple endocrine neoplasia type II)	10q11.21	Receptor protein tyrosine kinase (GDNF receptor)	caTg: thyroid cancer

finally confirmed that a tumor predisposition in the Eker rat is caused by a *Tsc2* germline mutation [9].

Genotype and phenotype of the *Tsc2* gene mutant (Eker) rat

We found that the homozygous mutant condition in the Eker rat is lethal at around day 13 of fetal life [10]. In heterozygotes, renal carcinomas (RCs) develop from early pre-neoplastic lesions (phenotypically altered renal tubules, which begin to appear at 2 months of age) to adenomas around the age of 1 year. Penetrance for this RC (*Tsc2*) gene is virtually 100%. Investigation of extrarenal primary tumors in Eker rats revealed hemangiomas/hemangiosarcomas of the spleen, and leiomyomas/leiomyosarcomas of the uterus [11], and pituitary adenomas [12–14]. The Eker rat thus bears a single gene mutation with a dominant predisposition to four different tumors, although predisposition for extrarenal tumors is not as complete as with RCs [15]. Brain lesions in the Eker rat, such as subependymal and subcortical hamartomas, were also reported [16]. In the Eker rat cerebrum, we identified two novel lesions, a cortical tuber and anaplastic ganglioglioma, in addition to the two types of hamartomas described above [17]. The presence of a cortical tuber is important, given that tubers are epileptogenic and presumably associated with autism and other neurological symptoms of human tuberous sclerosis.

Species-specific differences among human/rat/mouse *TSC*-related tumor phenotypes

Human tuberous sclerosis is an autosomal dominant multisystem disorder caused by a mutation either in the *TSC1* or the *TSC2* gene, characterized by phakomatosis with manifestations that include mental retardation and seizures. The phenotype in humans differs from that in the Eker rat, except for the occurrence of RCs (in humans, angiomyolipomas are more common), although subependymal, subcortical hamartomas and cortical tubers in the Eker rat have been reported [16, 17]. Thus, the same gene shows diverse phenotypes among species, although we do not have any good explanation for this difference. Similar phenomena have been observed in knockout mice for a number of tumor suppressor genes. To elicit insights into species-specific tumorigenesis caused by *Tsc2* gene inactivation, we generated a *Tsc2*-knockout mouse [18]. Mice heterozygous for *Tsc2* mutation developed RCs, but not angiomyolipoma, with complete penetrance, as seen in the Eker rat, although there is still a species-specific difference between the rat and mouse (liver hemangiomas are only observed in the mouse). We also produced a *Tsc1*-knockout mouse. The phenotype is similar to that of *Tsc2* knockout mice. However, it is noteworthy that renal carcinogenesis is slower and milder than that of *Tsc2* mutants [19].

Target gene-specific chemically induced tumorigenesis in the Eker rat

N-Ethyl-*N*-nitrosourea (ENU)-induced transplacental renal carcinogenesis in the rat results primarily in Wilms' tumors, not RCs, apparently because primitive nephroblasts are the preferred target. To examine this specificity of the tumor type further, we undertook studies of transplacental carcinogenesis in the Eker rat, which is heterozygous for a mutation that predisposes to RC with high penetrance, but does not predispose to Wilms' tumor [20]. Surprisingly, RCs, but not Wilms' tumors, began to appear as early as 1 week after birth [20]. Thus, the Eker rat is highly susceptible to induction of RCs (but not Wilms' tumors) by transplacental administration of ENU. Inheritance of the Eker mutation reduces the required number of carcinogenic events, as evidenced by the finding of renal lesions under conditions in which ENU produces no tumors in controls, and determines the specificity of tumor histology even with carcinogenesis *in utero*, where nephroblasts are the preferred target. Nephroblasts might be the targets for development of both kinds of tumor in ENU-induced transplacental renal carcinogenesis. Our data thus support the hypothesis that the Wilms' tumor gene may control differentiation of nephroblasts (metanephric stem cells) into renal tubular stem cells, whereas the Eker (*Tsc2*) gene may control terminal differentiation of these stem cells, but permit differentiation of nephroblasts into renal cells [20, 21].

Differences in the carcinogenicity of spontaneous *versus* chemicals *versus* radiation "second hits"

We have previously reported (using several DNA markers located on rat chromosome 10) that, of spontaneous RCs, 60% (6 of 10) showed loss of heterozygosity (LOH) covering more than 30 cM and, in contrast, 0% (0 of 9) had LOH in ENU-induced RCs [13]. We then characterized the second hit (intragenic mutations including point mutation) in LOH-negative lesions at the DNA-sequencing level of the predisposing *T*sc2 gene using methods such as polymerase chain reaction–single-strand conformations. The availability of the cDNA sequence of *Tsc2* permits comparative analysis of spontaneous and chemically induced tumors in the Eker rat. We showed that a qualitative difference in the second hit exists between spontaneous and ENU-induced mutations (e.g., deletion or duplication *versus* point mutation) [22]. We also investigated the second hit of radiation-induced RCs in the Eker rat and compared it with the former cases [23]. We also detected LOH in 4 of 11 uterine leiomyosarcomas (36%) and 11 of 31 pituitary adenomas (35%) from Eker rats, but in none of 9 pituitary adenomas from non-carrier rats, suggesting that inactivation of the *Tsc2* gene is also a critical event in the pathogenesis of these extrarenal tumors [14]. Our data indicate that there might be different pathways for tumorigenesis of pituitary adenomas between Eker and non-carrier rats [24]. However, none of the 5 investigated hemangiosarcomas of the spleen exhibited LOH, although one explanation might be contamination with an appreciable number of normal cellular components in such tumors and/or the small number of cases investigated.

Target genes in chemically induced renal carcinogenesis in non-Eker rat

We searched for mutations of the *Tsc2* gene in chemically induced non-Eker rat RCs by single-strand conformational polymorphism (SSCP) analysis [25, 26]. We simultaneously searched for mutation in the *Vhl* gene, a rat homolog of the von Hippel–Lindau (VHL) disease gene. Mutations in the *Vhl* gene were not detected in any spontaneous RCs of the Eker rat model, nor in chemically induced non-Eker rat tumors [25–27]. In contrast, *Tsc2* gene mutations were detected at a high frequency in non-Eker rat primary RCs induced with either *N*-ethyl-*N*-hydroxyethylnitrosamine or diethylnitrosamine [26]. We occasionally found mutations of the *Tsc1* gene, the rat homolog of the *TSC1* gene, in chemically induced rat RCs [27]. These findings call attention to other possible gene mutations (e.g., *Bhd*) in rat RCs.

Tsc2 gene mutation is the federal headship in multistep renal carcinogenesis

Successive stages in the development of RCs in *Tsc2* gene mutant (Eker) rat were observed, beginning with isolated phenotypically altered renal tubules (which begin to appear at 2months of age), characterized by partial or total replacement of the proximal tubular epithelium by large or weakly acidophilic cells with different degrees of nuclear atypia, or by basophilic cells [10, 21] These foci developed into atypical hyperplasia, then into adenomas of either eosinophilic or basophilic tubular type, and, finally, into fully developed carcinomas that were predominantly of one type, but also included carcinomas of mixed type with both basophilic and eosinophilic components (Fig. 1). These foci of atypical tubules were seen in all kidneys exhibiting tumors, and are presumed to give rise to the smaller, rounded early adenomas as well as to the later, larger carcinomas; they were not observed in normal rats [10]. Some tumors were cystic and had papillary projections. Microscopic invasion of perirenal tissue was detected infrequently.

The Eker rat provides a promising model for analyzing the essential events of carcinogenesis at different stages. We have previously reported that ionizing radiation induces additional tumors (large adenomas and carcinomas), with a linear dose-response relationship [10]. LOH at chromosome 10, where the predisposing *Tsc2* gene is located, was found in RCs that developed from hybrid F1 rats carrying the Eker mutation, indicating that in heterozygotes at least two events (one inherited, one somatic) are necessary to produce large adenomas and carcinomas. Using laser microscopic dissection, we found LOH

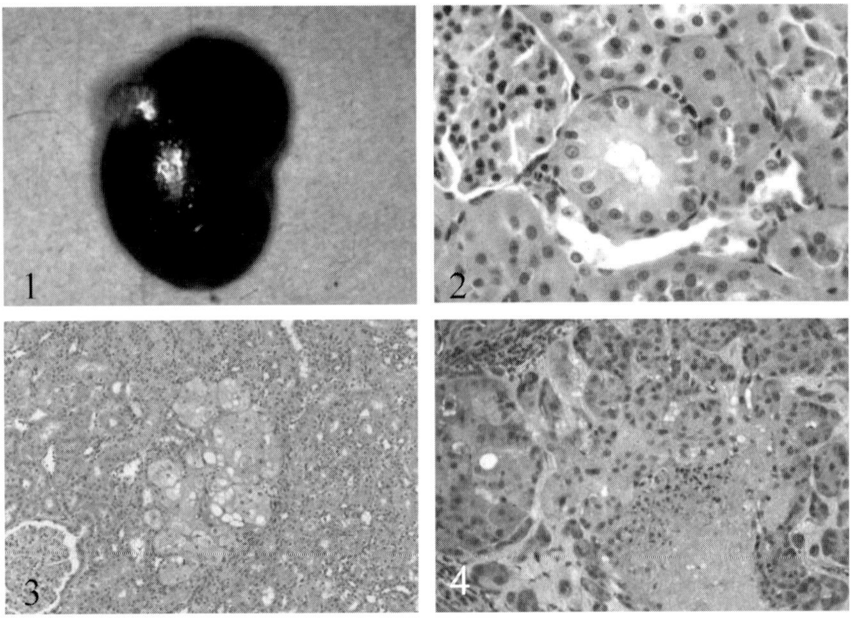

Figure 1. Multi-stage renal carcinogenesis in the Eker (*Tsc2* gene mutant) rat [32]. (1) Macroscopic renal tumor. (2) Small regions of phenotypically altered renal tubules, which develop via atypical hyperplasia into (3) adenomas, and finally, into (4) fully developed carcinomas

of the wild-type allele even in the earliest pre-neoplastic lesions, e.g., phenotypically altered renal tubules [28], supporting the hypothesis that a second, somatic mutation (second hit) might be a rate-limiting step for renal carcinogenesis in the Eker rat model of dominantly inherited cancer, as well as indicating a tumor-suppressor function for the *Tsc2* gene. Thus, heterozygosity is not by itself a sufficient condition for the development of cancer, but only one hit is enough to produce phenotypically altered renal tubules in the Eker rat. Such a lesion may initially be benign, but continued proliferation virtually ensures that other critical, though not rate-limiting, events will occur. Although the initial event that triggers Eker rat renal cancer is a somatic mutation of the *Tsc2* wild-type allele, other genetic or epigenetic modifications may also contribute to tumor progression in multistep renal carcinogenesis (Fig. 2).

Gene networks in Eker rat tumors

Carcinogenesis can be compared to an opened Japanese fan, because initiated cells grow in several directions and clinical tumors suggest that the edge of the fan has many gene abnormalities [29–32] (Fig. 2). To search for such alterations, we identified genes that were expressed more abundantly in Eker rat RC cells (RCC) than in the normal kidney [33].

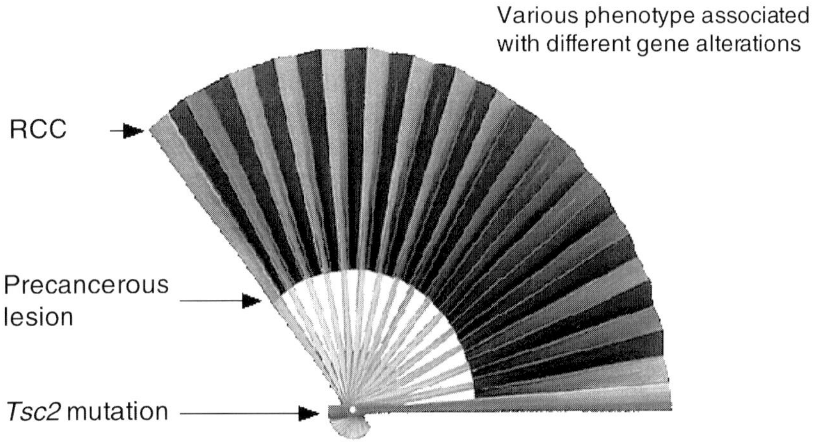

Figure 2. Fan model of renal carcinogenesis [29–32]. Carcinogenesis is like an opened Japanese fan. The primal force is "two hits" on the predisposing *Tsc2* gene. Initiated cells grow then in many directions and accumulate different clinical features at the periphery. The edge of the fan then corresponds to the diverse clinical features of the tumor cells.

AP-1 genes

We identified the highly expressed genes in Eker RCs as: the *C3* gene encoding the third component of complement, the *annexin II* gene encoding the calpactin 1 heavy-chain, the *Erc* (expressed in renal carcinoma) gene, and the *fra-1* gene encoding a transcriptional factor activator protein 1 (AP-1) [33]. We found that other members of the AP-1 transcriptional factors were involved in the renal carcinogenesis in the Eker rat model [34]. Interestingly, using immunohistochemistry, AP-1 proteins were shown to be highly expressed even in the earliest pre-neoplatic lesions (e.g., phenotypically altered tubules). Moreover, we transfected antisense oligonucleotides targeting *AP-1* genes into RCCs and demonstrated that their growth was strongly inhibited [34]. These data suggest that expression of *AP-1* genes might play a crucial role in renal carcinogenesis in the Eker rat model.

We also elucidated an underlying regulatory mechanism of *Tsc2* gene expression, occurring bidirectionally via two Ets-transcription factor binding sites, and found that Ets family proteins were highly expressed in the RCs [35]. These observations might be helpful in the search for the downstream molecular targets and the gene network of the *Tsc2* gene.

Erc gene/MPF/mesothelin

After we determined the complete primary structure of rat *Erc* cDNA, we showed that the putative rat Erc product has 56.1% identity with human

cinogenesis. *Oncogene* 23: 3495–3500
39 Hino O, Kobayashi E, Hirayama Y, Kobayashi T, Kubo Y, Tsuchiya H, Kikuchi Y, Mitani H (1995) Molecular genetic basis of renal carcinogenesis in the Eker rat model of tuberous sclerosis (Tsc2). *Mol Carcinog* 14: 23–27
40 Testa JR, Taguchi T, Knudson AG, Hino O (1992) Localization of interferon a gene cluster to rat chromosome bands 5q31–q33 by fluorescence *in situ* hybridization. *Cytogenet Cell Genet* 60: 247–249
41 Fukuda T, Hirayama Y, Mitani H, Maeda H, Tsutsumi M, Konishi Y, Hino O (1998) Generation of metastatic variants of Eker renal carcinoma cell lines for experimental investigation of renal cancer metastasis. *Jpn J Cancer Res* 89: 1104–1108
42 Miki S, Miki Y, Mitani H, Sato B, Yamamoto, M Hino O (1997) Hyperlipidemia enhancement of renal preneoplasia but not tumor formation is due to elevated apoptosis in the Eker rat. *Cancer Res* 57: 4673–4676
43 Hino O, Kobayashi T, Mitani H (2002) Prevention of hereditary carcinogenesis. *Proc Japan Acad* 78 (B): 31–33
44 Toyonaga T, Hino O, Sugai S, Wakasugi S, Abe K, Shichiri M, Yamamura K (1994) Chronic active hepatitis in transgenic mice expressing interferon-g in the liver. *Proc Natl Acad Sci USA* 91: 614–618
45 Okamoto T, Furuya M, Yamakawa T, Yamamura K, Hino O (1996) TNF-alpha gene expression in the liver of the IFN-gamma transgenic mouse with chronic active hepatitis. *Biochem Biophys Res Commun* 226: 762–768
46 Akbar SMF, Kajino K, Tanimoto K, Yamamura K, Onji M, Hino O (1999) Unique features of dendritic cells in IFN-g transgenic mice: relevance to cancer development and therapeutic implications. *Biochem Biophys Res Commun* 259: 294–299
47 Gao X, Zhang Y, Arrazola P, Hino O, Kobayashi T, Yeung RS, Ru B, Pan D (2002) Tsc tumor suppressor proteins antagonize amino-acid-TOR signalling. *Nat Cell Biol* 4: 699–704
48 Kobayashi T, Adachi H, Mitani H, Hirayama Y, Hino O (2003) Toward chemotherapy for Tsc2-mutant renal tumor. *Proc Japan Acad* 79B: 22–25
49 Momose S, Kobayashi T, Mitani H, Hirabayashi M, Ito K, Ueda M, Nabeshima Y, Hino O (2002) Identification of the coding sequences responsible for *Tsc2*-mediated tumor suppression using a transgenic rat system. *Hum Mol Genet* 11: 2997–3006
50 Kikuchi Y, Sudo A, Mitani H, Hino O (2004) Presence of a modifier gene(s) affecting early renal carcinogenesis in the Tsc2 mutant (Eker) rat model. *Int J Oncol* 24: 75–80
51 Okimoto K, Kouchi M, Kikawa E, Toyosawa K, Koujitani T, Tanaka K, Matsuoka N, Sakurai J, Hino O (2000) A novel "Nihon" rat model of a Mendelian dominantly inherited renal cell carcinoma. *Jpn J Cancer Res* 91: 1096–1099
52 Okimoto K, Kouchi M, Matsumoto I, Sakurai J, Kobayashi T, Hino O (2004) Natural history of the Nihon rat model of BHD. *Current Molecular Medicine* 4: 887–893
53 Hino O, Okimoto K, Kouchi M, Sakurai J (2001) A novel renal carcinoma predisposing gene of the Nihon rat maps on chromosome 10. *Jpn J Cancer Res* 92: 1147–1149
54 Okimoto K, Sakurai J, Kobayashi T, Mitani H, Hirayama Y, Nickerson ML, Warren MB, Zbar B, Schmidt LS, Hino O (2004) A germ-line insertion of the Birt-Hogg-Dube (BHD) gene gives rise to th Nihon rat model of inherited renal cancer. *Proc Natl Acad Sci USA* 101: 2023–2027
55 Bannasch P, Zerban H (1990) In: JN Eble (ed.): "Tumor and Tumor-like Conditions of the Kidneys and Ureters". Livingstone, New York, 1–34
56 Nickerson M, Warren MB, Toro JR, Matrosova V, Glenn G, Turner M, Duray P, Merino M, Choyke P, Pavlovich CP et al. (2003) *Cancer Cell* 2: 157–164
57 Birt AR, Hogg GR, Dube WJ (1977) Hereditary multiple fibro-folliculomas with trichodiscomas and acrochordons. *Arch Dermatol* 113: 1674–1677
58 Lane-Petter W, Pearson AEG (1971) The laboratory animal – principles and practice. Academic Press, London, New York
59 Russell WMS, Burch R (1959) The principles of humane experimental techniques. Mathuen and Co Ltd, London
60 Vogelstein B, Kinzier KW (1993) The multistep nature of cancer. *TIG* 9: 138–141
61 Frost P, Fidler IJ (1986) Biology of metastasis. *Cancer* 58: 550–553

Edited by Leon P. Bignold

Effects of ionizing radiation on cellular structures, induced instability and carcinogenesis

Marianne Sowa[1], Benjamin J. Arthurs[1], Brian J. Estes[1] and William F. Morgan[2]

[1] *Chemical Structure & Dynamics, Pacific Northwest National Laboratory, MS K8-88, 3335 Q Ave, Richland, WA 99352, USA*
[2] *Radiation Oncology Research Laboratory, BRB 7-011, and Greenebaum Cancer Center, University of Maryland, Baltimore, Baltimore, MD 21201-1551*

Abstract. Ionizing radiation is perhaps the most extensively studied human carcinogen. There have been a number of epidemiological studies on human populations exposed to radiation for medical or occupational reasons, as a result of protracted environmental exposures due to radiation accidents, or after atomic bombings. As a result of these studies exposure to ionizing radiation has been unambiguously linked to cancer causation. While cancer induction is the primary concern and the most important somatic effect of exposure to ionizing radiation, potential health risks do not only involve neoplastic diseases but also somatic mutations that might contribute to birth defects and ocular maladies, and heritable mutations that might impact on disease risks in future generations. Consequantly it is important we understand the long-term health risks associated with exposure to ionizing radiation.

Key words: Genomic instability, ionizing radiation, non-targeted effects, radiation carcinogenesis.

Introduction

According to the American Cancer Society, the United States can expect 1368030 new cases of cancer in 2004 [1]. Among the many carcinogens Americans are exposed to, ionizing radiation will contribute to this statistic. Humans live in a radiation environment. Ionizing radiation is in the air we breathe, the earth we live on, and the food we eat. Man-made radiation adds to this naturally occurring radiation level, thereby increasing the chance for human exposure. For many decades the scientific community, governmental regulatory bodies, and concerned citizens have struggled to estimate health risks associated with radiation exposures, particularly at low doses. While cancer induction is the primary concern and the most important somatic effect of exposure to ionizing radiation, potential health risks do not involve neoplastic diseases exclusively, but also include somatic mutations that might contribute to birth defects and ocular maladies, and heritable mutations that might impact on disease risks in future generations. Consequently, it is important we understand the effect of ionizing radiation on cellular structures and the subsequent long-term health risks associated with exposure to ionizing radiation.

Radiation carcinogenesis

Ionizing radiation is perhaps the most extensively studied human carcinogen. There have been a number of epidemiological studies on human populations exposed to radiation for medical or occupational reasons, as a result of protracted environmental exposures due to radiation accidents, or after atomic bombings (reviewed in [2]). As a result of these studies, exposure to ionizing radiation has been unambiguously linked to carcinogenesis. While many types of human cancer have been convincingly linked to radiation, there are a few notable exceptions including chronic lymphocytic leukemia, Hodgkin's disease, cervical cancer, and prostate cancer.

Cancer incidence is modified by the dose rate, the total dose of radiation delivered, and the quality of radiation, with high linear energy transfer radiation, e.g., radon α particles being more biologically effective than low linear energy transfer radiation, e.g., x- or γ-radiation. In general radiation carcinogenesis is a stochastic effect. That is, the probability of an effect increases as the dose increases, with no dose threshold. However, the severity of the effect is not dose related, such that a high dose of radiation does not induce a "worse" cancer than a low dose of radiation. It should be noted that there are also deterministic effects associated with radiation exposure. While these effects are comparatively rare relative to the well-documented stochastic effects, deterministic effects indicate a threshold of dose and the severity of the effect is dose related. Radiation induced cataracts are an example of a deterministic effect.

There are a number of biological modifiers of radiation-induced cancer risk. These include age at time of exposure, sex, and the target organ. In addition, the cancer risk can be modulated by potential genetic susceptibility factors such as polymorphisms in genes involved in cellular responses to DNA damage (reviewed in [3]). So while it is clear that radiation exposure can lead to cancer, what is not clear is how radiation causes cancer. Cancer appears to arise from the accumulation of multiple genetic abnormalities including gene mutations, deletions, rearrangements and/or alterations in gene expression, as well as chromosomal rearrangements and changes in chromosome number. Radiation-induced cancers have a long latency period between exposure and the appearance of the malignancy. The shortest period is for leukemia, with a peak 5–7 years, but solid tumors show a longer latency period, anything from 10 to 50 years and the excess risk appears to be a lifelong elevation of the natural age-specific cancer risk [4].

Unfortunately for those studying the mechanisms of radiation-induced carcinogenesis and attempting to understand radiation-induced cancer risk, there is no unique signature to cancers associated with radiation exposure. Instead, radiation-induced cancers are similar to those occurring spontaneously.

Radiation-induced genetic damage

Conventionally, radiobiologists have assumed radiation damage can only result from energy deposited directly within the cell nucleus. This deposition of energy results in single- and double-strand DNA cleavage, DNA-DNA and DNA-protein cross-links, and DNA base damages [5]. Failure to faithfully repair these induced damages can result in genetic recombination, deletions, mutations, chromosomal rearrangements and/or cell death [6]. There is no question that these forms of directly induced DNA damage contribute to radiation carcinogenesis. It is assumed that surviving cells can pass on this legacy of radiation to their progeny, thus initiating the carcinogenic process. According to this process, mutations activating oncogenes or deletions affecting tumor suppressor gene function are believed to lead to the accumulation of genomic changes associated with carcinogenesis. Cancer is, therefore, dependent upon the probability of an ionizing event occurring in the genome. This stochastic effect implies radiation causes cancer at random and higher doses only increase the chances of a direct DNA traversal. However, the long latency period between radiation-induced DNA damages and subsequent development of cancer begs the question that directly induced genomic damage might not be responsible for initiating the carcinogenic process, unless it can generate cellular processes leading to genomic instability and the subsequent accumulation of genetic abnormalities.

One of the hallmarks of cancer is its inherent genomic instability, e.g., [7, 8]. Because of the long latency period it is difficult to assign a causal relationship for a specific gene mutation to the development of radiation-induced cancer. Indeed, there is intriguing new evidence uncoupling directly induced DNA damages from the detrimental health effects of radiation exposure. Prakash Hande and colleagues [9] recently reported the detection and quantification of stable intra-chromosomal aberrations in lymphocytes of healthy former nuclear weapons workers exposed to plutonium. Even many years after occupational exposure, a high proportion, in some case more than 50%, of the blood cells of these healthy plutonium workers contained large (>6 Mb) intra-chromosomal rearrangements. The yield of these cytogenetic rearrangements was highly correlated with plutonium dose to the bone marrow, and the control, non-exposed group contained very few such intra-chromosomal aberrations. Thus, the radiation workers, despite their healthy and long-lived status, showed a significant chromosomal aberration burden many years after exposure. This indicates that directly induced radiation damage in this instance chromosomal rearrangements are a superb indicator of radiation exposure but do not necessarily initiate cancer risk. Nevertheless, it remains to be seen whether individuals with a high yield of intra-chromosomal rearrangements are more prone to developing cancer as this population continues to age.

At this stage it would be misleading to conclude that cellular responses to radiation-induced DNA damage have no implications for radiation carcinogenesis. Rearrangements involving *RET* are common in radiation-associated papillary thyroid cancer (*PTC*), e.g., in childhood thyroid cancers associated with the Chernobyl accident [10, 11], and in thyroid cancers from patients with a history of medical external irradiation [12, 13]. The *RET/PTC1* type of rearrangement is an inversion of chromosome 10 mediated by illegitimate recombination between *RET* and the *H4* gene, which is 30 Mb away from *RET*. Spatial contiguity of *RET* and *H4* might provide a structural basis for generation of *RET/PTC1* rearrangement presumably by allowing a singe track to produce a double-strand break in each gene at the same site in the nucleus [14]. Interestingly, such cytogenetic alterations can be detected 48 hours after exposing human fetal thyroid explants exposed to ionizing radiation [15]. It therefore appears reasonable to suggest that the induced *RET – H4* rearrangement facilitates formation of *RET/PTC1* in irradiated thyroid cells.

Radiation-induced genomic instability

Loss of genomic stability is becoming widely accepted as one of the most important processes in the development of cancer [16, 17]. There is now considerable evidence that exposure to ionizing radiation can result in induced genomic instability in the progeny of cells surviving irradiation. Radiation induced instability is a genome wide process, manifesting as the increased acquisition of chromosomal changes, mutation(s), micronuclei, gene amplifications, transformation, alterations in gene expression and/or cytotoxicity in the clonal descendants of an irradiated cell, and has been the subject of a number of recent reviews [18, 19]. The phenotype of radiation-induced instability suggests that genomic changes are not induced directly by the deposition of energy in the cell. Instead, it appears that the instability can manifest in the progeny of an irradiated cell some generations after the initial insult [18, 19]. In addition, recent evidence suggests that induced instability can occur in cells that were not actually irradiated. Instead, they may have been in a radiation environment but not traversed by the radiation [20–22], or have received medium from irradiated cells [23]. Thus, instability can also be a non-targeted or bystander-type consequence of radiation exposure [22, 24].

Mechanisms of radiation-induced genomic instability

Clonal expansion of cells surviving radiation exposure and subsequent cytogenetic analysis of progeny cells indicates chromosomal instability occurs at a very high frequency. For example, in their pioneering study of delayed effects of radiation, Kadhim et al. [20] found that up to 50% of surviving colonies showed chromosomal aberrations in the clonal descendants of an irradiated normal mouse bone marrow cells after exposure to α particles. This high frequency of induction has been confirmed by a number of other laboratories (reviewed in [25, 26]), and pooling the result from a large number of investigations in the Morgan laboratory, a frequency of 3% of surviving colonies exposed to low linear energy transfer x-rays [25] and 4% of surviving colonies exposed to high linear energy transfer radiation [27] displayed chromosomal instability. This high frequency event suggests that it is unlikely that a single mutation can account for the observed instability. Instead, more profound disruption of pathways controlling cellular homeostasis [28], alterations in gene expression [29], and/or modifications of the cell culture environment [30–32] are more likely to account for the observed phenotype.

Once initiated, instability can be perpetuated in clones by recombinational mechanisms involving interstitial repeat sequences within the genome [33], and/or the formation of dicentric chromosomes stimulating bridge breakage fusion cycles, which generate novel chromosomal rearrangements [34]. In many instances genomically unstable clones show persistently elevated levels of reactive oxygen species [35, 36] and a general failure to thrive, as measured by clonogenic survival as a function of time after the initial radiation exposure [37, 38]. This so called delayed reproductive cell death [37], or lethal mutation [38], results in increased numbers of apoptotic cells in unstable clones, which might in turn contribute to the perpetuation of instability by releasing lytic factors into the culture media. These lytic factors may provide a mechanism for DNA cleavage, resulting in DNA repair-mediated mutation induction or the interaction of induced breaks leading to chromosomal rearrangements. Whether or not lytic factors contribute, there is evidence for soluble factors either secreted by unstable cells or produced as byproducts of unstable cells contributing to the instability phenotype [32]. A schematic model for how radiation-induced genomic instability might be perpetuated *in vitro* is presented in Figure 1.

Radiation induced genomic instability and radiation carcinogenesis

Induced instability *in vitro* likely has an extracellular component whereby signal(s) from an unstable cell can elicit responses in non-irradiated cells [24]. These signals might be actively secreted by unstable cells or might be the result of lytic products from dead and dying cells characteristic of unstable clones [31, 39]. How this translates to the *in vivo* situation is not clear. Radiation-induced genomic instability has been described *in vivo*, although this is certainly not as straightforward or as convincing as the *in vitro* model systems (for discussion see [19]). Instability *in vivo* has a significant genetic component (reviewed in [40]) and is likely influenced by the instability endpoint being assayed [41]. A role for a transferable soluble secreted factor(s), so-called "clastogenic factor(s)" produced in blood plasma from irradiated humans and animals, capable of causing chromosomal damage in non-irradiated lymphocytes, has been described after a number of exposure situations (reviewed in [3]). Thus, there is the precedent for secreted factors eliciting a

8 Coleman WB, Tsongalis GJ (1999) The role of genomic instability in human carcinogenesis. *Anticancer Res* 19 (6A): 4645–4664
9 Hande MP, Azizova TV, Geard CR, Burak LE, Mitchell CR, Khokhryakov VF, Vasilenko EK, Brenner DJ (2003) Past exposure to densely ionizing radiation leaves a unique permanent signature in the genome. *Am J Hum Genet* 72: 1162–1170
10 Rabes HM, Demidchik EP, Sidorow JD, Lengfelder E, Beimfohr C, Hoelzel D, Klugbauer S (2000) Pattern of radiation-induced RET and NTRK1 rearrangements in 191 post-chernobyl papillary thyroid carcinomas: biological, phenotypic, and clinical implications. *Clin Cancer Res* 6: 1093–1103
11 Klugbauer S, Lengfelder E, Demidchik EP, Rabes HM (1995) High prevalence of RET rearrangement in thyroid tumors of children from Belarus after the Chernobyl reactor accident. *Oncogene* 11: 2459–2467
12 Bounacer A, Wicker R, Schlumberger M, Sarasin A, Suarez HG (1997) Oncogenic rearrangements of the ret proto-oncogene in thyroid tumors induced after exposure to ionizing radiation. *Biochimie* 79: 619–623
13 Bounacer A, Wicker R, Caillou B, Cailleux AF, Sarasin A, Schlumberger M, Suarez HG (1997) High prevalence of activating ret proto-oncogene rearrangements, in thyroid tumors from patients who had received external radiation. *Oncogene* 15: 1263–1273
14 Nikiforova MN, Stringer JR, Blough R, Medvedovic M, Fagin JA, Nikiforov YE (2000) Proximity of chromosomal loci that participate in radiation-induced rearrangements in human cells. *Science* 290: 138–141
15 Mizuno T, Kyoizumi S, Suzuki T, Iwamoto KS, Seyama T (1997) Continued expression of a tissue specific activated oncogene in the early steps of radiation-induced human thyroid carcinogenesis. *Oncogene* 15: 1455–1460
16 Loeb LA (2001) A mutator phenotype in cancer. *Cancer Res* 61: 3230–3239
17 Loeb LA, Loeb KR, Anderson JP (2003) Multiple mutations and cancer. *Proc Natl Acad Sci USA* 100: 776–781
18 Morgan WF (2003) Non-targeted and Delayed Effects of Exposure to Ionizing Radiation: I. Radiation-Induced Genomic Instability and Bystander Effects *In Vitro*. *Radiat Res* 159: 567–580
19 Morgan WF (2003) Non-targeted and delayed effects of exposure to ionizing radiation: II. Radiation-induced genomic instability and bystander effects *in vivo*, clastogenic factors and transgenerational effects. *Radiat Res* 159: 581–596
20 Kadhim MA, MacDonald DA, Goodhead DT, Lorimore SA, Marsden SJ, Wright EG (1992) Transmission of chromosomal instability after plutonium alpha-particle irradiation. *Nature* 355: 738–740
21 Lorimore SA, Kadhim MA, Pocock DA, Papworth D, Stevens DL, Goodhead DT, Wright EG (1998) Chromosomal instability in the descendants of unirradiated surviving cells after alpha-particle irradiation. *Proc Natl Acad Sci USA* 95: 5730–5733
22 Lorimore SA, Coates PJ, Wright EG (2003) Radiation-induced genomic instability and bystander effects: inter-related nontargeted effects of exposure to ionizing radiation. *Oncogene* 22: 7058–7069
23 Seymour CB, Mothersill C (1997) Delayed expression of lethal mutations and genomic instability in the progeny of human epithelial cells that survived in a bystander-killing environment. *Radiat Oncol Investig* 5: 106–110
24 Morgan WF (2003) Is there a common mechanism underlying genomic instability, bystander effects and other nontargeted effects of exposure to ionizing radiation. *Oncogene* 22: 7094–7099
25 Limoli CL, Corcoran JJ, Milligan JR, Ward JF Morgan WF (1999) Critical target and dose and dose-rate responses for the induction of chromosomal instability by ionizing radiation. *Radiat Res* 151: 677–685
26 Limoli CL, Ponnaiya B, Corcoran JJ, Giedzinski E, Kaplan MI, Hartmann A, Morgan WF (2000) Genomic instability induced by high and low LET ionizing radiation. *Adv Space Res* 25: 2107–2117
27 Limoli CL, Ponnaiya B, Corcoran JJ, Giedzinski E, Morgan WF (2000) Chromosomal instability induced by heavy ion irradiation. *Int J Radiat Biol* 76: 1599–1606
28 Baverstock K (2000) Radiation-induced genomic instability: a paradigm-breaking phenomenon and its relevance to environmentally induced cancer. *Mutat Res* 454: 89–109
29 Snyder AR, Morgan WF (2004) Gene expression profiling after irradiation: clues to understanding acute and persistent responses? *Cancer Metastasis Rev* 23: 259–268

30 Little JB, Azzam EI, de Toledo SM, Nagasawa H (2002) Bystander effects: intercellular transmission of radiation damage signals. *Radiat Prot Dosimetry* 99: 159–162
31 Nagar S, Smith LE, Morgan WF (2003) Characterization of a novel epigenetic effect of ionizing radiation: the death inducing effect. *Cancer Res* 63: 324–328
32 Sowa Resat M, Morgan WF (2004) Radiation-induced genomic instability: A role for secreted soluble factors in communicating the radiation response to non-irradiated cells. *J Cell Biochem* 92: 1013–1019
33 Day JP, Limoli CL, Morgan WF (1998) Recombination involving interstitial telomere repeat-like sequences promotes chromosomal instability in Chinese hamster cells. *Carcinogenesis* 19: 259–265
34 Marder BA, Morgan WF (1993) Delayed chromosomal instability induced by DNA damage. *Mol Cell Biol* 13: 6667–6677
35 Clutton SM, Townsend KM, Walker C, Ansell JD, Wright EG (1996) Radiation-induced genomic instability and persisting oxidative stress in primary bone marrow cultures. *Carcinogenesis* 17: 1633–1639
36 Limoli CL, Hartmann A, Shephard L, Yang CR, Boothman DA, Bartholomew J, Morgan WF (1998) Apoptosis, reproductive failure, and oxidative stress in Chinese hamster ovary cells with compromised genomic integrity. *Cancer Res* 58: 3712–3718
37 Chang WP, Little JB (1991) Delayed reproductive cell death in X-irradiated Chinese hamster ovary cells. *Int J Radiat Biol* 60: 483–496
38 Seymour CB, Mothersill C, Alper T (1986) High yields of lethal mutations in somatic mammalian cells that survive ionizing radiation. *Int J Radiat Biol Relat Stud Phys Chem Med* 50: 167–179
39 Nagar S, Smith LE, Morgan WF (2003) Mechanisms of cell death associated with death-inducing factors from genomically unstable cell lines. *Mutagenesis* 18: 549–560
40 Kadhim MA (2003) Role of genetic background in induced instability. *Oncogene* 22: 6994–6999
41 Morgan WF, Day JP, Kaplan MI, McGhee EM, Limoli CL (1996) Genomic instability induced by ionizing radiation. *Radiat Res* 146: 247–258
42 Lorimore SA, Coates PJ, Scobie GE, Milne G, Wright EG (2001) Inflammatory-type responses after exposure to ionizing radiation *in vivo*: a mechanism for radiation-induced bystander effects? *Oncogene* 20: 7085–7095
43 Neriishi K, Nakashima E, Delongchamp RR (2001) Persistent subclinical inflammation among A-bomb survivors. *Int J Radiat Biol* 77: 475–482
44 Holmberg K, Falt S, Johansson A, Lambert B (1993) Clonal chromosome aberrations and genomic instability in X-irradiated human T-lymphocyte cultures. *Mutat Res* 286: 321–330
45 Kadhim MA, Lorimore SA, Hepburn MD, Goodhead DT, Buckle VJ, Wright EG (1994) Alpha-particle-induced chromosomal instability in human bone marrow cells. *Lancet* 344: 987–988
46 Kadhim MA, Walker CA, Plumb MA, Wright EG (1996) No association between p53 status and alpha-particle-induced chromosomal instability in human lymphoblastoid cells. *Int J Radiat Biol* 69: 167–174
47 Sigurdson AJ, Jones IM (2003) Second cancers after radiotherapy: any evidence for radiation-induced genomic instability? *Oncogene* 22: 7018–7027
48 Goldberg Z (2003) Clinical implications of radiation-induced genomic instability. *Oncogene* 22: 7011–7017
49 Goldberg Z, Lehnert BE (2002) Radiation-induced effects in unirradiated cells: A review and implications in cancer. *Int J Oncol* 21: 337–349

Cancer: Cell Structures, Carcinogens and Genomic Instability
Edited by Leon P. Bignold

Genetic instability in human tumors

Stavroula Raptis[1, 2, 3] and Bharati Bapat[1, 2, 3]

[1] *Department of Pathology and Laboratory Medicine, Mount Sinai Hospital, Toronto, Ontario, M5G 1X5 Canada*
[2] *Samuel Lunenfeld Research Institute, Mount Sinai Hospital, 600 University Avenue, Toronto, Ontario, M5G 1X5 Canada*
[3] *Department of Laboratory Medicine and Pathobiology, Faculty of Medicine, University of Toronto, Toronto, ON, M5S 1A1, Canada*

Abstract. Genetic, or genomic, instability refers to a series of observed spontaneous genetic changes occurring at an accelerated rate in cell populations derived from the same ancestral precursor. This is far from a new finding, but is one that has increasingly gained more attention in the last decade due to its plausible role(s) in tumorigenesis. The majority of genetic alterations contributing to the malignant transformation are seen in growth regulatory genes, and in genes involved in cell cycle progression and arrest. Genomic instability may present itself through alterations in the length of short repeat stretches of coding and non-coding DNA, resulting in microsatellite instability. Tumors with such profiles are referred to as exhibiting a mutator phenotype, which is largely a consequence of inactivating mutations in DNA damage repair genes. Genomic instability may also, and most commonly, results from gross chromosomal changes, such as translocations or amplifications, which lead to chromosomal instability. Telomere length and telomerase activity, important in maintaining chromosomal structure and in regulating a normal cell's lifespan, have been shown to have a function in both suppressing and facilitating malignant transformation. In addition to such direct sequence and structural changes, gene silencing through the hypermethylation of promoter regions, or increased gene expression through the hypomethylation of such regions, together, form an alternative, epigenetic mechanism leading to instability. Emerging evidence also suggests that dietary and environmental agents can further modulate the contribution of genetic instability to tumorigenesis. Currently, there is still much debate over the distinct classes of genomic instability and their specific roles in the initiation of tumor formation, as well as in the progressive transition to a cancerous state. This review examines the various molecular mechanisms that result in this genomic instability and the potential contribution of the latter to human carcinogenesis.

Keywords: Cancer, CIN, epigenetics, genomic instability, MIN, telomeres.

Introduction

The last two decades have seen tremendous gains in promoting the understanding of the process of tumorigenesis. The identification of specific cancer susceptibility genes has provided evidence of an underlying genetic basis for cancer. As such, the advent of tumorigenesis requires the accumulation of multiple mutations in a single cell, thus rendering it with a selective advantage in the environment in which it is present. Clonal expansion of this more favorable cell follows, and, over time and through accrual of additional advantageous mutations, this cell proceeds to malignancy. Cancer results when the equilibrium between cell birth and cell death moves toward uncontrolled cell proliferation [1]. This shift is in part due to mutations in genes that regulate cell growth, differentiation and death, such as oncogenes and tumor suppressor genes. Mutations in genes that protect the integrity of the genome and mediate DNA repair processes also contribute to carcinogenesis.

The theory of the evolution of cancer has been one of long-standing debate. The predominant models of carcinogenesis include one of somatic evolution, where a cell acquires multiple somatic genetic mutations, which eventually lead to malignancy, or alternatively, a model based on the underlying instability in existence at the nucleotide or chromosomal level. The first argues that the normal spontaneous rate of mutation and selection of advantageous clones for expansion is sufficient to initiate the process of tumorigenesis [2, 3]. However, it fails to explain an important factor concerning cancer evolution, which involves the observance of a large number of mutations in tumors, despite a relatively low spontaneous mutation rate in normal cells [4]. The second proposed model of carcinogenesis argues for the role of an underlying genomic instability in cancer, which sees changes in the DNA sequence or chromosomal structure. This argument suggests that the mutator phenotype is necessary for cancer progression, and that tumorigenesis is often initiated by the inherent genomic instability within a cell, without which it is difficult to explain the multiple mutations observed in cancers [4, 5]. This instability may result from subtle mutations in DNA stability genes, leading to microsatellite instability (MIN), also known as MSI, or, more commonly, from structural or numerical changes in whole chromosomes, referred to as chromosomal instability (CIN). Another tumor instability category that confers neither the MIN nor CIN phenotype has been recently proposed, and results from defects in the base-excision repair (BER) pathway. A distinct mechanism contributing to the mutator phenotype involves epigenetic mechanisms, which together with subsequent genetic hits, facilitate the progression towards malignancy [6]. Although the presence of either CIN or MIN has been well documented in the majority of cancers, the speculation continues as to whether this instability is necessary for the initiation of cancer or whether it is the consequence of a cancer phenotype.

In the sections that follow, we review the types of alterations that have been observed in neoplastic cells, then examine the role of the two predominant types of instability described to date and some of the proposed dominant mechanisms leading to each of the instabilities, and their respective roles in tumorigenesis. We conclude with a look at the role of dietary and environmental factors in carcinogenesis.

Genetic alterations and tumors

By the time a tumor has been detected, the individual cells it is composed of have sustained many genetic alterations in key regions that contribute to the properties of the neoplasm. Several different types, or classes, of mutations have been found to be associated with cancer, and interestingly, the mutation class appears to dictate the cell type that will undergo malignant transformation. An example of this has been observed in thyroid carcinomas, whereby two different classes of mutations, chromosomal translocation and nucleotide substitution, in the RET kinase gene, result in different site-specific cancers–papillary thyroid and medullary thyroid carcinomas, respectively [7]. There are at least four main categories of genetic alterations that have been observed in cancer cells and these are described below.

Nucleotide sequence alterations

This type of alteration involves either substitution of a nucleotide, or insertion or deletion of one or more nucleotides. The resulting effects may be as mild as a silent mutation, where there is no change at the amino acid level, or a missense mutation, which still results in the production of a protein encoded by the gene, although the functionality of this protein may be altered from that of the normal. Other mutations, such as nonsense or frameshift mutations, usually have a more detrimental impact on the cell and, in most cases, lead to the production of a truncated protein that usually does not retain normal function. These types of changes to the sequence have been observed to arise both at the germline as well as the somatic levels, and cancer cells can harbor many such base substitutions in prominent cancer susceptibility genes.

Chromosomal aneuploidy

Upon examining cancer cells under a microscope, early cytogeneticists Hansemann and Boveri, made the salient observation that tumor cells often contained more than the normal complement of 46 chromosomes [8, 9]. Aneuploidy is frequently observed in the vast majority of human cancers and involves the gains and losses of whole chromosomes, as well as structural aberrations, such as inversions, deletions and duplications. Most tumors have actually been observed to lose more than half of their alleles. However in some instances, duplication of the remaining chromosome is able to maintain the chromosome number, albeit, retaining two copies of the same parental allele [10]. This could be potentially problematic if the duplicated copy harbors a detrimental mutation or is prone to genomic imprinting.

Chromosome translocations

The field of cytogenetics has enabled the visualization of two patterns of chromosomal translocations: simple translocations, which see discernable rearrangements of chromosomes that result in specific cancers, and complex translocations, which occur almost at random and are not common even among tumors of the same histology. The latter is observed to occur in many solid tumors and results in gains or losses of segments of chromosomes, while the former reproducible form of translocations displays itself mainly in leukemias and lymphomas [11].

Gene amplification

In some instances, gross chromosomal changes can lead to gene amplification. This has been observed to occur in the late-stages of cancer progression and usually manifests in oncogenes and genes involved in metabolism and inactivation of drugs, thus contributing to resistance [12]. Little is known about exactly how these multiple copies of a genomic segment are amplified, but they likely persist due to a defect in DNA damage signaling.

Mutator phenotype and DNA repair

The basis of genomic instability is the unfaithful transmission of genetic material from parent to daughter cell during cell division. Alterations occurring in genes controlling normal cellular functions, such as cell proliferation and DNA replication and repair, are commonly responsible for the failure of genomic integrity. These alterations can range from simple and subtle changes to the nucleotide sequence, to larger scale effects, such as those caused by chromosomal translocations and/or changes in chromosome number. Therefore, it is of no surprise that all organisms have developed systems that are responsible for recognition and repair of sustained DNA damage to protect the fidelity of their genomes. Loss of DNA repair mechanisms and/or loss of control of cell cycle checkpoints result in the loss of genomic stability, a feature presenting itself in most neoplasms. Cells of this type have an increased ability to sustain specific types of mutations, and are referred to as displaying a mutator phenotype. Three main DNA-repair pathways with partly overlapping functions confer this phenotype, and they include: mismatch repair (MMR), nucleotide-excision repair (NER), and base-excision repair (BER).

Mismatch repair and microsatellite instability

MMR is an evolutionarily conserved mechanism of repair that has evolved to correct changes introduced into the DNA during normal cell replication. At least seven proteins have been identified in humans to take part in this repair process, and they include MLH1, MSH2, MSH3, MSH6, PMS1, PMS2, and MLH3 [13, 14]. Heterodimers of MSH2/MSH6 or MSH2/MSH3 are involved in the recognition of mismatched base pairs or small insertion/deletion loops, respectively, with the former complex also having the ability to assist in the repair of insertion/deletions. Upon recognition and binding of one of these complexes, there is recruitment of MLH1-complexes to the site, which recruit additional proteins conducive to the repair [15].

Much of what is known about the development of cancer was discerned from the hereditary forms of colon cancer, the most common syndrome of which is hereditary non-polyposis colorectal cancer (HNPCC), or Lynch syndrome. Examination of colon tumors from HNPCC patients revealed a change in length in microsatellite sequences, which are short (<150 nucleotides), 1–5 nucleotide repeated sequences, scattered throughout the genome. These microsatellite regions appeared to have undergone a change in length due to the insertion or deletion of nucleotides. Further studies examining tumors from HNPCC patients led to a possible mechanism by which such widespread MIN evolved, implicating a defective DNA-MMR system. In 45–70% of HNPCC families, a germline defect in one of the key MMR genes, *MLH1* or *MSH2*, was identified [16–18]. This deficiency in MMR in HNPCC patients occurs through the inheritance of one mutated copy and subsequent loss or mutation of the second normal copy. Every somatic cell in such an individual harbors a mutation in one allele, therefore these individuals are more prone to developing tumors. Once the second allele is lost, the mutation rate of this cell increases by as much as 1000-fold compared with normal cells [19, 20]. This increased mutation rate allows for mutations to begin to accumulate in other important growth regulatory and cell cycle genes, particularly those containing repeat sequences, which are more prone to nucleotide misincorporation errors introduced by DNA replication polymerases. Examples of such genes include *TGFBR2*, *BAX*, *TCF4*, *AXIN2*, *PTEN* [15, 21]. Hence, MMR deficiency appears to exacerbate the number of mutations in a given cell, and therefore plays an important role in tumorigenesis.

Base-excision repair

BER is responsible for excision repair of damaged DNA bases sustained from endogenously formed products of normal cellular metabolic processes, such as reactive oxygen species, methylation, deamination, and hydroxylation [22, 23]. Similar to the other repair pathways, BER is a multi-step process requiring the activities and interactions of several proteins. In BER, the damaged or mispaired base is recognized and removed by DNA glycosylases, and, to date, ten such proteins have been identified in humans [24]. Three of the common glycosylases involved in the removal of 8-oxo-7,8-dihydro2'deoxyguanosine (8-oxoG), or oxidized guanine, the most stable product of oxidative DNA damage, are OGG1, MYH, and MTH1 [23]. When left unrepaired, 8-oxoG can readily mispair with adenines leading to G:C→T:A mutations [25], thus the proper functioning of these enzymes and their partial redundancy indicate their importance in protecting cells against the mutagenic effects of guanine oxidation. Each of these enzymes has a different function: OGG1 is involved in the removal of the oxidized guanine from its pairing to cytosine in the DNA helix, MYH excises misincorporated adenines found opposite unrepaired 8-oxoG, and MTH1 prevents oxidized nucleotide precursors from entering the nucleotide pool prior to the incorporation into DNA. Unlike MMR and NER, inherited BER deficiencies had not been documented in humans until recently [26], and this was believed to be due to the redundancy function between the different repair systems, as mouse knockout models with complete loss of various glycosylases presented with no apparent phenotype [27]. The discovery that biallelic mutations in the BER DNA glycosylase MYH resulted in an autosomal recessive form of adenomatous colorectal polyposis, a high-risk phenotype that may lead to cancer, has dispelled such long-held beliefs [26]. MYH is capable of removing misincorporated adenines from A/G mismatches, and with less efficiency from A/C mismatches, in replicated DNA, and has been shown to interact with the MSH2/MSH6 heterodimer, thereby indicating a possible partnership between MMR and BER [28]. The discovery of MYH involvement in a cancer phenotype occurred through examining the pattern of somatic *APC* mutations within a family having three siblings affected with colon cancer. Using DNA extracted from the tumors, the sequencing of *APC* revealed numerous somatic mutations of which the majority (~83%) were G:C→T:A transversions, mutations which were confirmed to be quite uncommon when compared with the available data on somatic *APC* mutations [26, 29]. Further studies of MYH-deficient colorectal tumors indicated that the two bases immediately following the somatically mutated guanine are almost always two adenine residues. This specificity of mutations at GAA sites is believed to reflect improper recognition and/or repair by the DNA glycosylases. Why defects in MYH seem to have a predilection for formation of tumors in the colon and not other sites is not yet clear, but there is speculation that the high incidence of GAA sites within *APC*, compared to other tumor suppressor genes, and the high levels of oxidative damage in the colon, are important contributing factors [23, 30]. Colorectal tumors arising due to *MYH* mutations do not exhibit either CIN or MIN.

Nucleotide-excision repair

NER is responsible for the removal of various DNA lesions that arise from exogenous agents, such as mutagens and carcinogens, as well as UV photoproducts, which include bulky DNA adducts [31]. There are at least three hereditary syndromes that are associated with defects in NER: xeroderma pigmentosum (XP), Cockayne syndrome and trichothiodystrophy (TTD). Each is characterized by neurological degeneration and sensitivity to sun exposure, the latter being the additional event that facilitates progression to cancer. Only patients with XP have been associated with the cancer phenotype but all have a hypersensitivity to killing by UV and exhibit defective DNA repair. NER utilizes over 30 proteins in humans and includes seven XP NER complementation groups (XP-A to XP-G), plus a variant form with normal excision repair. Patients with XP have an approximate 1000-fold increased risk of developing the skin cancers, basal cell carcinoma, squamous cell carcinoma and melanoma [27].

Chromosomal instability

CIN is apparent in the majority of cancers. In some instances it precedes the onset, and is in fact the underlying cause of certain cancers, such as in lymphomas and leukemias, but in other instances it appears that this instability is a result of previous inherent mutations in genes involved in cell division and differentiation and/or repair of chromosomal breaks occurring during the process. Here, we examine some of these genes and the evidence that exists for their involvement in this cancer phenotype.

CIN is more commonly observed in many human malignancies than MIN, and involves the gains and losses of whole chromosomes. Loss of a maternal or paternal allele, referred to as loss of heterozygosity (LOH), occurs quite frequently in tumors, and is often accompanied by a gain of the opposite allele through its duplication. The types of gene alterations that may lead to this observed phenotype have been extensively studied in yeast, and are involved in processes such as chromosome condensation, sister-chromatid cohesion, kinetochore structure and function, centrosome/microtubule formation and dynamics, and cell cycle checkpoints, whereas in humans to date, only genes involved in the latter process have been implicated in CIN [32, 33].

The mitotic spindle checkpoint, involved in the separation of chromatids only when proper alignment along the mitotic spindle has occurred, appears to play a role in the CIN observed in human cancers. Many studies have shown that genes involved in this process sometimes carry sequence alterations or are observed to have altered expression. Examples of such genes include, *MAD2, BUB1* and *BUBR1*, which appear to have decreased expression in breast and colon cancers respectively [10, 34].

In addition to the mitotic spindle checkpoint's role in CIN, another checkpoint, the DNA-damage checkpoint, plays a vital role in the observed chromosomal changes. This checkpoint is responsible for preventing cells containing DNA damage, sustained from various endogenous and exogenous sources, from entering mitosis. The genes involved in DNA-damage checkpoint control include *ATM, ATR, BRCA1, BRCA2,* and *TP53*, all of which are implicated in carcinogenesis [34].

Aneuploidy, or a change in chromosome number, is often seen in tumors with the CIN phenotype, and involves yet another component of cell division: centrosomes. The mitotic spindles involved in chromosome segregation appear to be multipolar in cancers and the centrosomes from which they are generated occur in greater than normal numbers. The basis for the latter remains a mystery, but may be related to a kinase, aurora2/STK15, which is involved in centrosome maturation and spindle fiber assembly in Drosophila, and is found to be highly expressed and sometimes amplified in human cancers [35]. CIN is a dominant phenotype, as fusion of tumor cells exhibiting CIN with tumor cells that do not yields cells that are chromosomally unstable [36]. Despite this, and the common observance of CIN in various cancers, there have been only a few examples of genetic alterations that can lead to the CIN phenotype. Therefore, alternative mechanisms likely exist that can contribute to the gross chromosomal changes leading to CIN tumors. One such mechanism may involve the telomeric sequences of chromosomes.

Telomere dysfunction and CIN

Although cancer is quite diverse, with extensive heterogeneity amongst tumors within different tissues, there are several characteristics that are commonly exhibited in all tumors, and these include unlimited replicative potential and widespread instability, both of which in part may be explained by telomeric dysfunction [37].

Telomeres serve an important role in maintaining chromosomal structure, provide the chromosome with protection from damage and degradation, and regulate the cell's lifespan. Telomeres consist of repetitive DNA sequences that are bound by specific proteins, some of which are able to form interactions with protein complexes involved in the cellular response to DNA damage, such as Mre11 and Ku, indicating a possible interplay between telomeres and DNA repair mechanisms [37]. A specialized enzyme called telomerase maintains the telomeric DNA sequences. Telomerase is a reverse transcriptase and is composed of two subunits: human telomerase RNA component (hTERC), and human telomerase catalytic component (hTERT). hTERC is ubiquitously expressed in mammalian cells and forms the RNA subunit that provides the template from which telomere synthesis can proceed [38], while hTERT is the catalytic subunit of telomerase and is expressed solely in cells that display telomerase activity [39]. Normally, most human and other mammalian cells do not demonstrate telomerase activity; exceptions of specific cells include those from the germline, including embryonic stem cells, and cells requiring ongoing proliferation, such as lymphocytes, basal keratinocytes, and intestinal crypt cells. Interestingly, the majority of cancer cell lines and >85% of tumors exhibit telomerase activity. In the absence of telomerase, and following each mitosis, an estimated 100 bases of human telomeric DNA is lost. It is precisely this shortening that determines a cell's finite lifespan. Experiments in which hTERT has been introduced into different types of primary human cells that lack telomerase have substantiated the regulatory role of the telomeres; these once mortal cells now see a halt to telomere shortening and seem to overcome one of the necessary proliferative barriers, known as replicative senescence, to become immortal [40, 41]. Consistent with this finding, inhibition of telomerase activity in immortal cancer cell lines results in telomere shortening and eventually leads to the end of cell proliferation. Further investigations indicate that human cells must not only bypass replicative senescence, but also a second barrier, called crisis, which is characterized by extremely short telomeres, increased chromosomal abnormalities, and widespread apoptosis. If cells reach this crisis state, approximately 1×10^{-7} cells survive, possess stable telomeres and even express telomerase, and hence have become immortal [42].

From these observations, it is obvious that the role of telomeres and telomerase is one of important paradox. In one sense, the progressive telomere shortening and absence of telomerase in normal cells leads to the arrest of their proliferation, and thus serves to function as a tumor suppressor. Whereas, the maintenance of telomeres by telomerase or another mechanism, called alternative lengthening of telomeres (ALT), is able to promote the development of cancer. When the telomere repeat sequences are critically shortened or the proteins that stabilize the telomeric structure are lost, telomere function is compromised, leaving it open for DNA-repair processes, such as end-to-end chromosome fusion and degradation by exonucleases. This chromosome end fusion resulting from the loss of telomeres can result in the formation of dicentric chromosomes, which can initiate ongoing CIN through chromosome breakage-fusion-bridge (BFB) cycles. During mitosis, the two centromeres of a dicentric chromosome may be pulled to opposite poles, thus causing the chromosome to break. These breaks can then contribute to genome rearrangements, LOH, and gene amplification, or even the formation of new dicentric chromosomes, which perpetuates the process of instability.

Epigenetics and instability

Epigenetic changes are modifications of the genome, leading to altered gene expression, that are heritable during cell division and do not involve a permanent change in the DNA sequence. Until now, three main types of epigenetic regulatory mechanisms have been described: DNA methylation, histone modifications and genomic imprinting, all of which have shown evidence of a role in carcinogenesis.

DNA methylation

DNA methylation is the most commonly studied form of epigenetic inheritance, and was thought to have first evolved in bacteria as a defense against foreign DNA. The equivalent methylation of cytosine residues in eukaryotes, however, served a far greater role in the maintenance of normal cellular functions and processes, involving the regulation of gene expression and the silencing of repeat elements in the genome [43]. In a normal cell, the DNA methylation pattern, which involves covalent modification of cytosine (carbon-5 position), is stably preserved following DNA replication and cell division by a maintenance DNA methyltransferase (DNMT1). CpG islands, which consist of short (0.5 to several kilobase) stretches of CG dinucleotides and are primarily located in the promoter region, most often observe this methylation that consequently silences normal gene expression [44]. This aberrant, or, more accurately, dysregulated DNA methylation has been implicated in tumorigenesis through mechanisms which include C→T transition mutations, global and focal hypomethylation, and hypermethylation [6].

Cytosine to thymine mutations

Cytosine residues located within CpG dinucleotide repeats exhibit a high transition rate to thymine residues as a consequence of hydrolytic deamination of 5-methylcytosine [45]. In addition to the high mutation rates of methylated cytosines, unmethylated cytosines may also undergo deamination, yielding uracil, which, if left unrepaired, may facilitate transition from cytosine→uracil→thymine. Such changes can assist oncogenesis when the mutations occur in critical genes, for example the tumor suppressor gene *TP53*. Mutations of this gene are found in more than 50% of human solid tumors [46], and almost half of these mutations are cytosine→thymine transitions within CpG sites, indicating the possible involvement of DNA methylation [47].

Hypomethylation and gene activation

The first epigenetic mechanism that was identified to contribute to the development of cancer was the loss of DNA methylation [48]. There has been evidence of age-related decreases in global DNA methylation in a tissue-specific manner, and tumor cells have also been found to display an overall hypomethylation of DNA [49]; these observations correspond to the apparent higher cancer incidence in aging populations [50, 51]. Hypomethylation is believed to occur early in tumorigenesis, and has been found to involve both benign and malignant lesions [52]. As a result of hypomethylation, the activation of specific oncogenes has been observed in tumors of certain types of cancers. For example, in B cell chronic lymphocytic leukemia the anti-apoptotic *BCL-2* [55] gene expression is increased, overexpression of cyclin D2 [53] and maspin [54] is seen in gastric carcinoma, and in lung and colon cancers, *K-RAS* proto-oncogene follows the same fate; all are attributed to decreased levels of methylation [49]. Hypomethylation appears to contribute to genomic instability in many cancers [56], and though the precise mechanism by which it is able to do so is unclear, it is possible that hypomethylation predisposes to chromosomal strand breakage and recombination within derepressed repetitive sequences [50, 56, 57]. Some evidence lending to this comes from studies of a developmental syndrome, ICF (immunodeficiency, chromosomal instability and facial anomalies), whereby loss-of-function mutations in DNMT3B, another DNA methyltransferase, results in unmethylated DNA-satellite repeats found on chromosomes 1, 9, and 16 in leukocytes, and leads to an overall destabilization of chromatin and formation of multi-radial chromosomes with arms from chromosomes 1 and 16 [58–60]. More direct evidence of genomic instability in cancer caused by under methylation is found in Wilms tumors, where specific unbalanced chromosomal translocations, t(1:16), have been observed and result in LOH for markers on chromosome 16 that are strongly correlated with tumor anaplasia [61, 62]. Recently, there has been evidence that mutations in the DNA methyltransferases that cause a decrease in the normal level of methylation could potentially cause this destabilization in the overall chromatin organization [58]. Other studies, analyzing hypomethylated tumors using comparative genome hybridization, indicate that hypomethylation is able to cause gains and losses of chromosomal regions, thus leading to a form of genomic instability, though the mechanism by which this occurs is not well understood. An example of this is seen in T cell lymphomas in mice, where gains in chromosomes 14 and 15 contributed to increased expression of c-myc due to the presence of multiple copies of chromosome 15 [63].

Hypermethylation and gene silencing

CpG island hypermethylation and its most evident contribution to cancer was first recognized in the retinoblastoma (*RB*) tumor suppressor gene [64]. Studies following this link of hypermethylation with cancer implicated many other tumor suppressor and cell cycle regulatory genes, including *VHL* [65], *p16/CDKN2A* [66], *CDH1* [67], *APC* [68], *p16/INK4b* [69] and *PTEN* [70]. Hypermethylation of tumor suppressor genes, as with hypomethylation of oncogenes, is believed to be an early event in carcinogenesis and may precede the neoplastic process [71, 72]. The collaborative effects of epigenetic silencing and genetic mutations in tumorigenesis have been well documented in colon cancer, whereby one allele of *MLH1* may be inactivated by mutation and the other allele is transcriptionally silenced, because of hypermethylation [73–75]. This example also lends to the idea of a role for hypermethylation in conferring a mutator phenotype, and consequently to genomic instability, as does the silencing (via hypermethylation) of the DNA repair gene coding for O6-methylguanine-DNA methyltransferase (MGMT), which has been found to be associated with specific mutations in *K-RAS* and *p53* [76, 77].

Subsequent studies involving genomic screening of different primary human tumors have indicated that, on average, there are approximately 600 aberrantly methylated CpG islands in each tumor [78], indicating that hypermethylation is affecting many genes which remain to be elucidated. Increased levels of methyltransferases have been observed in tumor cell lines, and contribute to up to a 3000-fold greater activity as compared to non-tumor cells [79]. Observations of increased methyltransferase activity in colon tumors *versus* normal colon mucosa from the same patient have also been reported [80], but the mechanism underlying this increased DNA methyltransferase activity is still unclear.

One possible explanation is due to the infidelity of DNMT1, which, similar to the DNA polymerases involved in genome replication, may be prone to methylation errors [81]. The steady, age-related increase in DNA methylation [82], as well as the increasing number of methylated CpG islands in cancer-related genes as tumor growth progresses [83], lends to this explanation. Another possible explanation involves the other DNA methyltransferases, DNMT3a and DNMT3b, which use unmethylated DNA as their template instead of the hemimethylated form used by DNMT1. The former enzymes may contribute to gene silencing possibly through incorrect identification of CpG islands, as it is not known exactly how the specificity of these enzymes is controlled. An alternative explanation involves faulty repair of aberrantly methylated CpG sequences by DNA demethylase, which is responsible for such repair [72, 84].

Histone modification

Histones may undergo modifications such as acetylation, methylation, and phosphorylation at their N-terminal tails, consequently affecting transcriptional regulation and chromatin stability [58]. Both aberrant methylation and deacetylation, via histone deacetylases (HDACs), have been observed to play a causal role in cancer development. Generally, acetylation of lysine residues in the histones leads to transcriptional activation. Deregulated HDACs have been observed to transcriptionally silence tumor suppressor genes in a variety of cancers. Histone methylation, on the other hand, is able to serve both functions, and as this modification is believed to be irreversible, with no demethylase described to date, histone methylation appears to serve an important function in long-term regulation [50]. Thus, histone modification may affect the regulation of expression of certain genes through chromatin remodeling. In addition, histone deacetylation, histone methylation, and DNA methylation may work together in accomplishing silenced gene expression [85, 86].

Genomic imprinting

Genomic imprinting and its contribution to cancer have recently attracted greater attention. Genomic imprinting refers to the epigenetic modification of maternal and paternal genomes during gametogenesis, resulting in the differential expression of the particular parental allele [6, 87]. In a variety of childhood and adult tumors, there is loss of imprinting (LOI), which leads to activation of growth-promoting imprinted genes, such as *IGF2* [6, 88], as well as silencing of potential tumor suppressor genes such as $p57^{\mathrm{KIP2}}$ and *ARHI* [89, 90]. LOI has been observed in colon cancer cases displaying MIN, and has also been found to occur even in the normal colonic mucosa of these patients, suggesting that such epigenetic dysregulation likely occurs early in neoplasia, and may be contributing to a form of genetic instability [91].

Environment shaping instability?

Exploration of the types of instability in human cancers, from epigenetic alterations to MIN and CIN, has indicated that the pathways of genomic instability are quite diverse. The changes that drive tumorigenesis, whether activation or amplification of oncogenes or loss of tumor suppressor genes, provide the cell with a number of different alterations, which are selected for, or against, based on the cell's environment and its exposure to specific carcinogens [92]. This idea is supported by the fact that CIN and MIN cancers tend to be mutually exclusive, and, depending on the type of instability, they may occur in anatomically different locations within the body. A prime example is in colorectal carcinogenesis, where MIN tumors are generally localized to the proximal segments of the bowel, while CIN tumors are more often found in the distal colon and rectum [10, 93]. Furthermore, the influence of dietary factors in various cancers, in particular colon cancer, have introduced the concept of how environment can possibly dictate the molecular pathway by which cancer can proceed [92]. Experiments performed by Bardelli et al. [94] using a colon cancer cell line indicated that exposure to cytotoxic levels of the carcinogens 2-amino-1-methyl-6-phenylimidazo [4,5-*b*]pyridine (PhIP), a common dietary heterocyclic amine resulting from well-cooked meat, and *N*-methyl-*N*'-nitro-*N*-nitrosoguanidine (MNNG) produced tumor cells characteristic of the CIN and MIN phenotypes, respectively. Thus, by these examples it is plausible that genetic instability occurs through somatic evolution, with environmental factors acting as selection pressures favoring the breakdown of DNA repair or the loss of cell cycle control [92]. However, is it possible that somatic mutations driven by environmental selection pressures will lead to cancer? As a number of mutations in a single cell are required to initiate tumorigenesis, and the likelihood that this phenomenon occurs by chance is not highly probable, we cannot ignore the role of genetics in this process, and the possibility that it supercedes the environment in certain instances.

Environmental and dietary factors thus likely do not act in isolation, rather, their effects vary according to the individual's genetic background, and it is this interaction that produces the phenotypic effects observed in cancer [95]. Hence genetic variation, or genetic polymorphisms, in nutrient metabolism genes and their potential role in carcinogenesis have received much attention in recent years. How can these alterations affect the type of instability that is observed in specific tumor types? A good example linking nutrient intake, nutrient metabolism and genetic instability may be seen in the well-studied methylenetetrahydrofolate reductase (*MTHFR*) gene. A common variant observed in *MTHFR* is the C677T polymorphism, which results in the disruption of the folate metabolic pathway. Folate is an important nutrient that is required for many methylation reactions, including methylation of cystosine at CpG sites, as well as the synthesis of thymidylate from dUMP. Folate deficiency can lead to misincorporation of uracil into DNA and inappropriate gene expression due to hypomethylation of gene regulatory regions, thereby affecting genomic stability. In addition, specific alleles of certain genes involved in cholesterol and lipid metabolism appear to influence the anatomical location of tumor formation in the colon. The apolipoprotein E4 (ApoE4) allele has been found to enhance cholesterol absorption and reduce bile acid output, offering a protective effect against the development of proximal tumors [96]. Therefore, the surrounding micro- or macro-environment is capable of modifying the risk of developing a tumor, but is likely to be dependent on underlying genetic variation.

Conclusion

The evolution of cancer is usually a long process, requiring multiple genetic mutations sustained over many years. Genomic instability, whether at the nucleotide or chromosomal level, is a common feature in the majority of tumors, yet its precise role in cancer initiation and progression is not completely understood. As cells are constantly exposed to an ever-changing microenvironment, their likelihood of coming into contact with potential DNA-damaging agents increases. Therefore, it is possible that genetic instability, or increased mutation rate, allows genetic and epigenetic alterations to accumulate during carcinogenesis to provide the tumor cells with a selective advantage that allow them to persist in their environment. Although there has been, and continues to be, much debate over whether an underlying genetic instability is necessary for cancer initiation, it is certain that without it, tumorigenesis would be a much more prolonged process.

Acknowledgements
B.B. is the recipient of CIHR Grant No. MOP_64225 and crCIHRt -43821 from the Canadian Institutes of Health Research. S.R. is the recipient of the Ontario Student Opportunity Trust Fund and the Samuel Lunenfeld Research Institute Fellowship.

References

1 Hanahan D, Weinberg RA (2000) The hallmarks of cancer. *Cell* 100: 57–70
2 Tomlinson IP, Novelli MR, Bodmer WF (1996) The mutation rate and cancer. *Proc Natl Acad Sci USA* 93: 14800–14803
3 Sieber OM, Heinimann K, Tomlinson IP (2003) Genomic instability—the engine of tumorigenesis? *Nat Rev Cancer* 3: 701–708
4 Loeb LA (2001) A mutator phenotype in cancer. *Cancer Res* 61: 3230–3239
5 Loeb LA (1991) Mutator phenotype may be required for multistage carcinogenesis. *Cancer Res* 51: 3075–3079
6 Feinberg AP (2004) The epigenetics of cancer etiology. *Semin Cancer Biol* 14: 427–432
7 Eng C, Mulligan LM (1997) Mutations of the RET proto-oncogene in the multiple endocrine neo-

plasia type 2 syndromes, related sporadic tumors, and hirschsprung disease. *Hum Mutat* 9: 97–109
8 von Hansemann D (1890) *Virchows Arch Pathol Anat* 119: 299–326
9 Boveri T (1914) *Zur Frage der Enstehung Maligner Tumoren*. Gustav Fischer, Jena. English translation: Boveri M (transl.) (1929) *The Origin of Malignant Tumors*. Williams and Wilkins, Baltimore
10 Lengauer C, Kinzler KW, Vogelstein B (1998) Genetic instabilities in human cancers. *Nature* 396: 643–649
11 Le Beau MM, Rowley JD (1986) Chromosomal abnormalities in leukemia and lymphoma: clinical and biological significance. *Adv Hum Genet* 15: 1–54
12 Schwab M (1999) Oncogene amplification in solid tumors. *Semin Cancer Biol* 9: 319–325
13 Marti TM, Kunz C, Fleck O (2002) DNA mismatch repair and mutation avoidance pathways. *J Cell Physiol* 191: 28–41
14 Jacob S, Praz F (2002) DNA mismatch repair defects: role in colorectal carcinogenesis. *Biochimie* 84: 27–47
15 Peltomaki P (2001) Deficient DNA mismatch repair: a common etiologic factor for colon cancer. *Hum Mol Genet* 10: 735–740
16 Liu B, Parsons R, Papadopoulos N, Nicolaides NC, Lynch HT, Watson P, Jass JR, Dunlop M, Wyllie A, Peltomaki P et al. (1996) Analysis of mismatch repair genes in hereditary non-polyposis colorectal cancer patients. *Nat Med* 2: 169–174
17 Marra G, Boland CR (1995) Hereditary nonpolyposis colorectal cancer: the syndrome, the genes, and historical perspectives. *J Natl Cancer Inst* 87: 1114–1125
18 Wijnen JT, Vasen HF, Khan PM, Zwinderman AH, van der Klift H, Mulder A, Tops C, Moller P, Fodde R (1998) Clinical findings with implications for genetic testing in families with clustering of colorectal cancer. *N Engl J Med* 339: 511–518
19 Parsons R, Li GM, Longley MJ, Fang WH, Papadopoulos N, Jen J, de la Chapelle A, Kinzler KW, Vogelstein B, Modrich P (1993) Hypermutability and mismatch repair deficiency in RER+ tumor cells. *Cell* 75: 1227–1236
20 Bhattacharyya NP, Skandalis A, Ganesh A, Groden J, Meuth M (1994) Mutator phenotypes in human colorectal carcinoma cell lines. *Proc Natl Acad Sci USA* 91: 6319–6323
21 Duval A, Hamelin R (2002) Genetic instability in human mismatch repair deficient cancers. *Ann Genet* 45: 71–75
22 Lindahl T, Wood RD (1999) Quality control by DNA repair. *Science* 286: 1897–1905
23 Cheadle JP, Sampson JR (2003) Exposing the MYtH about base excision repair and human inherited disease. *Hum Mol Genet* 12 Spec No 2: R159–R165
24 Wood RD, Mitchell M, Sgouros J, Lindahl T (2001) Human DNA repair genes. *Science* 291: 1284–1289
25 Shibutani S, Takeshita M, Grollman AP (1991) Insertion of specific bases during DNA synthesis past the oxidation-damaged base 8-oxodG. *Nature* 349: 431–434
26 Al-Tassan N, Chmiel NH, Maynard J, Fleming N, Livingston AL, Williams GT, Hodges AK, Davies DR, David SS, Sampson JR et al. (2002) Inherited variants of MYH associated with somatic G:C→T:A mutations in colorectal tumors. *Nat Genet* 30: 227–232
27 Hoeijmakers JH (2001) Genome maintenance mechanisms for preventing cancer. *Nature* 411: 366–374
28 Gu Y, Parker A, Wilson TM, Bai H, Chang DY, Lu AL (2002) Human MutY homolog, a DNA glycosylase involved in base excision repair, physically and functionally interacts with mismatch repair proteins human MutS homolog 2/human MutS homolog 6. *J Biol Chem* 277: 11135–11142
29 Cheadle JP, Krawczak M, Thomas MW, Hodges AK, Al-Tassan N, Fleming N, Sampson JR (2002) Different combinations of biallelic APC mutation confer different growth advantages in colorectal tumors. *Cancer Res* 62: 363–366
30 Babbs CF (1990) Free radicals and the etiology of colon cancer. *Free Radic Biol Med* 8: 191–200
31 Dixon K, Kopras E (2004) Genetic alterations and DNA repair in human carcinogenesis. *Semin Cancer Biol* 14: 441–448
32 Komarova NL, Sengupta A, Nowak MA (2003) Mutation-selection networks of cancer initiation: tumor suppressor genes and chromosomal instability. *J Theor Biol* 223: 433–450
33 Nowak MA, Komarova NL, Sengupta A, Jallepalli PV, Shih Ie M, Vogelstein B, Lengauer C (2002) The role of chromosomal instability in tumor initiation. *Proc Natl Acad Sci USA* 99: 16226–16231
34 Rajagopalan H, Lengauer C (2004) Aneuploidy and cancer. *Nature* 432: 338–341

35 Zhou H, Kuang J, Zhong L, Kuo WL, Gray JW, Sahin A, Brinkley BR, Sen S (1998) Tumor amplified kinase STK15/BTAK induces centrosome amplification, aneuploidy and transformation. *Nat Genet* 20: 189–193
36 Lengauer C, Kinzler KW, Vogelstein B (1997) Genetic instability in colorectal cancers. *Nature* 386: 623–627
37 Hahn WC (2003) Role of telomeres and telomerase in the pathogenesis of human cancer. *J Clin Oncol* 21: 2034–2043
38 Feng J, Funk WD, Wang SS, Weinrich SL, Avilion AA, Chiu CP, Adams RR, Chang E, Allsopp RC, Yu J et al. (1995) The RNA component of human telomerase. *Science* 269: 1236–1241
39 Nakamura TM, Cooper JP, Cech TR (1998) Two modes of survival of fission yeast without telomerase. *Science* 282: 493–496
40 Bodnar AG, Ouellette M, Frolkis M, Holt SE, Chiu CP, Morin GB, Harley CB, Shay JW, Lichtsteiner S, Wright WE (1998) Extension of life-span by introduction of telomerase into normal human cells. *Science* 279: 349–352
41 Vaziri H, Benchimol S (1998) Reconstitution of telomerase activity in normal human cells leads to elongation of telomeres and extended replicative life span. *Curr Biol* 8: 279–282
42 Lustig AJ (1999) Crisis intervention: the role of telomerase. *Proc Natl Acad Sci USA* 96: 3339–3341
43 Jacobsen SE (1999) Gene silencing: Maintaining methylation patterns. *Curr Biol* 9: R617–619
44 Bird A (2002) DNA methylation patterns and epigenetic memory. *Genes Dev* 16: 6–21
45 Laird PW, Jaenisch R (1996) The role of DNA methylation in cancer genetic and epigenetics. *Annu Rev Genet* 30: 441–464
46 Greenblatt MS, Bennett WP, Hollstein M, Harris CC (1994) Mutations in the p53 tumor suppressor gene: clues to cancer etiology and molecular pathogenesis. *Cancer Res* 54: 4855–4878
47 Magewu AN, Jones PA (1994) Ubiquitous and tenacious methylation of the CpG site in codon 248 of the p53 gene may explain its frequent appearance as a mutational hot spot in human cancer. *Mol Cell Biol* 14: 4225–4232
48 Feinberg AP, Tycko B (2004) The history of cancer epigenetics. *Nat Rev Cancer* 4: 143–153
49 Feinberg AP, Vogelstein B (1983) Hypomethylation distinguishes genes of some human cancers from their normal counterparts. *Nature* 301: 89–92
50 Lund AH, van Lohuizen M (2004) Epigenetics and cancer. *Genes Dev* 18: 2315–2335
51 Richardson B (2003) Impact of aging on DNA methylation. *Ageing Res Rev* 2: 245–261
52 Goelz SE, Vogelstein B, Hamilton SR, Feinberg AP (1985) Hypomethylation of DNA from benign and malignant human colon neoplasms. *Science* 228: 187–190
53 Oshimo Y, Nakayama H, Ito R, Kitadai Y, Yoshida K, Chayama K, Yasui W (2003) Promoter methylation of cyclin D2 gene in gastric carcinoma. *Int J Oncol* 23: 1663–1670
54 Akiyama Y, Maesawa C, Ogasawara S, Terashima M, Masuda T (2003) Cell-type-specific repression of the maspin gene is disrupted frequently by demethylation at the promoter region in gastric intestinal metaplasia and cancer cells. *Am J Pathol* 163: 1911–1919
55 Hanada M, Delia D, Aiello A, Stadtmauer E, Reed JC (1993) bcl-2 gene hypomethylation and high-level expression in B-cell chronic lymphocytic leukemia. *Blood* 82: 1820–1828
56 Schulz WA, Elo JP, Florl AR, Pennanen S, Santourlidis S, Engers R, Buchardt M, Seifert HH, Visakorpi T (2002) Genomewide DNA hypomethylation is associated with alterations on chromosome 8 in prostate carcinoma. *Genes Chromosomes Cancer* 35: 58–65
57 Tsuda H, Takarabe T, Kanai Y, Fukutomi T, Hirohashi S (2002) Correlation of DNA hypomethylation at pericentromeric heterochromatin regions of chromosomes 16 and 1 with histological features and chromosomal abnormalities of human breast carcinomas. *Am J Pathol* 161: 859–866
58 El-Osta A (2004) The rise and fall of genomic methylation in cancer. *Leukemia* 18: 233–237
59 Xu GL, Bestor TH, Bourc'his D, Hsieh CL, Tommerup N, Bugge M, Hulten M, Qu X, Russo JJ, Viegas-Pequignot E (1999) Chromosome instability and immunodeficiency syndrome caused by mutations in a DNA methyltransferase gene. *Nature* 402: 187–191
60 Carpenter NJ, Filipovich A, Blaese RM, Carey TL, Berkel AI (1988) Variable immunodeficiency with abnormal condensation of the heterochromatin of chromosomes 1, 9, and 16. *J Pediatr* 112: 757–760
61 Yeh A, Wei M, Golub SB, Yamashiro DJ, Murty VV, Tycko B (2002) Chromosome arm 16q in Wilms tumors: unbalanced chromosomal translocations, loss of heterozygosity, and assessment of the CTCF gene. *Genes Chromosomes Cancer* 35: 156–163
62 Qu GZ, Grundy PE, Narayan A, Ehrlich M (1999) Frequent hypomethylation in Wilms tumors of

pericentromeric DNA in chromosomes 1 and 16. *Cancer Genet Cytogenet* 109: 34–39
63 Gaudet F, Hodgson JG, Eden A, Jackson-Grusby L, Dausman J, Gray JW, Leonhardt H, Jaenisch R (2003) Induction of tumors in mice by genomic hypomethylation. *Science* 300: 489–492
64 Greger V, Passarge E, Hopping W, Messmer E, Horsthemke B (1989) Epigenetic changes may contribute to the formation and spontaneous regression of retinoblastoma. *Hum Genet* 83: 155–158
65 Herman JG, Latif F, Weng Y, Lerman MI, Zbar B, Liu S, Samid D, Duan DS, Gnarra JR, Linehan WM et al. (1994) Silencing of the VHL tumor-suppressor gene by DNA methylation in renal carcinoma. *Proc Natl Acad Sci USA* 91: 9700–9704
66 Gonzalez-Zulueta M, Bender CM, Yang AS, Nguyen T, Beart RW, Van Tornout JM, Jones PA (1995) Methylation of the 5' CpG island of the p16/CDKN2 tumor suppressor gene in normal and transformed human tissues correlates with gene silencing. *Cancer Res* 55: 4531–4535
67 Graff JR, Herman JG, Lapidus RG, Chopra H, Xu R, Jarrard DF, Isaacs WB, Pitha PM, Davidson NE, Baylin SB (1995) E-cadherin expression is silenced by DNA hypermethylation in human breast and prostate carcinomas. *Cancer Res* 55: 5195–5199
68 Hiltunen MO, Alhonen L, Koistinaho J, Myohanen S, Paakkonen M, Marin S, Kosma VM, Janne J (1997) Hypermethylation of the APC (adenomatous polyposis coli) gene promoter region in human colorectal carcinoma. *Int J Cancer* 70: 644–648
69 Baylin SB, Herman JG, Graff JR, Vertino PM, Issa JP (1998) Alterations in DNA methylation: a fundamental aspect of neoplasia. *Adv Cancer Res* 72: 141–196
70 Whang YE, Wu X, Suzuki H, Reiter RE, Tran C, Vessella RL, Said JW, Isaacs WB, Sawyers CL (1998) Inactivation of the tumor suppressor PTEN/MMAC1 in advanced human prostate cancer through loss of expression. *Proc Natl Acad Sci USA* 95: 5246–5250
71 Nephew KP, Huang TH (2003) Epigenetic gene silencing in cancer initiation and progression. *Cancer Lett* 190: 125–133
72 Momparler RL (2003) Cancer epigenetics. *Oncogene* 22: 6479–6483
73 Cunningham JM, Christensen ER, Tester DJ, Kim CY, Roche PC, Burgart LJ, Thibodeau SN (1998) Hypermethylation of the hMLH1 promoter in colon cancer with microsatellite instability. *Cancer Res* 58: 3455–3460
74 Veigl ML, Kasturi L, Olechnowicz J, Ma AH, Lutterbaugh JD, Periyasamy S, Li GM, Drummond J, Modrich PL, Sedwick WD et al. (1998) Biallelic inactivation of hMLH1 by epigenetic gene silencing, a novel mechanism causing human MSI cancers. *Proc Natl Acad Sci USA* 95: 8698–8702
75 Grady WM (2004) Genomic instability and colon cancer. *Cancer Metastasis Rev* 23: 11–27
76 Esteller M, Toyota M, Sanchez-Cespedes M, Capella G, Peinado MA, Watkins DN, Issa JP, Sidransky D, Baylin SB, Herman JG (2000) Inactivation of the DNA repair gene O6-methylguanine-DNA methyltransferase by promoter hypermethylation is associated with G to A mutations in K-ras in colorectal tumorigenesis. *Cancer Res* 60: 2368–2371
77 Nakamura M, Sakaki T, Hashimoto H, Nakase H, Ishida E, Shimada K, Konishi N (2001) Frequent alterations of the p14(ARF) and p16(INK4a) genes in primary central nervous system lymphomas. *Cancer Res* 61: 6335–6339
78 Costello JF, Fruhwald MC, Smiraglia DJ, Rush LJ, Robertson GP, Gao X, Wright FA, Feramisco JD, Peltomaki P, Lang JC et al. (2000) Aberrant CpG-island methylation has non-random and specific-type-specific patterns. *Nat Genet* 24: 132–138
79 Kautiainen TL, Jones PA (1986) DNA methyltransferase levels in tumorigenic and nontumorigenic cells in culture. *J Biol Chem* 261: 1594–1598
80 Issa JP, Vertino PM, Wu J, Sazawal S, Celano P, Nelkin BD, Hamilton SR, Baylin SB (1993) Increased cytosine DNA-methyltransferase activity during colon cancer progression. *J Natl Cancer Inst* 85: 1235–1240
81 Bebenek K, Kunkel TA (2000) Streisinger revisited: DNA synthesis errors mediated by substrate misalignments. *Cold Spring Harb Symp Quant Biol* 65: 81–91
82 Issa JP (1999) Aging, DNA methylation and cancer. *Crit Rev Oncol Hematol* 32: 31–43
83 Graff JR, Gabrielson E, Fujii H, Baylin SB, Herman JG (2000) Methylation patterns of the E-cadherin 5' CpG island are unstable and reflect the dynamic, heterogeneous loss of E-cadherin expression during metastatic progression. *J Biol Chem* 275: 2727–2732
84 Ramchandani S, Bhattacharya SK, Cervoni N, Szyf M (1999) DNA methylation is a reversible biological signal. *Proc Natl Acad Sci USA* 96: 6107–6112
85 Bird AP, Wolffe AP (1999) Methylation-induced repression—belts, braces, and chromatin. *Cell*

99: 451–454
86 Cameron EE, Bachman KE, Myohanen S, Herman JG, Baylin SB (1999) Synergy of demethylation and histone deacetylase inhibition in the re-expression of genes silenced in cancer. *Nat Genet* 21: 103–107
87 Feinberg AP (2001) Cancer epigenetics takes center stage. *Proc Natl Acad Sci USA* 98: 392–394
88 Ogawa O, Eccles MR, Szeto J, McNoe LA, Yun K, Maw MA, Smith PJ, Reeve AE (1993) Relaxation of insulin-like growth factor II gene imprinting implicated in Wilms' tumor. *Nature* 362: 749–751
89 Thompson JS, Reese KJ, DeBaun MR, Perlman EJ, Feinberg AP (1996) Reduced expression of the cyclin-dependent kinase inhibitor gene p57KIP2 in Wilms' tumor. *Cancer Res* 56: 5723–5727
90 Yu Y, Xu F, Peng H, Fang X, Zhao S, Li Y, Cuevas B, Kuo WL, Gray JW, Siciliano M et al. (1999) NOEY2 (ARHI), an imprinted putative tumor suppressor gene in ovarian and breast carcinomas. *Proc Natl Acad Sci USA* 96: 214–219
91 Cui H, Horon IL, Ohlsson R, Hamilton SR, Feinberg AP (1998) Loss of imprinting in normal tissue of colorectal cancer patients with microsatellite instability. *Nat Med* 4: 1276–1280
92 Breivik J, Gaudernack G (1999) Genomic instability, DNA methylation, and natural selection in colorectal carcinogenesis. *Semin Cancer Biol* 9: 245–254
93 Lindblom A (2001) Different mechanisms in the tumorigenesis of proximal and distal colon cancers. *Curr Opin Oncol* 13: 63–69
94 Bardelli A, Cahill DP, Lederer G, Speicher MR, Kinzler KW, Vogelstein B, Lengauer C (2001) Carcinogen-specific induction of genetic instability. *Proc Natl Acad Sci USA* 98: 5770–5775
95 Collins AR, Ferguson LR (2004) Nutrition and carcinogenesis. *Mutat Res* 551: 1–8
96 Kervinen K, Sodervik H, Makela J, Lehtola J, Niemi M, Kairaluoma MI, Kesaniemi YA (1996) Is the development of adenoma and carcinoma in proximal colon related to apolipoprotein E phenotype? *Gastroenterology* 110: 1785–1790

Cancer: Cell Structures, Carcinogens and Genomic Instability
Edited by Leon P. Bignold

Molecular mechanisms of human carcinogenesis

William B. Coleman[1] and Gregory J. Tsongalis[2]

[1] *Department of Pathology and Laboratory Medicine, Curriculum in Toxicology, UNC Lineberger Comprehensive Cancer Center, University of North Carolina School of Medicine, Chapel Hill NC, 27599, USA*
[2] *Laboratory of Molecular Pathology, Department of Pathology, Dartmouth-Hitchcock Medical Center, Lebanon, NH 03756, USA*

Abstract. Intensive research efforts during the last several decades have increased our understanding of carcinogenesis, and have identified a genetic basis for the multi-step process of cancer development. Tumors grow through a process of clonal expansion driven by mutation. Several forms of molecular alteration have been described in human cancers, and these can be generally classified as chromosomal abnormalities and nucleotide sequence abnormalities. Most cancer cells display a phenotype characterized by genomic hypermutability, suggesting that genomic instability may precede the acquisition of transforming mutations in critical target genes. Reduced to its essence, cancer is a disease of abnormal gene expression, and these genetic abnormalities contribute to cancer pathogenesis through inactivation of negative mediators of cell proliferation (including tumor suppressor genes) and activation of positive mediators of cell proliferation (including proto-oncogenes). In several human tumor systems, specific genetic alterations have been shown to correlate with well-defined histopathological stages of tumor development and progression. Although the significance of mutations to the etiological mechanisms of tumor development has been debated, a causal role for such genetic lesions is now commonly accepted for most human cancers. Thus, genetic lesions represent an integral part of the processes of neoplastic transformation, tumorigenesis, and tumor progression, and as such represent potentially valuable markers for cancer detection and staging.

Keywords: Chromosomal instability, genomic instability, microsatellite instability, tumor suppressor gene, proto-oncogene.

Cancer: a multi-step genetic disease

Cancer development is a multi-step process through which cells acquire increasingly abnormal proliferative and invasive behaviors. Cancer also represents a unique form of genetic disease, characterized by the accumulation of multiple somatic mutations in a population of cells undergoing neoplastic transformation [1, 2]. Genetic lesions represent an integral part of the processes of neoplastic transformation, tumorigenesis, and tumor progression, and as such represent potentially valuable markers for cancer detection and staging [3, 4]. Several forms of molecular alteration have been described in human cancers, including gene amplifications, deletions, insertions, rearrangements, and point mutations [2]. In many cases specific genetic lesions have been identified that are associated with neoplastic transformation and/or tumor progression in a particular tissue or cell type [1]. Statistical analyses of age-specific mortality rates for different forms of human cancer predict that multiple mutations in specific target genes are required for the genesis and outgrowth of most clinically diagnosable tumors [5]. In accordance with this prediction, it has been suggested that tumors grow through a process of clonal expansion driven by mutation [6], where the first mutation leads to limited expansion of progeny of a single cell, and each subsequent mutation gives rise to a new clonal outgrowth with greater proliferative potential. The idea that carcinogenesis is a multi-step process is supported by morphological observations of the transitions between pre-malignant (benign) cell growth and malignant tumors. In colorectal cancer (and some other tumor systems), the transition from a benign lesion to a malignant neoplasm can be easily documented and occurs in discernible stages, including benign adenoma, carcinoma *in situ*, invasive carcinoma, and eventually local and distant metastasis [7]. Moreover, specific genetic alterations have been shown to correlate with each of these well-defined histopathological stages of tumor development and progression [8]. However, it is important to recognize that it is the accumulation of multiple genetic alterations in affected cells, and not necessarily the order in which these changes accumulate, that determines tumor formation and progression. These observations suggest strongly that the molecular alterations observed in human cancers represent integral (necessary) components of the process of neoplastic transformation and tumor progression.

Mutations and cancer

Mutation is the ultimate source of variability for individual cells (and organisms), and is an essential component of the process of natural selection [9]. Tumorigenesis can be viewed simply as a process of natural selection in which cells develop a growth advantage that allows them to proliferate and invade under conditions where other (normal) cells cannot, and the acquisition of this ability is driven by mutation. In other words, tumor development and progression represents a form of somatic evolution, at the ultimate expense of the host organism [10]. The idea that somatic mutation could significantly contribute to cancer development was suggested by Boveri early in the 20th century [11]. At about the same time, De Vries proposed that certain forms of radiation (Röntgen rays) may be mutagenic [10], suggesting that mutation rates could be influenced by exogenous factors. Evidence from numerous investigations suggests that multiple somatic mutations contribute to the step-wise process of neoplastic transformation and tumorigenesis. In early studies, the nature of these mutations and their contributions to tumorigenesis were not at all clear. Nonetheless, the presence of multiple mutations in cancer cells could be observed in the form of karyotypic alterations and abnormal chromosome numbers in tumor cells. More recent studies utilizing comparative genomic hybridization extended these observations by identifying both gross (cytogenetically detectable) and subtle chromosomal abnormalities in different human neoplasms [12]. Subsequently, numerous positive and negative mediators (proto-oncogenes and tumor suppressor genes) of cell growth and differentiation have been identified and characterized, defining the basic role for these critical genetic elements in neoplastic transformation and tumorigenesis [1, 13]. Recently, microarray-based gene expression studies have provided definitive evidence that cancer is ultimately a disease of abnormal gene expression [14–16]. Somatic mutations occurring in developing cancers alter gene expression patterns, resulting in significant changes to cellular physiology, including unregulated (or abnormally regulated) cell proliferation and acquisition of invasive behaviors [17, 18]. The gene expression signature of a specific cancer can be used in differential diagnosis, prognostication, and prediction of responses to therapy [19, 20].

The exact number of critical mutations required for neoplastic transformation of normal cells is not known. Investigations involving the statistical analysis of human tumor incidence and natural history in sporadic and inherited human tumors formed the basis for the two-hit model of cancer development [21, 22]. In this model, genetic predisposition for a specific type of neoplasm is conferred on an individual that either inherits or acquires a germline mutation in one allele of a critical target (such as a tumor suppressor gene), constituting the first "hit", and the second "hit" represents an acquired somatic mutation in the remaining normal allele. Accumulation of two hits alters (or eliminates) normal gene function in affected cells, which proliferate to form a tumor. While the kinetics of tumor formation are consistent with this model for some neoplasms, it is now recognized that neoplastic transformation involves the mutational alteration or aberrant expression of multiple genes that function in cell proliferation or differentiation. Furthermore, epigenetic mechanisms can contribute to the multi-hit model of cancer induction through the silencing of critical genes [23–25]. In recent years, a re-examination of the number of critical mutations needed for cancer development led to the suggestion that six to eight mutations may be necessary for progression to an invasive tumor [5, 26]. These analyses provide estimates of the numbers of mutations involving genes that control proliferation and differentiation that are necessary for neoplastic transformation of a specific cell type. However, numerous lines of evidence suggest that tumors are mutation prone and/or accumulate large numbers of mutations [6, 27, 28], and some investigators have estimated that tumor cells may contain thousands or tens of thousands of mutations [29, 30].

Molecular alterations in cancer

The molecular alterations occurring in cancer typically reflect mutations, and can be categorized into two major groups: (i) chromosomal abnormalities, and (ii) nucleotide sequence abnormalities. There has been some debate in the literature as to which forms of mutation are more prevalent in cancer cells and/or constitute the foundations of the molecular mechanism of neoplastic transformation [31]. However, there is abundant evidence that representations of both of these major categories of genetic abnormalities exist in most tumor cells, and that both significantly contribute to neoplastic transformation.

Chromosomal abnormalities

Chromosomal alterations in cancer include the gain or loss of one or more chromosomes (aneuploidy), chromosomal rearrangements resulting from DNA strand breakage (translocations, inversions, and other rearrangements), and gain or loss of portions of chromosomes (amplification, large-scale deletion). The direct result of chromosomal translocation is the movement of some segment of DNA from its natural location into a new location within the genome, which can result in altered expression of the genes that are contained within the translocated region. If the chromosomal breakpoints utilized in a translocation are located within structural genes, then hybrid (chimeric) genes can be generated. The major consequence of chromosomal deletion (involving a whole chromosome or a large chromosomal region) is the loss of specific genes that are localized to the deleted chromosomal segment, resulting in changes in the copy number of the affected genes. Likewise, gain of chromosome number or amplification of chromosomal regions results in an increase in the copy numbers of genes found in these chromosomal locations.

Nucleotide sequence abnormalities

Nucleotide sequence alterations in cancer include changes in individual genes involving single nucleotide changes (missense and nonsense), and small insertions or deletions (some of which result in frameshift mutations). Single nucleotide alterations that involve a change in the normal coding sequence of the gene (point mutations) can give rise to an alteration in the amino acid sequence of the encoded protein. Missense mutations alter the translation of the affected codon, while nonsense mutations alter codons that encode amino acids to produce stop codons. This results in premature termination of translation and the synthesis of a truncated protein product. Small deletions and insertions are typically classified as frameshift mutations, because deletion or insertion of a single nucleotide (for instance) will alter the reading frame of the gene on the 3' side of the affected site. This alteration can result in the synthesis of a protein that bears very little resemblance to the normal gene product, or production of an abnormal/truncated protein due to the presence of a stop codon in the altered reading frame. In addition, deletion or insertion of one or more groups of three nucleotides will not alter the reading frame of the gene, but will alter the resulting polypeptide product, which will exhibit either loss of specific amino acids or the presence of additional amino acids within its primary structure.

Are cancer cells prone to mutation?

It is widely accepted that cancer cells accumulate numerous genetic abnormalities (consisting of chromosomal alterations and/or nucleotide sequence mutations) during the protracted interval between the initial carcinogenic insult and tumor outgrowth. At least a portion of the genetic changes occurring in neoplasia are related to the underlying molecular mechanism of neoplastic transformation [26, 32, 33]. Nonetheless, whether the myriad of genetic lesions found in cancer cells are the causes or consequences of neoplastic transformation continues to be the subject of debate [34]. Some investigators have suggested that the intrinsic mutation rate in mammalian cells is insufficient to account for the numerous genetic changes observed in cancer cells, leading to the suggestion that an early (essential?) step in neoplastic transformation is the development of a condition of hypermutability or genetic instability [35, 36]. In the past, increased rates of mutation in pre-neoplastic or neoplastic cells would have been attributed to exposure of these cells to exogenous mutagenic agents. However, more recent analyses of the nature and frequency of mutations occurring in human neoplasms suggests that a significant proportion result from spontaneous mutational mechanisms [37]. This observation strengthens the suggestion that cancer cells may exhibit diminished capacities for surveillance and repair of DNA lesions, leading to increased rates of spontaneous mutation and/or increased susceptibility to mutation following exposure to some exogenous carcinogenic agent. An alternative argument suggests that increased rates of mutation are not necessary for accumulation of large numbers of genetic lesions in cancer cells, but that selection of advantageous mutations is a more important feature of the process of tumorigenesis [36, 38].

Spontaneous mutation rates in normal cells

The measured spontaneous mutation rate of mammalian cells depends upon the exact experimental conditions employed and the nature of the cells and target sequence examined [39]. Somatic mutation rates have been determined for a variety of cultured cell types through examination of the spontaneous mutation frequency at one of several specific loci, such as the hypoxanthine-guanine phosphoribosyltransferase (*HPRT*) gene, the Na^{+}-K^{+}-*ATPase* gene, or the adenine phosphoribosyltransferase (*APRT*) gene. Using the results from several of these studies [40–42], the spontaneous mutation frequency at the HPRT locus can be estimated to be approximately $2.7 \times 10^{-10}–1 \times 10^{-9}$ mutations/nucleotide/cell generation in untransformed human cells. This mutation rate is sufficient to yield approximately three mutations per cell over the lifespan of an individual, which may be too low to account for the number of mutations thought to be required for carcinogenesis. This observation led to the hypothesis that an early event in neoplastic transformation may involve an increase in the spontaneous mutation rate in cells that are progressing through this multi-step pathway [42]. Cells expressing the "mutator phenotype" accumulate mutations more rapidly than normal cells, and would therefore be more likely to sustain mutations in critical genes required for enhanced growth and tumorigenesis [43, 44].

Mutation rates in cancer cells

In many studies, the measured mutation rate in malignant cells is significantly higher than that of corresponding normal cells. In some cases the elevated mutation rates were 100-fold higher than in untransformed cells [45, 46]. Tumor cell lines that are deficient for DNA repair exhibit mutation rates that are 750-fold higher than that displayed by DNA repair-proficient tumor cell lines [47]. In addition, the rate of gene amplification in malignant cells is much higher than in normal cells [48]. However, other studies find no difference in the spontaneous mutation rate between normal and malignant cells [40, 41], or suggest that selective pressures associated with clonal expansion of altered cells represent a much more important feature of carcinogenesis than a hypermutational phenotype [38]. Thus, some cancer cells may express a "mutator phenotype" and exhibit an enhanced mutation rate compared to normal cells [28], whereas other cancers may exhibit multiple mutations in the absence of any appreciable increase in mutation frequency. These observations suggest the possibility that multiple molecular mechanisms are needed to reconcile the occurrence of multiple mutations in human cancers and the expression of a mutator phenotype with elevated mutational frequency in only a subset of these tumors.

Genomic instability in human cancer

An appropriate definition of genomic instability is needed before a complete understanding of the interconnecting causes and consequences of genomic instability can be developed, and the contribution of this phenomenon to neoplastic transformation can be appreciated. The observation that most cancer cells contain discernible genetic abnormalities (chromosomal aberrations and/or DNA sequence abnormalities) suggests that all neoplastically transformed cells have sustained genetic damage and may have experienced some form of genomic instability. Normal human cells demonstrate a remarkable degree of genomic integrity, which reflects the combined contributions of high-fidelity DNA replication processes, and the expression of multiple mechanisms that recognize and repair DNA damage. Nonetheless, rare spontaneous mutations can occur in cells that are proficient for both DNA replication and repair. The observation that neoplastic cells contain variable numbers of mutations reflecting specific forms of DNA damage, and that tumors develop over widely variable periods of time, suggests the possible involvement of different pathogenic mechanisms that may reflect multiple distinct mutagenic pathways to neoplastic transformation. Tumors are highly variable with respect to their growth characteristics; some tumors become clinically evident early in the human lifespan, while others present later in life. This discrepancy could reflect individual differences among tumors with respect to the relative rapidity of their development and progression. Consistent with the proposal that tumors form through clonal expansion driven by mutation [49–51], tumors displaying early onset and rapid progression may accumulate a critical level of genetic damage more quickly than tumors with later onset and more indolent course.

The forms of genetic damage typically displayed by cancer cells (involving chromosomal alterations and/or DNA sequence alterations) are not mutually exclusive. However, different mutagenic mechanisms may be involved in the origins of these genetic abnormalities [52–54]. Nonetheless, it is likely that the same target genes might be involved in tumorigenesis driven by either form of genetic damage. Inactivation of the p53 tumor suppressor gene (loss of function) can be accomplished through point mutation at numerous nucleotide sites [13, 55] or through deletion of the locus on 17p [56]. Likewise, proto-oncogene activation can be accomplished by point mutation, as with the H-*ras* gene [57], or by chromosomal translocation, as with the c-*myc* gene [58].

Based upon these observations, a unifying hypothesis is required to describe the possible mechanisms of genomic instability that account for the disparate numbers of mutations (specific loci *versus* widespread mutation) and diverse nature of genetic damage (types of mutations) that characterize various human cancers. We have proposed that at least two broad categories of genomic instability may exist: (i) progressive (persistent) genomic instability, and (ii) episodic (transient) genomic instability [59]. Evidence supporting the existence of these forms of genetic instability has emerged from studies in bacteria [60], and examples of each form of genomic instability have been described in human cancers. Progressive instability defines an ongoing mutagenic process, with new mutations occurring in each cell generation, and is associated with cells that are compromised in their ability to safeguard genomic integrity. This form of genomic instability would be transmitted from cell generation to cell generation as a heritable trait [60]. For instance, tumor cells from patients with hereditary nonpolyposis colorectal cancer (HNPCC) exhibit progressive genomic instability, which is manifest as alterations in microsatellite sequences [61, 62]. In contrast to progressive instability, episodic instability describes sporadic genetic damage in cells that are proficient in the various pathways that govern genomic homeostasis. This form of instability is associated with tumors that contain specific mutations and/or chromosomal alterations, in the absence of wide-spread damage to the genome. The transient mutator state may account for a large portion of adaptive mutations occurring in cells [60]. For instance, cells exposed to high levels of oxidative stress may incur and accumulate adaptive mutations that enable the altered cells to thrive under highly selective conditions [63, 64]. These mutations can occur in the absence of cell proliferation [43, 44], but would facilitate clonal expansion of an altered clone in response to subsequent selection pressures [36]. Numerous sporadic tumor types exemplify this form of instability, including sporadic colorectal tumors of the tumor suppressor pathway [65], or the microsatellite mutator pathway [64, 66]. It can be envisioned that both chromosomal abnormalities and DNA sequence abnormalities could result from the expression of either of these forms of genomic instability during neoplastic transformation.

Chromosomal abnormalities in cancer

The majority of human cancers (including solid tumors, leukemias, and lymphomas) contain chromosomal abnormalities, consisting of either numerical changes (aneuploidy) and/or structural aberrations [67, 68]. These types of chromosomal damage may reflect two distinct mechanisms of chromosomal instability [2, 69]: (i) chromosome number instability, and (ii) chromosome structure instability. Recent evidence suggests a genetic basis for chromosomal instability in cancer, involving mutational inactivation of certain types of genes in aneuploid tumors [70].

Detailed karyotypic studies have been performed on a large number of tumor types; many of these studies have examined leukemia and lymphoma, partially reflecting the relative ease with which chromosomes can be prepared from these cancer cells. Traditional cytogenetic analyses of solid tumors are more difficult. Nonetheless, a substantial body of literature on the chromosomal aberrations of solid tumors has emerged [12]. Additional methods have also been applied to examination of chromosomal abnormalities in solid tumors [68]. Numerous studies have investigated allelic loss of heterozygosity (LOH) in various human solid tumors using Southern analysis or PCR [71–75]. While these methods do not provide the same information as karyotypic analysis, large-scale deletions can be inferred from the loss of multiple markers on a specific chromosomal arm [76]. In addition, flow cytometry is now widely employed for determination of tumor ploidy [77], and fluorescence *in situ* hybridization (or derivative methods) is used to examine specific chromosome numbers and alterations [78–80]. A detailed review of chromosomal alterations in human cancer is beyond the scope of this chapter. Several excellent reviews are available [12, 81].

Instability of chromosome number

Numerical alterations of chromosomes can involve both loss of entire chromosomes or allelic losses, which may be accompanied by duplication of the remaining allele. This phenomenon results in the generation of a tumor with normal karyotype, but an abnormal allelotype [71]. Several studies suggest that tumors arising in various tissues share a common chromosome number instability and may loose a significant number (25–50%) of alleles during neoplastic transformation and tumorigenesis [71, 73, 82, 83]. These large-scale genomic changes may be due to some form of progressive chromosomal instability [84, 85]. Supporting this suggestion, gains and losses of multiple chromosomes occur in aneuploid colorectal cancer cell lines 10- to 100-fold more frequently than in diploid cancer cell lines of the same histological subtype [53, 86]. In other studies, the rate of LOH at marker loci proximal to a selectable gene (*APRT*) was increased 10-fold in colorectal cancer cell lines that exhibit proficiency of mismatch repair (MMR) compared with cell lines that lack MMR [87, 88]. In addition, numerous studies combine to show that aneuploid cancers exhibit highly variable karyotypes [67, 89], suggesting that new chromosomal variations are produced in a progressive manner during tumor outgrowth and evolution.

The absence of chromosomal instability in diploid cancers and/or cancers that exhibit nucleotide sequence alterations, argues against a nonspecific mechanism for chromosomal instability related to abnormal properties of neoplastic cells [2]. Further, the high rates of numerical chromosomal alterations in aneuploid cells do not simply reflect the ability of these cells to survive changes in chromosome number [53]. Tetraploid cells resulting from the fusion of diploid cancer cells retain a stable tetraploid chromosome number [53], suggesting that the presence of a nondiploid chromosome number does not precipitate progressive chromosomal instability. Rather, the evidence supports the existence of a specific form of genetic instability in cancer cells that results from dysfunction of normal chromosomal homeostasis producing numerical chromosomal abnormalities. Several possibilities have been investigated, including the involvement of (i) mutant p53 protein, (ii) abnormal centrosomes, (iii) abnormal mitotic spindle checkpoint function, or (iv) abnormal DNA-damage checkpoint function [2, 85, 90].

Inactivation of the p53 tumor suppressor leads to abnormalities of chromosome number

The p53 tumor suppressor protein has long been suggested to play significant roles in cell cycle progression and cell cycle checkpoint function in response to DNA damage [91, 92]. The *p53* gene is commonly mutated in human cancers [37], and these same cancers frequently exhibit abnormalities of chromosome number [93–95]. Cells in culture often become aneuploid concurrent with mutation or inactivation of *p53* [96, 97], suggesting that loss of p53 function leads to abnormal regulation of mitosis and segregation of chromosomes [98]. However, other lines of evidence do not support a direct role for *p53* mutation in chromosomal instability. For example, aneuploidy occurs very early in the process of neoplastic transformation and tumorigenesis [99], and *p53* mutation typically occurs later in the process [100]. In addition, some diploid tumor cell lines that exhibit a stable karyotype also contain mutant *p53*

tribute to chromosomal instability through inactivation of the mitotic spindle checkpoint.

Abnormal DNA damage checkpoint function leads to aneuploidy

The DNA damage checkpoint represents the major cellular mechanism that guards against the replication of damaged DNA or entry of cells with DNA damage into mitosis. The types of DNA damage that elicit checkpoint activation include polymerase errors remaining after DNA replication and other forms of incompletely repaired DNA, damage resulting from exposure to exogenous genotoxins (ionizing radiation, chemical mutagens, and others), and damage related to endogenous genotoxic insult (such as reactive oxygen species). A number of genes have been implicated in the control of this checkpoint, including *p53* [92], *ATM* [123], *BRCA1* and *BRCA2* [124], and some others [2]. Functional inactivation of one or more of these genes through genetic or epigenetic mechanisms could result in a genomic instability related to the loss of the DNA damage checkpoint. Loss of this checkpoint might lead to aneuploidy directly resulting from abnormal segregation of damaged chromosomes [2].

Instability of chromosome structure

The majority of human cancers exhibit chromosomal abnormalities, including marker chromosomes with altered structure. It is generally accepted that many of the alterations of chromosome structure occurring in cancer cells confer some selective advantage to the evolving tumor. Thus, accumulation of a critical number of chromosomal aberrations or development of specific chromosomal abnormalities may represent essential steps in the process of neoplastic transformation. Three general forms of chromosomal alteration are observed in cancer cells: (i) gene amplifications, (ii) rearrangements and translocations, and (iii) large-scale deletions.

Gene amplification

The amplification of specific chromosomal segments or genes have been documented in some cancers and in many cancer cell lines [48, 125], some of which involve cellular proto-oncogenes, resulting in abnormal expression levels of the proto-oncogene products [126]. In general, gene amplification occurs late in tumorigenesis associated with tumor progression and is the recognized mechanism through which many tumors acquire resistance to chemotherapeutic agents. Thus, gene amplifications can profoundly affect tumor behavior, and can have prognostic significance for some cancers, but may not be involved with early genetic alterations in pre-neoplastic lesions. The mechanisms governing gene amplification have not been determined with any certainty. However, several studies suggest that gene amplification occurs at much higher rates in neoplastic cells than in normal cells [48]. A role for the *p53* tumor suppressor in gene amplification has been suggested by some investigators. Evidence supporting this suggestion includes the observation that gene amplification occurs more readily in cells following inactivation of *p53* function [127, 128]. However, gene amplification can also occur in cells with normal *p53* [127]. One possibility for the role of *p53* in this process is that amplification of a chromosomal segment in a normal cell may trigger apoptosis in response to perceived DNA damage [129], whereas in the absence of normal *p53* function cells would not undergo apoptosis, but would continue to accumulate amplicons in subsequent rounds of replication [2]. Thus, this form of chromosomal instability may involve a mechanism (or a mechanistic component) that increases the ability of an affected cell to survive the genetic alteration.

Chromosomal rearrangements and translocations

Chromosomal rearrangements can take on several different forms, the most common of which are translocations. Patterns of chromosomal translocation in human cancer can be classified as complex or simple [2]. In some human cancers no consistent pattern of chromosomal abnormality can be discerned (complex translocations). These tumors exhibit complex type translocations, which may appear to be random. Among individual tumors of one type, or individual cells of a single tumor, different chromosomal aberrations may be found. Very often, these rearrangements are accompanied by large-scale loss of chromosomal segments. While it is possible that some of these chromosomal alterations are not essential to tumorigenesis, it is unlikely that any chromosomal alteration that does not confer a proliferative or adaptive advantage would be preserved in an evolving tumor. In some human cancers specific chromosomal anomalies are consistently found in a high percentage of tumors (simple translocations). These recurrent chromosomal abnormalities may reflect molecular alterations that are essential and necessary to the molecular pathogenesis of the specific tumor type. The discovery of the Philadelphia chromosome [trans(9;22)(q34;q11)] in the cancer cells of patients with chronic myelogenous leukemia was the first report suggesting the involvement of nonrandom chromosomal changes in the molecular pathogenesis of the disease [130]. Subsequent studies suggest that the neoplastic cells of 80–90% of leukemia and lymphoma patients contain some sort of demonstrable karyotypic abnormality, and many of these are uniquely associated with morphologically or clinically defined subsets of these cancers [67, 89]. Similar relationships between chromosomal alterations and definable stages of tumor development and progression have been established for some human solid tumors [8, 26], and proposed for others [131, 132]. The role of chromosomal translocation in cancer pathogenesis is suggested to involve proto-oncogene activation by repositioning of the gene adjacent to a heterologous genetic control element. Evidence for this type of proto-oncogene activation includes studies of chromosome translocations in Burkitt's lymphoma [133]. In this cancer, the c-*myc* proto-oncogene is translocated from chromosome 8 to chromosome 14, proximal to the immunoglobulin enhancer sequences, resulting in abnormal constitutive expression of c-*myc* [58].

Large-scale chromosomal deletions

Large-scale deletions of whole chromosomes or chromosomal arms have been documented in many cancers. These deletions contribute to the abnormal allelotype of tumors, and may accompany chromosomal rearrangements and/or translocations. In most cases, such deletions are thought to be related to the presence of a tumor suppressor locus on the affected chromosomal arm. Large-scale deletions affecting several chromosomes have been documented in sporadic CRC, including deletions of 5q, 17p, and 18q [8]. Each of these chromosomal arms contains a known tumor suppressor locus; the adenomatous polyposis coli (*APC*) gene at 5q [134], the *p53* gene at 17p [56, 100], and the *DCC* (for "deleted in colorectal cancer") gene at 18q [135].

Microsatellite instability in human cancer

Microsatellite instability (MSI) is characterized by alterations to simple repeated sequences, including both expansions (insertions) and contractions (deletions), typically resulting in frameshift mutations. Microsatellites are repetitive sequences that consist of variable numbers of repeated units of one to four (or more) nucleotides. Such sequences are numerous and randomly distributed throughout the human genome. Mutational alterations of numerous adenine mononucleotide repeat motifs (polyA tracts) was the first characteristic used to define MSI in human tumors [136, 137]. These early studies of sporadic CRC suggested that 12% of all tumors harbor these mutations, with as many as 1×10^5 mutated polyA tracts per tumor [137]. Subsequent studies demonstrated frequent microsatellite alterations in hereditary CRC [61] and sporadic tumors [138] when higher order repeated units were examined.

Determination of microsatellite instability in human tumors

Tumors with MSI contain numerous altered microsatellite sequences, but not all microsatellite sequences are altered in tumors with MSI [59, 139]. In fact some studies have shown dramatic differences in susceptibility to mutation of individual microsatellite loci [140]. In addition, two distinct patterns of microsatellite alteration have been described in human cancers that display MSI, and specific microsatellite markers tend to be altered in a characteristic pattern [141, 142]. The pattern of alteration observed at a specific microsatellite locus may reflect the nature of the genomic instability displayed by a tumor. Several factors influence the probability of mutation at a specific microsatellite locus: (i) the type of repeated sequence (mononucleotide, dinucleotide, etc.), (ii) the length of the microsatellite sequence (number of repeated units), (iii) the location of the microsatellite sequence within the genome, and (iv) the underlying molecular lesion. Thus, no single type of microsatellite will be diagnostic for MSI in all tumors. This is supported by the observation that numerous polyA repeats are altered in various human cancers [143, 144], but not all neoplasms that exhibit MSI demonstrate alterations in polyA sequences, and may only show alterations in higher order repeats [145]. A direct relationship has been observed between the length of polyA tracts and their mutation frequency among genetically unstable tumors [146], consistent with the suggestion that the probability of sustaining a mutation in an individual microsatellite sequence is proportional to the length of its sequence [147]. Extensive comparison of the mutation of dinucleotide *versus* higher order repeat units (trinucleotide or tetranucleotide) in human tumors suggests that larger alleles are more susceptible to mutation in genetically unstable tumors [148]. Studies with cancer cell lines that harbor MMR gene mutations demonstrate instability of specific classes of microsatellites. Cells possessing a defect in *hPMS2* exhibit instability of trinucleotide repeats [149], while cells deficient for *hMSH3* or *hMSH6* demonstrate an inability to correct mismatches in dinucleotide (or higher order) repeats [150]. Furthermore, cells lacking *hMSH* demonstrate minimal levels of dinucleotide instability, while cell lines lacking *hMSH2* or *hMLH1* demonstrate profound dinucleotide instability [151]. In addition, specific MMR gene mutations can affect the extent of hypermutability at microsatellite sequences [140]. The microsatellite mutation rate in cells lacking *hMLH1* and *hMSH3* is tenfold greater than that of cells lacking *hPMS2* and *hMSH6* [140]. These observations suggest that individual MMR complexes exhibit specificity for certain types of mismatches, and that the MSI displayed by cancer cells may be directly related to the number [152] and nature [140] of MMR gene mutations.

Frequency of microsatellite instability in human cancer

A large number of studies have documented the occurrence of MSI in human cancer [59]. Tumors from patients diagnosed with HNPCC frequently exhibit MSI (141/159 tumors, 89%), while only 15% (887/5724 tumors) sporadic colorectal cancers demonstrate MSI [59]. However, sporadic CRC occurring in young patients (<35 years of age) and in patients with multiple primary tumors exhibit MSI at higher frequency (64% and 71%, respectively) than sporadic

CRC in general [153]. Sporadic gastric cancers exhibit MSI in 19% (276/1485) of tumors [59], while gastric carcinomas from patients with multiple primary tumors or familial predisposition exhibit an elevated frequency of MSI (61% and 32%, respectively) compared to sporadic tumors [154, 155]. Several other cancers exhibit MSI in 15–35% of tumors examined. Sporadic breast cancers demonstrate MSI in 17% (64/372 tumors) of cases, but this percentage varies widely between studies [59]. The combined results of six studies failed to detect MSI at even one locus among 522 tumors examined [59], suggesting that the actual frequency of occurrence of MSI among breast cancers is very low. MSI has been documented in 24% (168/713) of tumors of the endometrium, 13% (16/123) of ovarian cancers, 27% (25/92) of esophageal tumors, 28% (25/88) of liver tumors, 29% (78/272) of non-small cell lung cancers, and 32% (79/247) of prostate cancers [59]. Hodgkin's disease and some forms of leukemia exhibit MSI in a high percentage of cases. However, additional studies will be needed to determine more precisely the prevalence of this genetic abnormality in these neoplasms, particularly among the various forms of leukemia. MSI is rare (<10% tumors) among gliomas, neuroblastomas, and cancers of the testicles, thyroid, and uterine-cervix. Evidence for the involvement of MSI in some other tumors has been produced, although the numbers of tumors examined and the numbers of studies conducted are limited.

Mismatch repair defects lead to nicrosatellite instability

The molecular defects responsible for the MSI in human tumors involve the genes that encode proteins required for normal MMR [156]. These include *hMSH2* [157, 158], *hMSH3* [159], and *hMSH6/GTBP* [160, 161], which are human homologs of the bacterial *MutS* gene, and *hMLH1* [162, 163], *hPMS1* [164], *hPMS2* [164], and *hMLH3* [165], which are human homologs of the bacterial *MutL* gene. One or more of these genes are mutated in the germline of the majority of individuals with HNPCC [166], and somatic mutations have been identified in sporadic CRC that display MSI [167]. MMR gene defects have also been identified in other cancers that exhibit MSI [52, 59, 168, 169].

The proteins involved with MMR operate in concert to recognize mispaired or unpaired nucleotides, and facilitate their removal and repair [170]. This mechanism differs from nucleotide-excision repair, which recognizes and repairs abnormal (adducts) nucleotides [168]. The observation that microsatellite mutations consist of expansion or contraction of the repeated sequence through insertion or deletion of variable numbers of repeat units suggests that such mutations arise through a slippage mechanism during replication of these simple repeat sequences [171]. Strand slippage of the primer at a repetitive sequence during replication generates a misaligned intermediate that is stabilized by correct base pairing between discrete repeat units on the misaligned strand. Such a misaligned intermediate is normally repaired through the proofreading function of the polymerase complex, or by post-replication repair mechanisms [172]. If the intermediate is not repaired, subsequent rounds of replication will generate insertion or deletion mutations in the newly synthesized DNA strands. The relative location of the unpaired repeat sequence in the replication intermediate determines whether an insertion or deletion will result (contraction or expansion of the microsatellite).

Microsatellite mutation rates in DNA repair-proficient and -deficient cells

The spontaneous mutation rate of a dinucleotide microsatellite repeat sequence in normal human fibroblasts has been estimated to be 12.7×10^{-8} mutations/cell/generation [173], suggesting that dinucleotide repeat sequences are remarkably stable in normal human cells. MMR-proficient cancer cells exhibit a microsatellite mutation rate of 9.8×10^{-6} mutations/cell/generation, whereas MMR-deficient cells exhibited mutation rates of 1.6×10^{-4} and 3.3×10^{-3} mutations/cell/generation, respectively [174]. The estimated mutation rates at the dinucleotide repeat in MMR-deficient tumor cells were 16-fold and 337-fold higher than that of the MMR-proficient tumor cells, and 1260-fold and 25984-fold higher than that of normal fibroblasts [173]. These results highlight the propensity for spontaneous mutation at microsatellite repeat sequences of tumor cells that are deficient for MMR, and support the notion that cells displaying MSI harbor sustained lesions in MMR.

Specific mismatch repair genes are responsible for specific forms of microsatellite instability

Genetic complementation studies have produced direct evidence for the involvement of specific chromosomal loci or specific genes in MMR-deficient tumor cells that exhibit MSI. Transfer of human chromosome 2, which contains the *hMSH2* and *hMSH6* genes, restores genetic stability and MMR-proficiency to *hMSH2*-mutant cancer cells [175]. Furthermore, cells containing chromosome 2 demonstrate microsatellite stability at a trinucleotide repeat (D7S1794) and a dinucleotide repeat (D14S73), whereas cells containing other transferred chromosomes (such as chromosome 17) continue to exhibit instability at these loci [175]. Furthermore, transfer of chromosome 2 restores genetic stability to cancer cells that carry mutations of both *hMSH6* and DNA polymerase δ [161, 176], suggesting that the DNA polymerase δ defect is not the primary determinant of genetic instability in these cells [175]. However, other studies question whether *hMSH6* plays a major role in MSI [177]. Transfer of chromosome 3, which contains *hMLH1*, into tumor cells that are homozygous for *hMLH1* mutation, restores MMR and stability to the D5S107 dinucleotide microsatellite repeat [178]. In similar studies, transfer of human chromosome 5 (containing *hMSH3*) or human chromosome 2, into tumor cells resulted in partial correction of the MMR defect [159]. More recent studies have utilized single gene transfer to correct MMR deficiency. Cancer cells that harbor an *hPMS2* mutation and display MMR deficiency [149] show increased microsatellite stability and reduced mutation rate at the *HPRT* locus, and cell extracts can perform strand-specific MMR following transfection with a wild-type *hPMS2* gene [179]. Likewise, transfection of tumor cells with *hMSH6* resulted in restoration of MMR, increased stability of the BAT26 polyA tract, and reduction in the mutation rate at the *HPRT* locus [180].

Epigenetic silencing of mismatch repair genes leads to microsatellite instability

Mutational inactivation of MMR genes has been documented in numerous human tumors that display MSI. However, in many cases the underlying molecular defect in MMR cannot be identified, suggesting that additional MMR genes exist, or that alternative mechanisms for microsatellite mutation are operational in these tumors. Several studies have produced strong evidence that epigenetic regulation of MMR gene expression may be responsible for loss of MMR function in tumors that display MSI. Initially, a strong correlation between general methylation status and MMR proficiency in CRC cell lines was noted [181]. Cell lines that were deficient for MMR and showed MSI demonstrated hypermethylation of endogenous and exogenous DNA sequences [181]. Subsequently, several laboratories examined expression of *hMLH1*, methylation of the *hMLH1* promoter, and MSI status among sporadic CRC [182–184]. Tumors exhibiting high level MSI, no detectable expression of *hMLH1*, and no *hMLH1* point mutation, also showed hypermethylation of the *hMLH1* promoter region [182–184]. In cell lines that exhibit loss of *hMLH1* and hypermethylation of the *hMLH1* promoter, treatment with 5-aza-2'-deoxycytidine resulted in re-expression of *hMLH1* and restoration of MMR capacity [183]. These results suggest that inactivation of *hMLH1* through hypermethylation of its promoter may represent the principle mechanism of gene inactivation in sporadic CRC characterized by widespread MSI. Consistent with this suggestion, the *hMLH1* promoter has been shown to be hypermethylated in 122/167 (73%) CRC with MSI, but in only 20/138 (14%) of microsatellite stable CRC [59]. Similar relationships between *hMLH1* promoter hypermethylation and MSI have been observed in gastric and endometrial cancers [185–187]. In contrast to the relationship observed in sporadic cancers with MSI, tumors from HNPCC patients that harbor mutations in MMR genes do not show *hMLH1* promoter hypermethylation.

Oxidative stress and loss of mismatch repair function

Chronic inflammation is known to contribute to DNA damage related to excess levels of free radicals. MMR-proficient cells are protected from mutational alteration of microsatellite sequences after exposure to low levels of hydrogen peroxide [66]. However, MMR-deficient cells demonstrate numerous microsatellite alterations in response to oxidative stress, and oxygen radical scavengers diminish the damaging effects of free radicals in these cells [188]. These observations suggest that DNA damage related to free radical exposure can contribute to MSI. Many forms of cancer are closely associated with chronic inflammation, leading to the suggestion that oxidative stress may significantly contribute to DNA damage elevating the risk for neoplastic transformation in affected tissues. Ulcerative colitis is an inflammatory bowel disease associated with increased risk for colorectal cancer [189]. MSI is one of the features of ulcerative colitis and related pre-neoplastic lesions [190], suggesting a role for MMR defects in this condition. The accumulation of microsatellite mutations in ulcerative colitis could be related to a failure of the MMR pathways to correct the excess damage resulting from elevated levels of free radicals [191]. Alternatively, MMR function may be disabled through DNA methylation [192] or directly by oxidative stress [193]. The inactivation of MMR function in response to oxidative stress is mediated by oxidative damage to MMR complexes, possibly involving hMutSα, hMutSβ, and hMutLα [193]. This type of mechanism could account for MSI in chronically inflamed non-neoplastic tissues [194], as well as cancers associated with inflammatory processes [195, 196].

Molecular targets of microsatellite instability

Numerous simple repeat sequences are found in the human genome. Some of these occur within the coding regions of structural genes. These genes may be targets for mutation in cells that display MSI [197, 198]. The *TGFβRII* gene contains two simple repeat sequences: (i) a 10-bp adenine mononucleotide tract and (ii) a 6-bp GT repeat [199]. This gene represents the first recognized target for inactivation due to microsatellite mutations in human tumors and cell lines, and both simple repeat sequences are subject to mutation [200]. Mutation of the $(GT)_3$ repeat region in one tumor by insertion of an additional GT repeat unit resulted in a frameshift, which was predicted to significantly alter the C terminus of the receptor protein [200]. Additional mutations were documented in the $(A)_{10}$ repeat region of the *TGFβRII* gene (deletion of one or two bases), resulting in frameshifts that were predicted to give rise to truncated receptor proteins [200]. Inactivating *TGFβRII* mutations involving these simple repeat regions have now been identified in a significant number of human tumors that exhibit MSI, including sporadic and hereditary CRC, as well as cancers of the stomach, endometrium, and acute lymphoblastic leukemia [59]. However, cancers of the esophagus [201] and gliomas [202] display no microsatellite mutations involving the *TGFβRII* gene.

A number of other genes that function in various aspects of normal cellular homeostasis (growth control and DNA repair) exhibit frameshift mutations at microsatellite loci, including *APC*, *BAX*, *E2F-4*, *IGFIIR*, *hMSH3*, *hMSH6*, *TCF-4*, *BLM*, and others [59, 198, 203, 204]. Mutation in these genes have been identified in a significant percentage (as high as 50–55%) of gastrointestinal cancers (HNPCC, sporadic CRC, stomach) that exhibit MSI [59]. However, some other cancers that display MSI do not contain these mutations [205], suggesting that these genes may be preferential targets in tumors of the gastrointestinal tract. Other tumors with MSI may mutate different genes from those that have been identified to be susceptible to this form of genetic event [198, 204].

Conclusions

A large amount of evidence has now accumulated suggesting a genetic basis for the development of neoplastic disease in humans. However, the genetic damage documented in human cancers includes both large-scale alterations (chromosomal aberrations and ploidy changes) and DNA sequence alterations (single nucleotide changes or alterations in short segments of DNA). In addition, the patterns of genetic damage within a single tumor can vary from a few molecular alterations at specific loci to genome-wide mutations involving a large number of loci. Several distinct forms of genomic instability may provide the molecular basis for neoplastic transformation in humans. Cells undergoing neoplastic transformation may accumulate genetic damage related to progressive genomic instability, or due to episodic genomic instability. Transforming mutations could arise through either of these mechanisms, involving chromosomal alterations or sequence alterations (point mutations and/or MSI). Although the significance of mutations to the etiological mechanisms of tumor development has been debated, a causal role for genetic lesions in the genesis of cancer is commonly accepted. Thus, genetic lesions represent an integral part of the processes of neoplastic transformation, tumorigenesis, and tumor progression, and as such represent potentially valuable markers for cancer detection, diagnosis, staging, and prediction of clinical outcome [3, 4].

References

1 Bishop JM (1991) Molecular themes in oncogenesis. *Cell* 64: 235–248
2 Lengauer C, Kinzler KW, Vogelstein B (1998) Genetic instabilities in human cancers. *Nature* 396: 643–649
3 Mao L, Sidransky D (1994) Cancer screening based on genetic alterations in human tumors. *Cancer Res* 54: 1939s–1940s
4 Sidransky D (1995) Molecular markers in cancer: can we make better predictions? *Int J Cancer* 64: 1–2
5 Renan MJ (1993) How many mutations are required for tumorigenesis? Implications from human cancer data. *Mol Carcinog* 7: 139–146
6 Loeb KR, Loeb LA (2000) Significance of multiple mutations in cancer. *Carcinogenesis* 21: 379–385

7 Cohen AM, Minsky BD, Schilsky RL. (1997) Cancer of the colon. In: VT DeVita, S Hellman, SA Rosenberg (eds): *Cancer: Principles and Practice of Oncology, 5th ed.* Lippincott-Raven, Philadelphia, 1144–1197

8 Kinzler KW, Vogelstein B (2001) Colorectal tumors. In: CR Scriver, AL Beaudet, WS Sly, D Valle (eds): *The Metabolic and Molecular Bases of Inherited Disease*, 8th ed. McGraw-Hill. New York 1033–1062

9 Crow JF (1997) The high spontaneous mutation rate: is it a health risk? *Proc Natl Acad Sci USA* 94: 8380–8386

10 Cairns J (1998) Mutation and cancer: the antecedents to our studies of adaptive mutation. *Genetics* 148: 1433–1440

11 Ruddon RW (1995) *Cancer Biology, 3rd ed.* Oxford University Press, New York

12 Albertson DG, Collins C, McCormick F, Gray JW (2003) Chromosome aberrations in solid tumors. *Nat Genet* 34: 369–376

13 Hussain SP, Harris CC (1998) Molecular epidemiology of human cancer: contribution of mutation spectra studies of tumor suppressor genes. *Cancer Res* 58: 4023–4037

14 Hedenfalk I, Ringner M, Ben-Dor A, Yakhini Z, Chen Y, Chebil G, Ach R, Loman N, Olsson H, Meltzer P et al. (2003) Molecular classification of familial non-BRCA1/BRCA2 breast cancer. *Proc Natl Acad Sci USA* 100: 2532–2537

15 Chung CH, Parker JS, Karaca G, Wu J, Funkhouser WK, Moore D, Butterfoss D, Xiang D, Zanation A, Yin X et al. (2004) Molecular classification of head and neck squamous cell carcinomas using patterns of gene expression. *Cancer Cell* 5: 489–500

16 Meyerson M, Franklin WA, Kelley MJ (2004) Molecular classification and molecular genetics of human lung cancers. *Semin Oncol* 31: 4–19

17 Hoang CD, D'Cunha J, Tawfic SH, Gruessner AC, Kratzke RA, Maddaus MA (2004) Expression profiling of non-small cell lung carcinoma identifies metastatic genotypes based on lymph node tumor burden. *J Thorac Cardiovasc Surg* 127: 1332–1341; discussion 1342

18 Warner GC, Reis PP, Jurisica I, Sultan M, Arora S, Macmillan C, Makitie AA, Grenman R, Reid N, Sukhai M et al. (2004) Molecular classification of oral cancer by cDNA microarrays identifies overexpressed genes correlated with nodal metastasis. *Int J Cancer* 110: 857–868

19 Cleator S, Ashworth A (2004) Molecular profiling of breast cancer: clinical implications. *Br J Cancer* 90: 1120–1124

20 Troester MA, Hoadley KA, Sorlie T, Herbert BS, Borresen-Dale AL, Lonning PE, Shay JW, Kaufmann WK, Perou CM (2004) Cell-type-specific responses to chemotherapeutics in breast cancer. *Cancer Res* 64: 4218–4226

21 Armitage P, Doll R (1957) A two-stage theory of carcinogenesis in relation to the age distribution of human cancer. *Br J Cancer* 11: 161–169

22 Knudson AG Jr, (1971) Mutation and cancer: statistical study of retinoblastoma. *Proc Natl Acad Sci USA* 68: 820–823

23 Baylin SB (2002) Mechanisms underlying epigenetically mediated gene silencing in cancer. *Semin Cancer Biol* 12: 331–337

24 Herman JG, Baylin SB (2003) Gene silencing in cancer in association with promoter hypermethylation. *N Engl J Med* 349: 2042–2054

25 Fazzari MJ, Greally JM (2004) Epigenomics: beyond CpG islands. *Nat Rev Genet* 5: 446–455

26 Fearon ER, Vogelstein B (1990) A genetic model for colorectal tumorigenesis. *Cell* 61: 759–767

27 Jackson AL, Loeb LA (1998) The mutation rate and cancer. *Genetics* 148: 1483–1490

28 Loeb LA (2001) A mutator phenotype in cancer. *Cancer Res* 61: 3230–3239

29 Loeb LA, Christians FC (1996) Multiple mutations in human cancers. *Mutat Res* 350: 279–286

30 Perucho M (1996) Cancer of the microsatellite mutator phenotype. *Biol Chem* 377: 675–684

31 Orr-Weaver TL, Weinberg RA (1998) A checkpoint on the road to cancer. *Nature* 392: 223–224

32 Goyette MC, Cho K, Fasching CL, Levy DB, Kinzler KW, Paraskeva C, Vogelstein B, Stanbridge EJ (1992) Progression of colorectal cancer is associated with multiple tumor suppressor gene defects but inhibition of tumorigenicity is accomplished by correction of any single defect via chromosome transfer. *Mol Cell Biol* 12: 1387–1395

33 Vogelstein B, Fearon ER, Hamilton SR, Kern SE, Preisinger AC, Leppert M, Nakamura Y, White R, Smits AM, Bos JL (1988) Genetic alterations during colorectal-tumor development. *N Engl J Med* 319: 525–532

34 Duesberg P, Rausch C, Rasnick D, Hehlmann R (1998) Genetic instability of cancer cells is proportional to their degree of aneuploidy. *Proc Natl Acad Sci USA* 95: 13692–13697

35 Strauss BS (1998) Hypermutability in carcinogenesis. *Genetics* 148: 1619–1626
36 Sieber OM, Heinimann K, Tomlinson IP (2003) Genomic instability—the engine of tumorigenesis? *Nat Rev Cancer* 3: 701–708
37 Hollstein M, Shomer B, Greenblatt M, Soussi T, Hovig E, Montesano R, Harris CC (1996) Somatic point mutations in the p53 gene of human tumors and cell lines: updated compilation. *Nucleic Acids Res* 24: 141–146
38 Moolgavkar SH, Luebeck EG (2003) Multistage carcinogenesis and the incidence of human cancer. *Genes Chromosomes Cancer* 38: 302–306
39 Boesen JJ, Niericker MJ, Dieteren N, Simons JW (1994) How variable is a spontaneous mutation rate in cultured mammalian cells? *Mutat Res* 307: 121–129
40 Eldridge SR, Gould MN (1992) Comparison of spontaneous mutagenesis in early-passage human mammary cells from normal and malignant tissues. *Int J Cancer* 50: 321–324
41 Wittenkeller JL, Storer B, Bittner G, Schiller JH (1997) Comparison of spontaneous and induced mutation rates in an immortalized human bronchial epithelial cell line and its tumorigenic derivative. *Oncology* 54: 335–341
42 Loeb LA (1991) Mutator phenotype may be required for multistage carcinogenesis. *Cancer Res* 51: 3075–3079
43 Loeb LA (1997) Transient expression of a mutator phenotype in cancer cells. *Science* 277: 1449–1450
44 Richards B, Zhang H, Phear G, Meuth M (1997) Conditional mutator phenotypes in hMSH2-deficient tumor cell lines. *Science* 277: 1523–1526
45 Bhattacharyya NP, Skandalis A, Ganesh A, Groden J, Meuth M (1994) Mutator phenotypes in human colorectal carcinoma cell lines. *Proc Natl Acad Sci USA* 91: 6319–6323
46 Eshleman JR, Lang EZ, Bowerfind GK, Parsons R, Vogelstein B, Willson JK, Veigl ML, Sedwick WD, Markowitz SD (1995) Increased mutation rate at the hprt locus accompanies microsatellite instability in colon cancer. *Oncogene* 10: 33–37
47 Glaab WE, Tindall KR (1997) Mutation rate at the hprt locus in human cancer cell lines with specific mismatch repair-gene defects. *Carcinogenesis* 18: 1–8
48 Tlsty TD, Margolin BH, Lum K (1989) Differences in the rates of gene amplification in nontumorigenic and tumorigenic cell lines as measured by Luria-Delbruck fluctuation analysis. *Proc Natl Acad Sci USA* 86: 9441–9445
49 Cairns J (1975) Mutation selection and the natural history of cancer. *Nature* 255: 197–200
50 Foulds L (1958) The natural history of cancer. *J Chronic Dis* 8: 2–37
51 Nowell PC (1976) The clonal evolution of tumor cell populations. *Science* 194: 23–28
52 Eshleman JR, Markowitz SD (1996) Mismatch repair defects in human carcinogenesis. *Hum Mol Genet* 5 Spec No: 1489–1494
53 Lengauer C, Kinzler KW, Vogelstein B (1997) Genetic instability in colorectal cancers. *Nature* 386: 623–627
54 Cahill DP, Lengauer C, Yu J, Riggins GJ, Willson JK, Markowitz SD, Kinzler KW, Vogelstein B (1998) Mutations of mitotic checkpoint genes in human cancers. *Nature* 392: 300–303
55 Greenblatt MS, Bennett WP, Hollstein M, Harris CC (1994) Mutations in the p53 tumor suppressor gene: clues to cancer etiology and molecular pathogenesis. *Cancer Res* 54: 4855–4878
56 Baker SJ, Fearon ER, Nigro JM, Hamilton SR, Preisinger AC, Jessup JM, vanTuinen P, Ledbetter DH, Barker DF, Nakamura Y et al. (1989) Chromosome 17 deletions and p53 gene mutations in colorectal carcinomas. *Science* 244: 217–221
57 Bos JL (1989) ras oncogenes in human cancer: a review. *Cancer Res* 49: 4682–4689
58 Dalla-Favera R, Martinotti S, Gallo RC, Erikson J, Croce CM (1983) Translocation and rearrangements of the c-myc oncogene locus in human undifferentiated B-cell lymphomas. *Science* 219: 963–967
59 Coleman WB, Tsongalis GJ (2002) The role of genomic instability in the development of human cancer. In: WB Coleman, GJ Tsongalis (eds): *The Molecular Basis of Human Cancer*. Humana Press, Totowa, 115–142
60 Rosenberg SM, Thulin C, Harris RS (1998) Transient and heritable mutators in adaptive evolution in the lab and in nature. *Genetics* 148: 1559–1566
61 Aaltonen LA, Peltomaki P, Leach FS, Sistonen P, Pylkkanen L, Mecklin JP, Jarvinen H, Powell SM, Jen J, Hamilton SR et al. (1993) Clues to the pathogenesis of familial colorectal cancer. *Science* 260: 812–816
62 Aaltonen LA, Peltomaki P, Mecklin JP, Jarvinen H, Jass JR, Green JS, Lynch HT, Watson P,

Tallqvist G, Juhola M et al. (1994) Replication errors in benign and malignant tumors from hereditary nonpolyposis colorectal cancer patients. *Cancer Res* 54: 1645–1648

63 Feig DI, Reid TM, Loeb LA (1994) Reactive oxygen species in tumorigenesis. *Cancer Res* 54: 1890s–1894s

64 Jackson AL, Chen R, Loeb LA (1998) Induction of microsatellite instability by oxidative DNA damage. *Proc Natl Acad Sci USA* 95: 12468–12473

65 Baba S (1997) Recent advances in molecular genetics of colorectal cancer. *World J Surg* 21: 678–687

66 Yamada NA, Parker JM, Farber RA (2003) Mutation frequency analysis of mononucleotide and dinucleotide repeats after oxidative stress. *Environ Mol Mutagen* 42: 75–84

67 Mitelman F (1994) *Catalog of Chromosome Aberrations in Cancer, 5th ed.* Wiley-Liss Publishers, New York

68 Meltzer PS, Kallioniemi A, Trent JM (2001) Chromosome alterations in human solid tumors. In: CR Scriver, AL Beaudet, WS Sly, D Valle (eds): *The Metabolic and Molecular Bases of Inhierited Disease, 8th ed.* McGraw-Hill, New York, 575–596

69 Cahill DP, Lengauer C (2001) Tumor genome instabilities. In: CR Scriver, AL Beaudet, WS Sly, D Valle (eds): *The Metabolic and Molecular Bases of Inherited Disease, 8th ed.* McGraw-Hill, New York 611–612

70 Wang Z, Cummins JM, Shen D, Cahill DP, Jallepalli PV, Wang TL, Parsons DW, Traverso G, Awad M, Silliman N et al. (2004) Three classes of genes mutated in colorectal cancers with chromosomal instability. *Cancer Res* 64: 2998–3001

71 Vogelstein B, Fearon ER, Kern SE, Hamilton SR, Preisinger AC, Nakamura Y, White R (1989) Allelotype of colorectal carcinomas. *Science* 244: 207–211

72 Tsuchiya E, Nakamura Y, Weng SY, Nakagawa K, Tsuchiya S, Sugano H, Kitagawa T (1992) Allelotype of non-small cell lung carcinoma—comparison between loss of heterozygosity in squamous cell carcinoma and adenocarcinoma. *Cancer Res* 52: 2478–2481

73 Seymour AB, Hruban RH, Redston M, Caldas C, Powell SM, Kinzler KW, Yeo CJ, Kern SE (1994) Allelotype of pancreatic adenocarcinoma. *Cancer Res* 54: 2761–2764

74 Roncalli M, Borzio M, Bianchi P, Laghi L (2000) Comprehensive allelotype study of hepatocellular carcinoma. *Hepatology* 32: 876

75 Miller BJ, Wang D, Krahe R, Wright FA (2003) Pooled analysis of loss of heterozygosity in breast cancer: a genome scan provides comparative evidence for multiple tumor suppressors and identifies novel candidate regions. *Am J Hum Genet* 73: 748–767

76 Thiagalingam S, Laken S, Willson JK, Markowitz SD, Kinzler KW, Vogelstein B, Lengauer C (2001) Mechanisms underlying losses of heterozygosity in human colorectal cancers. *Proc Natl Acad Sci USA* 98: 2698–2702

77 El-Naggar AK, Vielh P (2004) Solid tumor DNA content analysis. *Methods Mol Biol* 263: 355–370

78 Kallioniemi OP, Kallioniemi A, Piper J, Isola J, Waldman FM, Gray JW, Pinkel D (1994) Optimizing comparative genomic hybridization for analysis of DNA sequence copy number changes in solid tumors. *Genes Chromosomes Cancer* 10: 231–243

79 Kallioniemi OP, Kallioniemi A, Sudar D, Rutovitz D, Gray JW, Waldman F, Pinkel D (1993) Comparative genomic hybridization: a rapid new method for detecting and mapping DNA amplification in tumors. *Semin Cancer Biol* 4: 41–46

80 Bayani JM, Squire JA (2002) Applications of SKY in cancer cytogenetics. *Cancer Invest* 20: 373–386

81 Patel AS, Hawkins AL, Griffin CA (2000) Cytogenetics and cancer. *Curr Opin Oncol* 12: 62–67

82 Radford DM, Fair KL, Phillips NJ, Ritter JH, Steinbrueck T, Holt MS, Donis-Keller H (1995) Allelotyping of ductal carcinoma *in situ* of the breast: deletion of loci on 8p, 13q, 16q, 17p and 17q. *Cancer Res* 55: 3399–3405

83 Boige V, Laurent-Puig P, Fouchet P, Flejou JF, Monges G, Bedossa P, Bioulac-Sage P, Capron F, Schmitz A, Olschwang S, Thomas G (1997) Concerted nonsyntenic allelic losses in hyperploid hepatocellular carcinoma as determined by a high-resolution allelotype. *Cancer Res* 57: 1986–1990

84 Nowak MA, Komarova NL, Sengupta A, Jallepalli PV, Shih Ie M, Vogelstein B, Lengauer C (2002) The role of chromosomal instability in tumor initiation. *Proc Natl Acad Sci USA* 99: 16226–16231

85 Rajagopalan H, Nowak MA, Vogelstein B, Lengauer C (2003) The significance of unstable chromosomes in colorectal cancer. *Nat Rev Cancer* 3: 695–701

86 Ghadimi BM, Sackett DL, Difilippantonio MJ, Schrock E, Neumann T, Jauho A, Auer G, Ried T (2000) Centrosome amplification and instability occurs exclusively in aneuploid, but not in diploid colorectal cancer cell lines, and correlates with numerical chromosomal aberrations. *Genes Chromosomes Cancer* 27: 183–190
87 Harwood J, Tachibana A, Davis R, Bhattacharyya NP, Meuth M (1993) High rate of multilocus deletion in a human tumor cell line. *Hum Mol Genet* 2: 165–171
88 Phear G, Bhattacharyya NP, Meuth M (1996) Loss of heterozygosity and base substitution at the APRT locus in mismatch-repair-proficient and -deficient colorectal carcinoma cell lines. *Mol Cell Biol* 16: 6516–6523
89 Le Beau MM (1997) Molecular biology of cancer: Cytogenetics. In: VT DeVita, S Hellman, SA Rosenberg (eds): *Cancer: Principles and Practice of Oncology, 5th ed.* Lippincott-Raven, Philadelphia 103–119
90 Gollin SM (2004) Chromosomal instability. *Curr Opin Oncol* 16: 25–31
91 Kastan MB, Kuerbitz SJ (1993) Control of G1 arrest after DNA damage. *Environ Health Perspect* 101, Suppl 5: 55–58
92 Kastan MB, Onyekwere O, Sidransky D, Vogelstein B, Craig RW (1991) Participation of p53 protein in the cellular response to DNA damage. *Cancer Res* 51: 6304–6311
93 Solomon E, Borrow J, Goddard AD (1991) Chromosome aberrations and cancer. *Science* 254: 1153–1160
94 Jong YJ, Li LH, Tsou MH, Chen YJ, Cheng SH, Wang-Wuu S, Tsai SF, Chen CM, Huang AT, Hsu MT, Lin CH (2004) Chromosomal comparative genomic hybridization abnormalities in early- and late-onset human breast cancers: correlation with disease progression and TP53 mutations. *Cancer Genet Cytogenet* 148: 55–65
95 Sugai T, Takahashi H, Habano W, Nakamura S, Sato K, Orii S, Suzuki K (2003) Analysis of genetic alterations, classified according to their DNA ploidy pattern, in the progression of colorectal adenomas and early colorectal carcinomas. *J Pathol* 200: 168–176
96 Filatov L, Golubovskaya V, Hurt JC, Byrd LL, Phillips JM, Kaufmann WK (1998) Chromosomal instability is correlated with telomere erosion and inactivation of G2 checkpoint function in human fibroblasts expressing human papillomavirus type 16 E6 oncoprotein. *Oncogene* 16: 1825–1838
97 Honma M, Momose M, Tanabe H, Sakamoto H, Yu Y, Little JB, Sofuni T, Hayashi M (2000) Requirement of wild-type p53 protein for maintenance of chromosomal integrity. *Mol Carcinog* 28: 203–214
98 Tarapore P, Fukasawa K (2000) p53 mutation and mitotic infidelity. *Cancer Invest* 18: 148–155
99 Shih IM, Zhou W, Goodman SN, Lengauer C, Kinzler KW, Vogelstein B (2001) Evidence that genetic instability occurs at an early stage of colorectal tumorigenesis. *Cancer Res* 61: 818–822
100 Baker SJ, Preisinger AC, Jessup JM, Paraskeva C, Markowitz S, Willson JK, Hamilton S, Vogelstein B (1990) p53 gene mutations occur in combination with 17p allelic deletions as late events in colorectal tumorigenesis. *Cancer Res* 50: 7717–7722
101 Eshleman JR, Casey G, Kochera ME, Sedwick WD, Swinler SE, Veigl ML, Willson JK, Schwartz S, Markowitz SD (1998) Chromosome number and structure both are markedly stable in RER colorectal cancers and are not destabilized by mutation of p53. *Oncogene* 17: 719–725
102 Kramer A, Neben K, Ho AD (2002) Centrosome replication, genomic instability and cancer. *Leukemia* 16: 767–775
103 Lingle WL, Barrett SL, Negron VC, D'Assoro AB, Boeneman K, Liu W, Whitehead CM, Reynolds C, Salisbury JL (2002) Centrosome amplification drives chromosomal instability in breast tumor development. *Proc Natl Acad Sci USA* 99: 1978–1983
104 Bischoff JR, Anderson L, Zhu Y, Mossie K, Ng L, Souza B, Schryver B, Flanagan P, Clairvoyant F, Ginther C et al. (1998) A homologue of Drosophila aurora kinase is oncogenic and amplified in human colorectal cancers. *Embo J* 17: 3052–3065
105 Zhou H, Kuang J, Zhong L, Kuo WL, Gray JW, Sahin A, Brinkley BR, Sen S (1998) Tumour amplified kinase STK15/BTAK induces centrosome amplification, aneuploidy and transformation. *Nat Genet* 20: 189–193
106 Wolf G, Elez R, Doermer A, Holtrich U, Ackermann H, Stutte HJ, Altmannsberger HM, Rubsamen-Waigmann H, Strebhardt K (1997) Prognostic significance of polo-like kinase (PLK) expression in non-small cell lung cancer. *Oncogene* 14: 543–549
107 Tanaka T, Kimura M, Matsunaga K, Fukada D, Mori H, Okano Y (1999) Centrosomal kinase AIK1 is overexpressed in invasive ductal carcinoma of the breast. *Cancer Res* 59: 2041–2044

108 Carroll PE, Okuda M, Horn HF, Biddinger P, Stambrook PJ, Gleich LL, Li YQ, Tarapore P, Fukasawa K (1999) Centrosome hyperamplification in human cancer: chromosome instability induced by p53 mutation and/or Mdm2 overexpression. *Oncogene* 18: 1935–1944
109 Tarapore P, Fukasawa K (2002) Loss of p53 and centrosome hyperamplification. *Oncogene* 21: 6234–6240
110 Tutt A, Gabriel A, Bertwistle D, Connor F, Paterson H, Peacock J, Ross G, Ashworth A (1999) Absence of Brca2 causes genome instability by chromosome breakage and loss associated with centrosome amplification. *Curr Biol* 9: 1107–1110
111 Deng CX (2002) Roles of BRCA1 in centrosome duplication. *Oncogene* 21: 6222–6227
112 Bharadwaj R, Yu H (2004) The spindle checkpoint, aneuploidy, and cancer. *Oncogene* 23: 2016–2027
113 Hartwell LH, Kastan MB (1994) Cell cycle control and cancer. *Science* 266: 1821–1828
114 Cahill DP, da Costa LT, Carson-Walter EB, Kinzler KW, Vogelstein B, Lengauer C (1999) Characterization of MAD2B and other mitotic spindle checkpoint genes. *Genomics* 58: 181–187
115 Li Y, Benezra R (1996) Identification of a human mitotic checkpoint gene: hsMAD2. *Science* 274: 246–248
116 Jaffrey RG, Pritchard SC, Clark C, Murray GI, Cassidy J, Kerr KM, Nicolson MC, McLeod HL (2000) Genomic instability at the BUB1 locus in colorectal cancer, but not in non-small cell lung cancer. *Cancer Res* 60: 4349–4352
117 Myrie KA, Percy MJ, Azim JN, Neeley CK, Petty EM (2000) Mutation and expression analysis of human BUB1 and BUB1B in aneuploid breast cancer cell lines. *Cancer Lett* 152: 193–199
118 Sato M, Sekido Y, Horio Y, Takahashi M, Saito H, Minna JD, Shimokata K, Hasegawa Y (2000) Infrequent mutation of the hBUB1 and hBUBR1 genes in human lung cancer. *Jpn J Cancer Res* 91: 504–509
119 Yamaguchi K, Okami K, Hibi K, Wehage SL, Jen J, Sidransky D (1999) Mutation analysis of hBUB1 in aneuploid HNSCC and lung cancer cell lines. *Cancer Lett* 139: 183–187
120 Imai Y, Shiratori Y, Kato N, Inoue T, Omata M (1999) Mutational inactivation of mitotic checkpoint genes, hsMAD2 and hBUB1, is rare in sporadic digestive tract cancers. *Jpn J Cancer Res* 90: 837–840
121 Gualberto A, Aldape K, Kozakiewicz K, Tlsty TD (1998) An oncogenic form of p53 confers a dominant, gain-of-function phenotype that disrupts spindle checkpoint control. *Proc Natl Acad Sci USA* 95: 5166–5171
122 Shigeta T, Takagi M, Delia D, Chessa L, Iwata S, Kanke Y, Asada M, Eguchi M, Mizutani S (1999) Defective control of apoptosis and mitotic spindle checkpoint in heterozygous carriers of ATM mutations. *Cancer Res* 59: 2602–2607
123 Canman CE, Lim DS (1998) The role of ATM in DNA damage responses and cancer. *Oncogene* 17: 3301–3308
124 Zhang H, Tombline G, Weber BL (1998) BRCA1, BRCA2, and DNA damage response: collision or collusion? *Cell* 92: 433–436
125 Tlsty TD, Briot A, Gualberto A, Hall I, Hess S, Hixon M, Kuppuswamy D, Romanov S, Sage M, White A (1995) Genomic instability and cancer. *Mutat Res* 337: 1–7
126 Hogarty MD, Brodeur GM (2001) Gene amplification in human cancers: Biological and clinical significance. In: CR Scriver, AL Beaudet, WS Sly, D Valle (eds): *The Metabolic and Molecular Bases of Inherited Disease, 8th ed.* McGraw-Hill, New York, 597–610
127 Livingstone LR, White A, Sprouse J, Livanos E, Jacks T, Tlsty TD (1992) Altered cell cycle arrest and gene amplification potential accompany loss of wild-type p53. *Cell* 70: 923–935
128 Yin Y, Tainsky MA, Bischoff FZ, Strong LC, Wahl GM (1992) Wild-type p53 restores cell cycle control and inhibits gene amplification in cells with mutant p53 alleles. *Cell* 70: 937–948
129 Oren M (1994) Relationship of p53 to the control of apoptotic cell death. *Semin Cancer Biol* 5: 221–227
130 Nowell PC, Hungerford DA (1960) Chromosome studies on normal and leukemic human leukocytes. *J Natl Cancer Inst* 25: 85–109
131 Dubeau L. (2001) Ovarian cancer. In: CR Scriver, AL Beaudet, WS Sly, D Valle (eds): *The Metabolic and Molecular Bases of Inherited Disease, 8th ed.* McGraw-Hill, New York, 1091–1096
132 Hruban RH, Wilentz RE, Kern SE (2000) Genetic progression in the pancreatic ducts. *Am J Pathol* 156: 1821–1825
133 Croce CM (1986) Chromosome translocations and human cancer. *Cancer Res* 46: 6019–6023

134 Kinzler KW, Nilbert MC, Su LK, Vogelstein B, Bryan TM, Levy DB, Smith KJ, Preisinger AC, Hedge P, McKechnie D et al. (1991) Identification of FAP locus genes from chromosome 5q21. *Science* 253: 661–665
135 Fearon ER, Cho KR, Nigro JM, Kern SE, Simons JW, Ruppert JM, Hamilton SR, Preisinger AC, Thomas G, Kinzler KW et al. (1990) Identification of a chromosome 18q gene that is altered in colorectal cancers. *Science* 247: 49–56
136 Peinado MA, Malkhosyan S, Velazquez A, Perucho M (1992) Isolation and characterization of allelic losses and gains in colorectal tumors by arbitrarily primed polymerase chain reaction. *Proc Natl Acad Sci USA* 89: 10065–10069
137 Ionov Y, Peinado MA, Malkhosyan S, Shibata D, Perucho M (1993) Ubiquitous somatic mutations in simple repeated sequences reveal a new mechanism for colonic carcinogenesis. *Nature* 363: 558–561
138 Thibodeau SN, Bren G, Schaid D (1993) Microsatellite instability in cancer of the proximal colon. *Science* 260: 816–819
139 Zhang L, Yu J, Willson JK, Markowitz SD, Kinzler KW, Vogelstein B (2001) Short mononucleotide repeat sequence variability in mismatch repair-deficient cancers. *Cancer Res* 61: 3801–3805
140 Yamada NA, Castro A, Farber RA (2003) Variation in the extent of microsatellite instability in human cell lines with defects in different mismatch repair genes. *Mutagenesis* 18: 277–282
141 Honchel R, Halling KC, Thibodeau SN (1995) Genomic instability in neoplasia. *Semin Cell Biol* 6: 45–52
142 Thibodeau SN, French AJ, Cunningham JM, Tester D, Burgart LJ, Roche PC, McDonnell SK, Schaid DJ, Vockley CW, Michels VV et al. (1998) Microsatellite instability in colorectal cancer: different mutator phenotypes and the principal involvement of hMLH1. *Cancer Res* 58: 1713–1718
143 Hoang JM, Cottu PH, Thuille B, Salmon RJ, Thomas G, Hamelin R (1997) BAT-26, an indicator of the replication error phenotype in colorectal cancers and cell lines. *Cancer Res* 57: 300–303
144 Zhou XP, Hoang JM, Li YJ, Seruca R, Carneiro F, Sobrinho-Simoes M, Lothe RA, Gleeson CM, Russell SE, Muzeau F et al. (1998) Determination of the replication error phenotype in human tumors without the requirement for matching normal DNA by analysis of mononucleotide repeat microsatellites. *Genes Chromosomes Cancer* 21: 101–107
145 Dietmaier W, Wallinger S, Bocker T, Kullmann F, Fishel R, Ruschoff J (1997) Diagnostic microsatellite instability: definition and correlation with mismatch repair protein expression. *Cancer Res* 57: 4749–4756
146 Parsons R, Myeroff LL, Liu B, Willson JK, Markowitz SD, Kinzler KW, Vogelstein B (1995) Microsatellite instability and mutations of the transforming growth factor beta type II receptor gene in colorectal cancer. *Cancer Res* 55: 5548–5550
147 Eichler EE, Holden JJ, Popovich BW, Reiss AL, Snow K, Thibodeau SN, Richards CS, Ward PA, Nelson DL (1994) Length of uninterrupted CGG repeats determines instability in the FMR1 gene. *Nat Genet* 8: 88–94
148 Mao L, Lee DJ, Tockman MS, Erozan YS, Askin F, Sidransky D (1994) Microsatellite alterations as clonal markers for the detection of human cancer. *Proc Natl Acad Sci USA* 91: 9871–9875
149 Risinger JI, Umar A, Barrett JC, Kunkel TA (1995) A hPMS2 mutant cell line is defective in strand-specific mismatch repair. *J Biol Chem* 270: 18183–18186
150 Risinger JI, Umar A, Boyd J, Berchuck A, Kunkel TA, Barrett JC (1996) Mutation of MSH3 in endometrial cancer and evidence for its functional role in heteroduplex repair. *Nat Genet* 14: 102–105
151 Oki E, Oda S, Maehara Y, Sugimachi K (1999) Mutated gene-specific phenotypes of dinucleotide repeat instability in human colorectal carcinoma cell lines deficient in DNA mismatch repair. *Oncogene* 18: 2143–2147
152 Baranovskaya S, Soto JL, Perucho M, Malkhosyan SR (2001) Functional significance of concomitant inactivation of hMLH1 and hMSH6 in tumor cells of the microsatellite mutator phenotype. *Proc Natl Acad Sci USA* 98: 15107–15112
153 Horii A, Han HJ, Shimada M, Yanagisawa A, Kato Y, Ohta H, Yasui W, Tahara E, Nakamura Y (1994) Frequent replication errors at microsatellite loci in tumors of patients with multiple primary cancers. *Cancer Res* 54: 3373–3375
154 Akiyama Y, Nakasaki H, Nihei Z, Iwama T, Nomizu T, Utsunomiya J, Yuasa Y (1996) Frequent microsatellite instabilities and analyses of the related genes in familial gastric cancers. *Jpn J*

Cancer Res 87: 595–601
155 Nakashima H, Honda M, Inoue H, Shibuta K, Arinaga S, Era S, Ueo H, Mori M, Akiyoshi T (1995) Microsatellite instability in multiple gastric cancers. *Int J Cancer* 64: 239–242
156 Jiricny J, Nystrom-Lahti M (2000) Mismatch repair defects in cancer. *Curr Opin Genet Dev* 10: 157–161
157 Fishel R, Lescoe MK, Rao MR, Copeland NG, Jenkins NA, Garber J, Kane M, Kolodner R (1993) The human mutator gene homolog MSH2 and its association with hereditary nonpolyposis colon cancer. *Cell* 75: 1027–1038
158 Leach FS, Nicolaides NC, Papadopoulos N, Liu B, Jen J, Parsons R, Peltomaki P, Sistonen P, Aaltonen LA, Nystrom-Lahti M et al. (1993) Mutations of a mutS homolog in hereditary nonpolyposis colorectal cancer. *Cell* 75: 1215–1225
159 Umar A, Risinger JI, Glaab WE, Tindall KR, Barrett JC, Kunkel TA (1998) Functional overlap in mismatch repair by human MSH3 and MSH6. *Genetics* 148: 1637–1646
160 Palombo F, Gallinari P, Iaccarino I, Lettieri T, Hughes M, D'Arrigo A, Truong O, Hsuan JJ, Jiricny J (1995) GTBP, a 160-kilodalton protein essential for mismatch-binding activity in human cells. *Science* 268: 1912–1914
161 Papadopoulos N, Nicolaides NC, Liu B, Parsons R, Lengauer C, Palombo F, D'Arrigo A, Markowitz S, Willson JK, Kinzler KW et al. (1995) Mutations of GTBP in genetically unstable cells. *Science* 268: 1915–1917
162 Bronner CE, Baker SM, Morrison PT, Warren G, Smith LG, Lescoe MK, Kane M, Earabino C, Lipford J, Lindblom A et al. (1994) Mutation in the DNA mismatch repair gene homologue hMLH1 is associated with hereditary non-polyposis colon cancer. *Nature* 368: 258–261
163 Papadopoulos N, Nicolaides NC, Wei YF, Ruben SM, Carter KC, Rosen CA, Haseltine WA, Fleischmann RD, Fraser CM, Adams MD et al. (1994) Mutation of a mutL homolog in hereditary colon cancer. *Science* 263: 1625–1629
164 Nicolaides NC, Papadopoulos N, Liu B, Wei YF, Carter KC, Ruben SM, Rosen CA, Haseltine WA, Fleischmann RD, Fraser CM et al. (1994) Mutations of two PMS homologues in hereditary nonpolyposis colon cancer. *Nature* 371: 75–80
165 Lipkin SM, Wang V, Jacoby R, Banerjee-Basu S, Baxevanis AD, Lynch HT, Elliott RM, Collins FS (2000) MLH3: a DNA mismatch repair gene associated with mammalian microsatellite instability. *Nat Genet* 24: 27–35
166 Muller A, Fishel R (2002) Mismatch repair and the hereditary non-polyposis colorectal cancer syndrome (HNPCC). *Cancer Invest* 20: 102–109
167 Liu B, Nicolaides NC, Markowitz S, Willson JK, Parsons RE, Jen J, Papadopolous N, Peltomaki P, de la Chapelle A, Hamilton SR et al. (1995) Mismatch repair gene defects in sporadic colorectal cancers with microsatellite instability. *Nat Genet* 9: 48–55
168 Schmutte C, Fishel R (1999) Genomic instability: first step to carcinogenesis. *Anticancer Res* 19: 4665–4696
169 Arzimanoglou II, Gilbert F, Barber HR (1998) Microsatellite instability in human solid tumors. *Cancer* 82: 1808–1820
170 Lothe RA (1997) Microsatellite instability in human solid tumors. *Mol Med Today* 3: 61–68
171 Schlotterer C, Tautz D (1992) Slippage synthesis of simple sequence DNA. *Nucleic Acids Res* 20: 211–215
172 Thomas DC, Umar A, Kunkel TA (1996) Microsatellite instability and mismatch repair defects in cancer. *Mutat Res* 350: 201–205
173 Boyer JC, Farber RA (1998) Mutation rate of a microsatellite sequence in normal human fibroblasts. *Cancer Res* 58: 3946–3949
174 Hanford MG, Rushton BC, Gowen LC, Farber RA (1998) Microsatellite mutation rates in cancer cell lines deficient or proficient in mismatch repair. *Oncogene* 16: 2389–2393
175 Umar A, Koi M, Risinger JI, Glaab WE, Tindall KR, Kolodner RD, Boland CR, Barrett JC, Kunkel TA (1997) Correction of hypermutability, N-methyl-N'-nitro-N-nitrosoguanidine resistance, and defective DNA mismatch repair by introducing chromosome 2 into human tumor cells with mutations in MSH2 and MSH6. *Cancer Res* 57: 3949–3955
176 da Costa LT, Liu B, el-Deiry W, Hamilton SR, Kinzler KW, Vogelstein B, Markowitz S, Willson JK, de la Chapelle A, Downey KM et al. (1995) Polymerase delta variants in RER colorectal tumours. *Nat Genet* 9: 10–11
177 Parc YR, Halling KC, Wang L, Christensen ER, Cunningham JM, French AJ, Burgart LJ, Price-Troska TL, Roche PC, Thibodeau SN (2000) HMSH6 alterations in patients with microsatellite

instability-low colorectal cancer. *Cancer Res* 60: 2225–2231
178 Koi M, Umar A, Chauhan DP, Cherian SP, Carethers JM, Kunkel TA, Boland CR (1994) Human chromosome 3 corrects mismatch repair deficiency and microsatellite instability and reduces N-methyl-N'-nitro-N-nitrosoguanidine tolerance in colon tumor cells with homozygous hMLH1 mutation. *Cancer Res* 54: 4308–4312
179 Risinger JI, Umar A, Glaab WE, Tindall KR, Kunkel TA, Barrett JC (1998) Single gene complementation of the hPMS2 defect in HEC-1-A endometrial carcinoma cells. *Cancer Res* 58: 2978–2981
180 Lettieri T, Marra G, Aquilina G, Bignami M, Crompton NE, Palombo F, Jiricny J (1999) Effect of hMSH6 cDNA expression on the phenotype of mismatch repair-deficient colon cancer cell line HCT15. *Carcinogenesis* 20: 373–382
181 Lengauer C, Kinzler KW, Vogelstein B (1997) DNA methylation and genetic instability in colorectal cancer cells. *Proc Natl Acad Sci USA* 94: 2545–2550
182 Cunningham JM, Christensen ER, Tester DJ, Kim CY, Roche PC, Burgart LJ, Thibodeau SN (1998) Hypermethylation of the hMLH1 promoter in colon cancer with microsatellite instability. *Cancer Res* 58: 3455–3460
183 Herman JG, Umar A, Polyak K, Graff JR, Ahuja N, Issa JP, Markowitz S, Willson JK, Hamilton SR, Kinzler KW et al. (1998) Incidence and functional consequences of hMLH1 promoter hypermethylation in colorectal carcinoma. *Proc Natl Acad Sci USA* 95: 6870–6875
184 Kane MF, Loda M, Gaida GM, Lipman J, Mishra R, Goldman H, Jessup JM, Kolodner R (1997) Methylation of the hMLH1 promoter correlates with lack of expression of hMLH1 in sporadic colon tumors and mismatch repair-defective human tumor cell lines. *Cancer Res* 57: 808–811
185 Bevilacqua RA, Simpson AJ (2000) Methylation of the hMLH1 promoter but no hMLH1 mutations in sporadic gastric carcinomas with high-level microsatellite instability. *Int J Cancer* 87: 200–203
186 Endoh Y, Tamura G, Ajioka Y, Watanabe H, Motoyama T (2000) Frequent hypermethylation of the hMLH1 gene promoter in differentiated-type tumors of the stomach with the gastric foveolar phenotype. *Am J Pathol* 157: 717–722
187 Simpkins SB, Bocker T, Swisher EM, Mutch DG, Gersell DJ, Kovatich AJ, Palazzo JP, Fishel R, Goodfellow PJ (1999) MLH1 promoter methylation and gene silencing is the primary cause of microsatellite instability in sporadic endometrial cancers. *Hum Mol Genet* 8: 661–666
188 Glaab WE, Hill RB, Skopek TR (2001) Suppression of spontaneous and hydrogen peroxide-induced mutagenesis by the antioxidant ascorbate in mismatch repair-deficient human colon cancer cells. *Carcinogenesis* 22: 1709–1713
189 Ekbom A, Helmick C, Zack M, Adami HO (1990) Ulcerative colitis and colorectal cancer. A population-based study. *N Engl J Med* 323: 1228–1233
190 Willenbucher RF, Aust DE, Chang CG, Zelman SJ, Ferrell LD, Moore DH, 2nd, Waldman FM (1999) Genomic instability is an early event during the progression pathway of ulcerative-colitis-related neoplasia. *Am J Pathol* 154: 1825–1830
191 Loeb KR, Loeb LA (1999) Genetic instability and the mutator phenotype. Studies in ulcerative colitis. *Am J Pathol* 154: 1621–1626
192 Fleisher AS, Esteller M, Tamura G, Rashid A, Stine OC, Yin J, Zou TT, Abraham JM, Kong D, Nishizuka S et al. (2001) Hypermethylation of the hMLH1 gene promoter is associated with microsatellite instability in early human gastric neoplasia. *Oncogene* 20: 329–335
193 Chang CL, Marra G, Chauhan DP, Ha HT, Chang DK, Ricciardiello L, Randolph A, Carethers JM, Boland CR (2002) Oxidative stress inactivates the human DNA mismatch repair system. *Am J Physiol Cell Physiol* 283: C148–154
194 Brentnall TA, Crispin DA, Bronner MP, Cherian SP, Hueffed M, Rabinovitch PS, Rubin CE, Haggitt RC, Boland CR (1996) Microsatellite instability in nonneoplastic mucosa from patients with chronic ulcerative colitis. *Cancer Res* 56: 1237–1240
195 Iwaya T, Maesawa C, Nishizuka S, Suzuki Y, Sakata K, Sato N, Ikeda K, Koeda K, Ogasawara S, Otsuka K et al. (1998) Infrequent frameshift mutations of polynucleotide repeats in multiple primary cancers affecting the esophagus and other organs. *Genes Chromosomes Cancer* 23: 317–322
196 Suzuki H, Harpaz N, Tarmin L, Yin J, Jiang HY, Bell JD, Hontanosas M, Groisman GM, Abraham JM, Meltzer SJ (1994) Microsatellite instability in ulcerative colitis-associated colorectal dysplasias and cancers. *Cancer Res* 54: 4841–4844
197 Perucho M (2003) Tumors with microsatellite instability: many mutations, targets and paradox-

es. *Oncogene* 22: 2223–2225
198 Duval A, Hamelin R (2002) Genetic instability in human mismatch repair deficient cancers. *Ann Genet* 45: 71–75
199 Lin HY, Wang XF, Ng-Eaton E, Weinberg RA, Lodish HF (1992) Expression cloning of the TGF-beta type II receptor, a functional transmembrane serine/threonine kinase. *Cell* 68: 775–785
200 Markowitz S, Wang J, Myeroff L, Parsons R, Sun L, Lutterbaugh J, Fan RS, Zborowska E, Kinzler KW, Vogelstein B et al. (1995) Inactivation of the type II TGF-beta receptor in colon cancer cells with microsatellite instability. *Science* 268: 1336–1338
201 Tomita S, Miyazato H, Tamai O, Muto Y, Toda T (1999) Analyses of microsatellite instability and the transforming growth factor-beta receptor type II gene mutation in sporadic human gastrointestinal cancer. *Cancer Genet Cytogenet* 115: 23–27
202 Leung SY, Chan TL, Chung LP, Chan AS, Fan YW, Hung KN, Kwong WK, Ho JW, Yuen ST (1998) Microsatellite instability and mutation of DNA mismatch repair genes in gliomas. *Am J Pathol* 153: 1181–1188
203 Woerner SM, Benner A, Sutter C, Schiller M, Yuan YP, Keller G, Bork P, Doeberitz MK, Gebert JF (2003) Pathogenesis of DNA repair-deficient cancers: a statistical meta-analysis of putative Real Common Target genes. *Oncogene* 22: 2226–2235
204 Duval A, Hamelin R (2002) Mutations at coding repeat sequences in mismatch repair-deficient human cancers: toward a new concept of target genes for instability. *Cancer Res* 62: 2447–2454
205 Malkhosyan S, Rampino N, Yamamoto H, Perucho M (1996) Frameshift mutator mutations. *Nature* 382: 499–500

Edited by Leon P. Bignold

Mechanisms of abnormal gene expression in tumor cells

Assam El-Osta

The Alfred Medical Research and Education Precinct (AMREP), Baker Medical Research Institute, Epigenetics in Human Health and Disease Laboratory, Second Floor, Commercial Road, Prahran, Victoria 3181, Australia

Abstract. Epigenetic mechanisms are involved in critical nuclear processes such as transcriptional control, genome stability, replication and repair. Recent evidence suggests that changes in the epigenetic repertoire can drive tumorigenesis. This review examines the latest experimental evidence that questions the mechanisms underlying the consequence of epigenetic changes in gene regulation and cancer development.

Key words: Cancer, chromatin, DNA methyltransferase, methylation, transcriptional silencing, tumor suppressor gene.

Introduction

There are many ways in which genes are regulated, and the field of epigenetics has seen a recent surge of interest in the study of modifications of the genome and histone tails to explain transcriptional competence. The term epigenetics refers to heritable changes in gene expression that are not the result of changes in the DNA code. DNA methylation is the best studied of these mechanisms with CpG methylation recognized as a major component of gene silencing in cancer [1]. Microinjection experiments using methylated gene constructs indicate that transcriptional repression occurs once chromatin is assembled [2]. Nuclease resistance in mammalian nuclei is due to CpG methylation, and this correlated with transcriptional repression mediated by methyl-CpG binding (MBD) proteins [3, 4]. It is not coincidental then that MeCP2, a global transcriptional repressor, silences gene activity and binds to chromatin in a methylation-dependent manner [5]. Before focusing on the impact of DNA methylation in tumorigenesis, the relevance of epigenetic mechanisms and transcriptional control is discussed.

DNA methylation influences chromatin function

Recent studies are beginning to provide a molecular explanation as to how chromatin assembly on methylated DNA can repress transcription. It is well

Table 1. DNMT associated binding partners that modify chromatin

Binding partner	Proposed function	Refs
DNMT1		
HDAC1	Chromatin remodeling, transcriptional silencing	[54]
HDAC2	Chromatin remodeling, transcriptional silencing	[55]
DMAP1	Histone deacetylation following DNA replication, transcriptional silencing	[55]
pRB	Chromatin remodeling, transcriptional silencing	[56]
MBD3	Binds hemi-methylated DNA, transcriptional silencing	[57]
PCNA	Targeting to replication foci, maintain DNA methylation	[58–60]
RUNX1/MTG8	Targeted recruitment and silencing in acute myeloid leukemia	[61]
p53	Transcriptional silencing	[62]
RGS6	Cooperates with DMAP1 complex, transcriptional silencing	[63]
SuV39H1	Histone tail modification at H3K9	[64]
p33ING1	Cooperates with DMAP1 and co-repressor complex, histone modification	[65]
DNMT3a/DNMT3b		
RP58	Maintain transcriptionally repressive chromatin in late S-phase	[28, 29]
Condensin	Mitotic chromosome condensation	[66]
hSNF2H	Epigenetic regulation	[67]
DNMT3L		
HDAC1	Transcriptional silencing	[68, 69]

reveal that the DNMT3L protein can mediate transcriptional repression by its biochemical interaction with histone deacetylase. These observations suggest the methylation machinery are connected with chromatin remodeling; however, the biggest challenge in the area is to determine the mechanisms by which the determinants are localized and segregated on target genes. In the next section, I discuss possible mechanisms that could explain aberrant DNA methylation patterns in cancer.

Mistargeting of the DNMT co-repressor complex

A question that has long caused confusion in the cancer-epigenetics field is the specificity of genomic methylation patterns. Recent studies in the area have revealed interesting exceptions to the belief that hypermethylation of tumor suppressor genes is the primary mechanism of cancer development [30]. Indeed, hypomethylation events have been described and attributed to genomic instability in cancer [31, 32]. Almost two decades ago, studies demonstrated that reductions in genomic methylation are associated with cancer progression [33, 34]. One of the best-studied models of cancer development is tumor suppressor gene silencing and has been studied in different contexts and diseases. For example, the retinoblastoma tumor suppressor gene is silenced by CpG methylation [35]. Alternatively, demethylating agents such as azacytidine have been used to induce promoter sequence hypomethylation and derepress gene silencing [36]. Clearly, experimental evidence suggests that hypomethylation and hypermethylation events can be associated with tumor development. However, hypermethylation of tumor suppressor genes and transcriptional repression do not explain how determinants could be mistargeted in cancer when hypomethylation is believed to be the primary cause of tumorigenesis. In this section I consider recent advances to our knowledge of methylation-mediated mechanisms in cancer and examine both hyper- and hypo- methylation events in cancer.

Gene silencing, DNMT recruitment and chromatin disruption

DNA hypermethylation has been described in a number of cancer types including retinoblastoma, breast cancer, colorectal carcinoma, melanoma, leukemia and renal carcinoma [37–43]. Histone deactylase inhibitors such as TSA are not effective to derepress hypermethylated promoters [12]. The mechanism of repression is believed to involve the recruitment of a co-repressor complex that belong to the MBD protein family. Epigenetic modifiers such as 5adC and TSA reactivate gene activity by promoting DNA demethylation and increased in histone tail acetylation (see Fig. 1) [44, 45].

Disruption of the DNMT1 gene in colorectal carcinoma cells (DNMT1$^{-/-}$) significantly decreases methylatransferase activity, and is correlated with changes in DNA methylation [27]. By contrast, the tumor suppressor gene *p16INK4A* and *Alu* repeats retained characteristic hypermethylation pattern and remained transcriptionally repressed. However, when DNMT1$^{-/-}$ cells were exposed to the demethylating agent 5adC, *p16INK4A* showed demethylation and derepression, suggesting other methyltransferase activities could cooperate with silencing. Recent studies have brought to light additional enzymes that participate with cancer progression in carcinoma. Depletion of DNMT1 and DNMT3b show marked reductions in methylation content at repetitive sequences and derepression of tumor suppressor genes *p16INK4a* and *TIMP3* (see Tab. 2) [46]. The findings of these experiments suggest that DNMT cooperativity, transcriptional silencing and methylation could contribute to tumorigenesis. The results do not explain how the DNMT methyltransferases and associated co-repressors are specified on focal areas of promoters to silence transcription while the genome experiences global hypomethylation events. To understand changes in methylation regulating transcription, it is often useful to examine different cancer models. PML-RAR is a

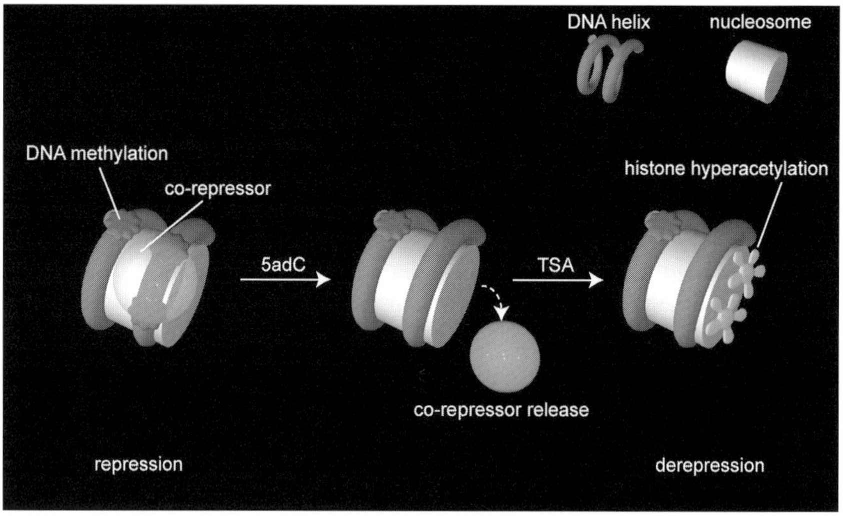

Figure 1. Model of methylation-mediated transcriptional regulation. Hypermethylation of the promoter sequence is dominant in silencing gene transcription. Methylated CpG sequences become recruitment sites for methyl-CpG-specific proteins and are associated with HDAC and Sin3 co-repressors. Demethylation by 5adC reduces the silencing potential mediated by methylation and the robust release of the co-repressor complex. Hyperacetylation of histone tails can be induced using HDAC inhibitors such as TSA, thereby decondensing chromatin and allowing assembly of activator complexes that drive gene expression.

mutant oncogenic transcription factor caused by translocation between promyelocytic leukaemia (PML) and retinoic acid receptor (RAR). This fusion protein recruits histone deacetylase and is thought to remodel chromatin and regulate transcription [47]. Evidence suggests that PML-RAR can recruit DNMTs to RA target genes with consequential promoter hypermethylation and transcriptional repression [48]. Chromatin immunoprecipitation experiments show enrichment of DNMT1 and DNMT3a on the RARβ2 promoter. Interestingly, TSA and 5adC could partially restore transcriptional competence, and this was correlated with changes in the methylation status of the RARβ2 promoter (see Fig. 2). A surprising result is that RA treatment could reduce promoter methylation, suggesting that cooperation of the DNMT methylases

Table 2. Consequence of DNMT disruption

Gene	Enzyme activity	Reduction in methylation content	Gene expression
DNMT1$^{-/-}$	96%	20%	None
DNMT3b$^{-/-}$	87%	3%	None
DNMT1/3b$^{-/-}$	99.9%	95%	Expressed

are central to carcinogenesis. Taken together, these results suggest a leukemia-promoting protein is directly associated with carcinogenesis by inducing gene hypermethylation and the recruitment of DNMTs. These observations clearly identify that DNA hypermethylation is associated with silencing of tumor susceptibility genes in several forms of cancer. However, direct proof that CpG hypermethylation and transcriptional silencing are the primary mechanisms of cellular transformation is currently lacking.

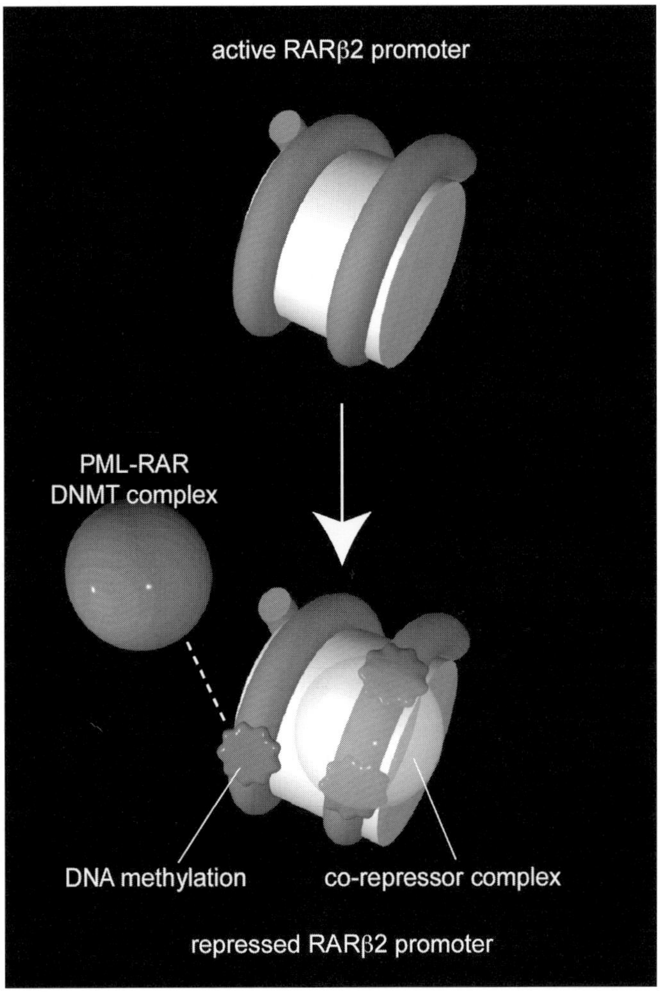

Figure 2. Recruitment model of PML-RAR/methyltransferase silencing on retinoic acid target genes. The active promoter of RARβ2 gene is targeted by the PML-RAR/DNMT methyltransferase associated complex and undergoes endogenous CpG methylation before recruitment of a methylation-dependent co-repressor complex and transcriptional silencing. Epigenetic modification induced by 5adC or RAR can reverse silencing by DNA demethylation.

If DNA methylation is inversely correlated to transcriptional repression, then recent findings that chromatin remodelling can change genomic methylation events pose some interesting questions on the antithetical nature of epigenetic modification [19, 34]. Lymphoid specific helicase (Lsh) belongs to the SNF2 subfamily of ATPase-dependent chromatin remodelling proteins [49, 50]. Results with $Lsh^{-/-}$ mice reveal substantial changes in genomic methylation levels, suggesting a role in regulating DNA methylation, histone tail modification and genetic instability during tumor progression [50–52]. In *Arabidopsis thaliana* the ddm1 (decrease in DNA methylation) gene is responsible for significant reductions in *de novo* methylation [53]. The models discussed in this review are by no means meant to represent the mechanistic riposte, nevertheless, the experimental findings expand our understanding of DNA methylation and highlights the diverse biological nature at a molecular level.

Conclusion

It is clear that the study of epigenetics continues to attract widespread interest, both within basic and medical research. The future holds great promise and, given these recent research findings, may lead to the development of new therapeutic tools based on the pharmaceutical reversal of the methylation signal and/or regulation of the machinery responsible for methylation.

References

1 Jones PA, Baylin SB (2002) The fundamental role of epigenetic events in cancer. *Nat Rev Genet* 3: 415–428
2 Buschhausen G, Wittig B, Graessmann M, Graessmann A (1987) Chromatin structure is required to block transcription of the methylated herpes simplex virus thymidine kinase gene. *Proc Natl Acad Sci USA* 84: 1177–1181
3 Antequera F, Macleod D, Bird AP (1989) Specific protection of methylated CpGs in mammalian nuclei. *Cell* 58: 509–517
4 Meehan RR, Lewis JD, McKay S, Kleiner EL, Bird AP (1989) Identification of a mammalian protein that binds specifically to DNA containing methylated CpGs. *Cell* 58: 499–507
5 Nan X, Campoy FJ, Bird A (1997) MeCP2 is a transcriptional repressor with abundant binding sites in genomic chromatin. *Cell* 88: 471–481
6 Kass SU, Landsberger N, Wolffe AP (1997) DNA methylation directs a time-dependent repression of transcription initiation. *Curr Biol* 7: 157–165
7 Boyes J, Bird A (1991) DNA methylation inhibits transcription indirectly via a methyl-CpG binding protein. *Cell* 64: 1123–1134
8 Buschhausen G, Graessmann M, and Graessmann A (1985) Inhibition of herpes simplex thymidine kinase gene expression by DNA methylation is an indirect effect. *Nucleic Acids Res* 13: 5503–5513
9 Prendergast GC, Ziff EB (1991) Methylation-sensitive sequence-specific DNA binding by the c-Myc basic region. *Science* 251: 186–189
10 Clark SJ, Harrison J, Molloy PL (1997) Sp1 binding is inhibited by (m)Cp(m)CpG methylation. *Gene* 195: 67–71
11 Kass SU, Pruss D, Wolffe AP (1997) How does DNA methylation repress transcription? *Trends Genet* 13: 444–449

12 Cameron EE, Bachman KE, Myohanen S, Herman JG, Baylin SB (1999) Synergy of demethylation and histone deacetylase inhibition in the re- expression of genes silenced in cancer. *Nat Genet* 21: 103–107
13 Jones PL, Veenstra GJ, Wade PA, Vermaak D, Kass SU, Landsberger N, Strouboulis J, Wolffe AP (1998) Methylated DNA and MeCP2 recruit histone deacetylase to repress transcription. *Nat Genet* 19: 187–191
14 Nan X, Ng HH, Johnson CA, Laherty CD, Turner BM, Eisenman RN, Bird A (1998) Transcriptional repression by the methyl-CpG-binding protein MeCP2 involves a histone deacetylase complex. *Nature* 393: 386–389
15 Wade PA (2001) Methyl CpG-binding proteins and transcriptional repression. *Bioessays* 23: 1131–1137
16 Hendrich B, Guy J, Ramsahoye B, Wilson VA, Bird A (2001) Closely related proteins MBD2 and MBD3 play distinctive but interacting roles in mouse development. *Genes Dev* 15: 710–723
17 Wade PA, Gegonne A, Jones PL, Ballestar E, Aubry F, Wolffe AP (1999) Mi-2 complex couples DNA methylation to chromatin remodelling and histone deacetylation. *Nat Genet* 23: 62–66
18 Zhang Y, Ng HH, Erdjument-Bromage H, Tempst P, Bird A, Reinberg D (1999) Analysis of the NuRD subunits reveals a histone deacetylase core complex and a connection with DNA methylation. *Genes Dev* 13: 1924–1935
19 El-Osta A (2003) DNMT cooperativity—the developing links between methylation, chromatin structure and cancer. *Bioessays* 25: 1071–1084
20 Bestor T, Laudano A, Mattaliano R, Ingram V (1988) Cloning and sequencing of a cDNA encoding DNA methyltransferase of mouse cells. The carboxyl-terminal domain of the mammalian enzymes is related to bacterial restriction methyltransferases. *J Mol Biol* 203: 971–983
21 Okano M, Xie S, Li E (1998) Cloning and characterization of a family of novel mammalian DNA (cytosine-5) methyltransferases. *Nat Genet* 19: 219–220
22 Bestor TH (2000) The DNA methyltransferases of mammals. *Hum Mol Genet* 9: 2395–2402
23 Kumar S, Cheng X, Klimasauskas S, Mi S, Posfai J, Roberts RJ, Wilson GG (1994) The DNA (cytosine-5) methyltransferases. *Nucleic Acids Res* 22: 1–10
24 Pradhan S, Roberts RJ (2000) Hybrid mouse-prokaryotic DNA (cytosine-5) methyltransferases retain the specificity of the parental C-terminal domain. *Embo J* 19: 2103–2114
25 Gowher H, Jeltsch A (2002) Molecular enzymology of the catalytic domains of the Dnmt3a and Dnmt3b DNA methyltransferases. *J Biol Chem* 277: 20409–20414
26 Lei H, Oh SP, Okano M, Juttermann R, Goss KA, Jaenisch R, Li E (1996) *De novo* DNA cytosine methyltransferase activities in mouse embryonic stem cells. *Development* 122: 3195–3205
27 Rhee I, Jair KW, Yen RW, Lengauer C, Herman JG, Kinzler KW, Vogelstein B, Baylin SB, Schuebel KE (2000) CpG methylation is maintained in human cancer cells lacking DNMT1. *Nature* 404: 1003–1007
28 Bachman KE, Rountree MR, and Baylin SB (2001) Dnmt3a and Dnmt3b are transcriptional repressors that exhibit unique localization properties to heterochromatin. *J Biol Chem* 276: 32282–32287
29 Fuks F, Burgers WA, Godin N, Kasai M, Kouzarides T (2001) Dnmt3a binds deacetylases and is recruited by a sequence-specific repressor to silence transcription. *Embo J* 20: 2536–2544
30 Belinsky SA (2004) Gene-promoter hypermethylation as a biomarker in lung cancer. *Nat Rev Cancer* 4: 707–717
31 Eden A, Gaudet F, Waghmare A, Jaenisch R (2003) Chromosomal instability and tumors promoted by DNA hypomethylation. *Science* 300: 455
32 Gaudet F, Hodgson JG, Eden A, Jackson-Grusby L, Dausman J, Gray JW, Leonhardt H, Jaenisch R (2003) Induction of tumors in mice by genomic hypomethylation. *Science* 300: 489–492
33 Feinberg AP, Vogelstein B (1983) Hypomethylation distinguishes genes of some human cancers from their normal counterparts. *Nature* 301: 89–92
34 El-Osta A (2004) The rise and fall of genomic methylation in cancer. *Leukemia* 18: 233–237
35 Ohtani-Fujita N, Fujita T, Aoike A, Osifchin NE, Robbins PD, Sakai T (1993) CpG methylation inactivates the promoter activity of the human retinoblastoma tumor-suppressor gene. *Oncogene* 8: 1063–1067
36 Merlo A, Herman JG, Mao L, Lee DJ, Gabrielson E, Burger PC, Baylin SB, Sidransky D (1995) 5' CpG island methylation is associated with transcriptional silencing of the tumour suppressor p16/CDKN2/MTS1 in human cancers. *Nat Med* 1: 686–692
37 Sakai T, Toguchida J, Ohtani N, Yandell DW, Rapaport JM, Dryja TP (1991) Allele-specific

hypermethylation of the retinoblastoma tumor-suppressor gene. *Am J Hum Genet* 48: 880–888
38 Rice JC, Ozcelik H, Maxeiner P, Andrulis I, Futscher BW (2000) Methylation of the BRCA1 promoter is associated with decreased BRCA1 mRNA levels in clinical breast cancer specimens. *Carcinogenesis* 21: 1761–1765
39 Herman JG, Umar A, Polyak K, Graff JR, Ahuja N, Issa JP, Markowitz S, Willson JK, Hamilton SR, Kinzler KW et al. (1998) Incidence and functional consequences of hMLH1 promoter hypermethylation in colorectal carcinoma. *Proc Natl Acad Sci USA* 95: 6870–6875
40 Hiltunen MO, Alhonen L, Koistinaho J, Myohanen S, Paakkonen M, Marin S, Kosma VM, Janne J (1997) Hypermethylation of the APC (adenomatous polyposis coli) gene promoter region in human colorectal carcinoma. *Int J Cancer* 70: 644–648
41 van dVPA, Metzelaar-Blok JA, Bergman W, Monique H, Hurks H, Frants RR, Gruis NA, Jager MJ (2001) Promoter hypermethylation: a common cause of reduced p16(INK4a) expression in uveal melanoma. *Cancer Res* 61: 5303–5306
42 Wang JC, Chen W, Nallusamy S, Chen C, Novetsky AD (2002) Hypermethylation of the P15INK4b and P16INK4a in agnogenic myeloid metaplasia (AMM) and AMM in leukaemic transformation. *Br J Haematol* 116: 582–586
43 Herman JG, Latif F, Weng Y, Lerman MI, Zbar B, Liu S, Samid D, Duan DS, Gnarra JR, Linehan WM et al. (1994) Silencing of the VHL tumor-suppressor gene by DNA methylation in renal carcinoma. *Proc Natl Acad Sci USA* 91: 9700–9704
44 El-Osta A, Kantharidis P, Zalcberg JR, Wolffe AP (2002) Precipitous release of methyl-CpG binding protein 2 and histone deacetylase 1 from the methylated human multidrug resistance gene (MDR1) on activation. *Mol Cell Biol* 22: 1844–1857
45 Magdinier F, and Wolffe AP (2001) Selective association of the methyl-CpG binding protein MBD2 with the silent p14/p16 locus in human neoplasia. *Proc Natl Acad Sci USA* 98: 4990–4995
46 Rhee I, Bachman KE, Park BH, Jair KW, Yen RW, Schuebel KE, Cui H, Feinberg AP, Lengauer C, Kinzler KW et al. (2002) DNMT1 and DNMT3b cooperate to silence genes in human cancer cells. *Nature* 416: 552–556
47 Grignani F, De MS, Nervi C, Tomassoni L, Gelmetti V, Cioce M, Fanelli M, Ruthardt M, Ferrara FF, Zamir I et al. (1998) Fusion proteins of the retinoic acid receptor-alpha recruit histone deacetylase in promyelocytic leukaemia. *Nature* 391: 815–818
48 Di CL, Raker VA, Corsaro M, Fazi F, Fanelli M, Faretta M, Fuks F, Lo CF, Kouzarides T, Nervi C et al. (2002) Methyltransferase recruitment and DNA hypermethylation of target promoters by an oncogenic transcription factor. *Science* 295: 1079–1082
49 Dennis K, Fan T, Geiman T, Yan Q, Muegge K (2001) Lsh, a member of the SNF2 family, is required for genome-wide methylation. *Genes Dev* 15: 2940–2944
50 Geiman TM, Tessarollo L, Anver MR, Kopp JB, Ward JM, Muegge K (2001) Lsh, a SNF2 family member, is required for normal murine development. *Biochim Biophys Acta* 1526: 211–220
51 Fan T, Yan Q, Huang J, Austin S, Cho E, Ferris D, Muegge K (2003) Lsh-deficient murine embryonal fibroblasts show reduced proliferation with signs of abnormal mitosis. *Cancer Res* 63: 4677–4683
52 Yan Q, Huang J, Fan T, Zhu H, Muegge K (2003) Lsh, a modulator of CpG methylation, is crucial for normal histone methylation. *Embo J* 22: 5154–5162
53 Vongs A, Kakutani T, Martienssen RA, Richards EJ (1993) *Arabidopsis thaliana* DNA methylation mutants. *Science* 260: 1926–1928
54 Fuks F, Burgers WA, Brehm A, Hughes-Davies L, Kouzarides T (2000) DNA methyltransferase Dnmt1 associates with histone deacetylase activity. *Nat Genet* 24: 88–91
55 Rountree MR, Bachman KE, Baylin SB (2000) DNMT1 binds HDAC2 and a new co-repressor, DMAP1, to form a complex at replication foci. *Nat Genet* 25: 269–277
56 Robertson KD, Ait-Si-Ali S, Yokochi T, Wade PA, Jones PL, Wolffe AP (2000) DNMT1 forms a complex with Rb, E2F1 and HDAC1 and represses transcription from E2F-responsive promoters. *Nat Genet* 25: 338–342
57 Tatematsu KI, Yamazaki T, Ishikawa F (2000) MBD2-MBD3 complex binds to hemi-methylated DNA and forms a complex containing DNMT1 at the replication foci in late S phase. *Genes Cells* 5: 677–688
58 Chuang LS, Ian HI, Koh TW, Ng HH, Xu G, Li BF (1997) Human DNA-(cytosine-5) methyltransferase-PCNA complex as a target for p21WAF1. *Science* 277: 1996–2000
59 Leonhardt H, Page AW, Weier HU, Bestor TH (1992) A targeting sequence directs DNA methyltransferase to sites of DNA replication in mammalian nuclei. *Cell* 71: 865–873

60 Vertino PM, Sekowski JA, Coll JM, Applegren N, Han S, Hickey RJ, Malkas LH (2002) DNMT1 is a component of a multiprotein DNA replication complex. *Cell Cycle* 1: 416–423
61 Liu S, Shen T, Huynh L, Klisovic MI, Rush LJ, Ford JL, Yu J, Becknell B, Li Y, Liu C et al. (2005) Interplay of RUNX1/MTG8 and DNA methyltransferase 1 in acute myeloid leukemia. *Cancer Res* 65: 1277–1284
62 Esteve PO, Chin HG, Pradhan S (2005) Human maintenance DNA (cytosine-5)-methyltransferase and p53 modulate expression of p53-repressed promoters. *Proc Natl Acad Sci USA* 102: 1000–1005
63 Liu Z, Fisher RA (2004) RGS6 interacts with DMAP1 and DNMT1 and inhibits DMAP1 transcriptional repressor activity. *J Biol Chem* 279: 14120–14128
64 Fuks F, Hurd PJ, Deplus R, Kouzarides T (2003) The DNA methyltransferases associate with HP1 and the SUV39H1 histone methyltransferase. *Nucleic Acids Res* 31: 2305–2312
65 Xin H, Yoon HG, Singh PB, Wong J, Qin J (2004) Components of a pathway maintaining histone modification and heterochromatin protein 1 binding at the pericentric heterochromatin in Mammalian cells. *J Biol Chem* 279: 9539–9546
66 Geiman TM, Sankpal UT, Robertson AK, Chen Y, Mazumdar M, Heale JT, Schmiesing JA, Kim W, Yokomori K, Zhao Y, Robertson KD (2004) Isolation and characterization of a novel DNA methyltransferase complex linking DNMT3B with components of the mitotic chromosome condensation machinery. *Nucleic Acids Res* 32: 2716–2729
67 Geiman TM, Sankpal UT, Robertson AK, Zhao Y, Zhao Y, Robertson KD (2004) DNMT3B interacts with hSNF2H chromatin remodeling enzyme, HDACs 1 and 2, and components of the histone methylation system. *Biochem Biophys Res Commun* 318: 544–555
68 Deplus R, Brenner C, Burgers WA, Putmans P, Kouzarides T, de Launoit Y, Fuks F (2002) Dnmt3L is a transcriptional repressor that recruits histone deacetylase. *Nucleic Acids Res* 30: 3831–3838
69 Aapola U, Liiv I, Peterson P (2002) Imprinting regulator DNMT3L is a transcriptional repressor associated with histone deacetylase activity. *Nucleic Acids Res* 30: 3602–3608

Index

The EXS-Series
Experientia Supplementa

Experientia Supplementa (EXS) is a multidisciplinary book series originally created as supplement to the journal *Experientia* which appears now under the cover of *Cellular and Molecular Life Sciences*. The multi-authored volumes focus on selected topics of biological or biomedical research, discussing current methodologies, technological innovations, novel tools and applications, new developments and recent findings.

The series is a valuable source of information not only for scientists and graduate students in medical, pharmacological and biological research, but also for physicians as well as practitioners in industry.

Forthcoming titles:

NPY Family of Peptides in Neurobiology, Cardiovascular and Metabolic Disorders: from Genes to Therapeutics, EXS 95, G.Z. Feuerstein, Z. Zukowska (Editors), 2005
Neurotransmitter Interactions and Cognitive Function, EXS 97, E.D. Levin (Editor), 2006
Plant Systems Biology, EXS 98, S. Baginsky, A.R. Fernie (Editors), 2006